APPLIED GEOSTATISTICS

Applied Geostatistics

EDWARD H. ISAAKS
Department of Applied Earth Sciences,
Stanford University

R. MOHAN SRIVASTAVA
FSS International,
Vancouver, British Columbia

New York   Oxford   OXFORD UNIVERSITY PRESS   1989

Oxford University Press

Oxford   New York   Toronto
Delhi   Bombay   Calcutta   Madras   Karachi
Petaling Jaya   Singapore   Hong Kong   Tokyo
Nairobi   Dar es Salaam   Cape Town
Melbourne   Auckland

and associated companies in
Berlin   Ibadan

Copyright © 1989 by Oxford University Press, Inc.

Published by Oxford University Press, Inc.,
200 Madison Avenue, New York, New York 10016

Oxford is a registered trademark of Oxford University Press

Library of Congress Cataloging-in-Publication Data
Isaaks, Edward H.
Applied geostatistics / Edward H. Isaaks and R. Mohan Srivastava.
p. cm. Bibliography: p. Includes index.
ISBN 0-19-505012-6—ISBN 0-19-505013-4 (pbk.) :
1. Geology—Statistical methods. I. Srivastava, R. Mohan. II. Title.
QE33.2.M3I83   1989   551'.72—dc20   89-34891   CIP

9 8 7 6 5 4 3 2
Printed in the United States of America
on acid-free paper

To my son James Reid and my daughter Carolyn Lee

E.I.

ACKNOWLEDGMENTS

This began as an attempt to write the book that we wish we had read when we were trying to learn geostatistics, a task that turned out to be much more difficult than we originally envisaged. To the many people who provided encouragement, support, and advice throughout the writing of this book, we are very grateful.

We owe a lot to André Journel, without whom this book would never have been written. In addition to providing the support necessary for this project, he has been an insightful technical reviewer and an energetic cheerleader.

We are indebted to Gordon Luster for his thorough and thoughtful proofreading; at times it seemed as if Gordon was spending as much energy reviewing our material as we were in writing it. Harry Parker also spent considerable time in reviewing the book and his comments have helped us to keep the book focused on its goal of a simplified practical introduction. The book has also benefitted from the comments and criticisms of the following people who have read all or parts of it: Kadri Dagdelen, Bruce Davis, Doug Hartzell, Young C. Kim, Neil Schofield, and Andy Solow.

Though the preparation of camera-ready copy has taught us how hard it is to produce a book, it has been a very satisfying and rewarding experience. We would like to express our thanks to Oxford University Press for their tolerance and patience as we struggled with the aesthetic detail that is normally their domain, and to Stanford University for making available the hardware and software tools necessary for the preparation of camera-ready copy.

FOREWORD

This is a book that few would have dared to write, a book that presents the fundamentals of an applied discipline without resorting to abstract concepts. There is a traditional approach to the definition of probability and random variables—one that dwells on the clean ergodic properties of Gaussian, isofactorial, and factorable random processes. Had the authors wanted simply to add to their list of publications, they could have followed this safe and well-worn path, and produced yet one more book on random fields, novel for its terminology, traditional and unassailable, but not really useful. Instead, by questioning supposedly unquestionable dogma and by placing practicality above mathematical elegance, they have chosen a more difficult path—one that risks the scorn of the self-ordained Keepers of the Tablets.

Geostatistics owes much to practice. It has evolved through what appeared initially as inconsistent applications or ad hoc adaptations of well-established models. As these adaptations established their practical utility through several successful applications, theoreticians belatedly granted them respectiblity and established their theoretical pedigree. Despite having sprung from practice that was once dismissed as theoretically inconsistent, many of these ideas are now presented as clean and logical derivations from the basic principles of random function theory. Two most enlightening examples are:

- The practice introduced by Michel David of the general relative variogram (Chapter 7 of this book) whereby the traditional experimental variogram $\gamma(h)$ is divided by the squared mean $[m(h)]^2$ of the data used for each lag h. Though inconsistent with the stationarity hypothesis, this practice proved very successful in cleaning up experimental variograms and in revealing features of spatial continuity that were later confirmed by additional data. It was much later understood that the theoretical objections to David's proposal do not hold since all variogram estimators are conditional to the available data locations and are therefore nonstationary. Moreover, his general relative variogram can be shown theoretically to filter biases due to preferential data clusters, a feature commonly encountered in earth science data.

- The practice of using a moving data neighborhood for ordinary kriging (OK), with a rescaling of the kriging variance by some function of the local mean data value. Though strictly inconsistent with the stationarity hypothesis underlying OK, this practice is the single most important reason for the practical success of the OK algorithm that drives geostatistics as a whole. It was later understood that OK with moving data neighborhoods is in fact a nonstationary estimation algorithm that allows for local fluctuations of the mean while assuming a stationary variogram. Rather than being motivated by theoretical considerations of robustness, the now common practice of OK with moving data neighborhoods was motivated in the 1960s by trite considerations of computer memory and CPU time.

Geostatistics with Mo and Ed, as this book is known at Stanford, is remarkable in the statistical literature and unique in geostatistics in that concepts and models are introduced from the needs of data analysis rather than from axioms or through formal derivations. This presentation of geostatistics is centered around the analysis of a real data set with "distressing" complexity. The availability of both sparse sampling and the exhaustive reference allows assumptions and their consequences to be checked through actual hindsight comparisons rather than through checking some theoretical property of the elusive random function generator. One may argue that the results presented could be too specific to the particular data set used. My immediate answer would be that a real data set with true complexity represents as much generality as a simplistic random function model, most often neatly stationary and Gaussian-related, on which supposedly general results can be established.

Applied geostatistics, or for that matter any applied statistics, is an art in the best sense of the term and, as such, is neither completely automatable nor purely objective. In a recent experiment conducted by the U.S. Environmental Protection Agency, 12 independent reputable geostatisticians were given the same sample data set and asked to perform the same straightforward block estimation. The 12 results were widely different due to widely different data analysis conclusions, variogram models, choices of kriging type, and search strategy. In the face of such an experiment, the illusion of objectivity can be maintained only by imposing one's decisions upon others by what I liken to scientific bullying in which laymen are dismissed as incapable of understanding the theory and are therefore disqualified from questioning the universal expertise written into some cryptic software package that delivers the correct and objective answer.

It bears repeating that there is no accepted universal algorithm for determining a variogram/covariance model, whether generalized or not, that cross-validation is no guarantee that an estimation procedure will actually produce good estimates of unsampled values, that kriging need not be the most appropriate estimation method, and that the most consequential decisions of any geostatistical study are taken early in the exploratory data analysis phase. *An Introduction to Applied Geostatistics* delivers such messages in plain terms yet with a rigor that would please both practitioners and mature theoreticians (i.e., from well-interpreted observations and comparative studies rather than from theoretical concepts whose practical relevance is obscure).

This book is sown with eye-opening remarks leading to the most recent developments in geostatistical methodology. Though academics will be rewarded with multiple challenges and seed ideas for new research work, the main public for this book will be undergraduates and practitioners who want to add geostatistics to their own toolbox. This book demonstrates that geostatistics can be learned and used properly without graduate-level courses in stochastic processes. Mo and Ed came to geostatistics not directly from academia but from the harsh reality of the practice of resource estimation within producing companies. They returned to university to better understand the tools that they found useful and are now back solving problems, sometimes using geostatistics. Their book puts geostatistics back where it belongs, in the hands of practitioners mastering both the tools and the material. Listen to their unassuming experience and remember: you are in command!

May, 1989   *André G. Journel*

CONTENTS

APPLIED GEOSTATISTICS

1

INTRODUCTION

This book presents an introduction to the set of tools that has become known commonly as *geostatistics*. Many statistical tools are useful in developing qualitative insights into a wide variety of natural phenomena; many others can be used to develop quantitative answers to specific questions. Unfortunately, most classical statistical methods make no use of the spatial information in earth science data sets. Geostatistics offers a way of describing the spatial continuity that is an essential feature of many natural phenomena and provides adaptations of classical regression techniques to take advantage of this continuity.

The presentation of geostatistics in this book is not heavily mathematical. Few theoretical derivations or formal proofs are given; instead, references are provided to more rigorous treatments of the material. The reader should be able to recall basic calculus and be comfortable with finding the minimum of a function by using the first derivative and representing a spatial average as an integral. Matrix notation is used in some of the later chapters since it offers a compact way of writing systems of simultaneous equations. The reader should also have some familiarity with the statistical concepts presented in Chapters 2 and 3.

Though we have avoided mathematical formalism, the presentation is not simplistic. The book is built around a series of case studies on a distressingly real data set. As we soon shall see, analysis of earth science data can be both frustrating and fraught with difficulty. We intend to trudge through the muddy spots, stumble into the pitfalls, and wander into some of the dead ends. Anyone who has already tackled a geostatistical study will sympathize with us in our many dilemmas.

Our case studies different from those that practitioners encounter in only one aspect; throughout our study we will have access to the correct answers. The data set with which we perform the studies is in fact a subset of a much larger, completely known data set. This gives us a yardstick by which we can measure the success of several different approaches.

A warning is appropriate here. The solutions we propose in the various case studies are particular to the data set we use. It is not our intention to propose these as general recipes. The hallmark of a good geostatistical study is customization of the approach to the problem at hand. All we intend in these studies is to cultivate an understanding of what various geostatistical tools can do and, more importantly, what their limitations are.

The Walker Lake Data Set

The focus of this book is a data set that was derived from a digital elevation model from the western United States; the Walker Lake area in Nevada.

We will not be using the original elevation values as variables in our case studies. The variables we do use, however, are related to the elevation and, as we shall see, their maps exhibit features which are related to the topographic features in Figure 1.1. For this reason, we will be referring to specific sub areas within the Walker Lake area by the geographic names given in Figure 1.1.

The original digital elevation model contained elevations for about 2 million points on a regular grid. These elevations have been transformed to produce a data set consisting of three variables measured at each of 78,000 points on a 260 x 300 rectangular grid. The first two variables are continuous and their values range from zero to several thousands. The third variable is discrete and its value is either one or two. Details on how to obtain the digital elevation model and reproduce this data set are given in Appendix A.

We have tried to avoid writing a book that is too specific to one field of application. For this reason the variables in the Walker Lake data set are referred to anonymously as V, U and T. Unfortunately, a bias toward mining applications will occasionally creep

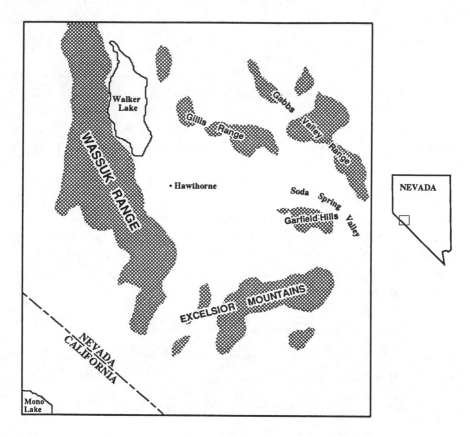

Figure 1.1 A location map of the Walker Lake area in Nevada. The small rectangle on the outline of Nevada shows the relative location of the area within the state. The larger rectangle shows the major topographic features within the area.

in; this reflects both the historical roots of geostatistics as well as the experience of the authors. The methods discussed here, however, are quite generally applicable to any data set in which the values are spatially continuous.

The continuous variables, V and U, could be thicknesses of a geologic horizon or the concentration of some pollutant; they could be soil strength measurements or permeabilities; they could be rainfall measurements or the diameters of trees. The discrete variable, T, can be viewed as a number that assigns each point to one of two possible categories; it could record some important color difference or two different species; it could separate different rock types or different soil lithologies; it could record some chemical difference such as the presence or absence of a particular element.

For the sake of convenience and consistency we will refer to V and U as concentrations of some material and will give both of them units of parts per million (ppm). We will treat T as an indicator of two types that will be referred to as *type 1* and *type 2*. Finally, we will assign units of meters to our grid even though its original dimensions are much larger than 260 x 300 m^2.

The Walker Lake data set consists of V, U and T measurements at each of 78,000 points on a 1 x 1 m^2 grid. From this extremely dense data set a subset of 470 sample points has been chosen to represent a typical sample data set. To distinguish between these two data sets, the complete set of all information for the 78,000 points is called the *exhaustive* data set, while the smaller subset of 470 points is called the *sample* data set.

Goals of the Case Studies

Using the 470 samples in the sample data set we will address the following problems:

1. The description of the important features of the data.

2. The estimation of an average value over a large area.

3. The estimation of an unknown value at a particular location.

4. The estimation of an average value over small areas.

5. The use of the available sampling to check the performance of an estimation methodology.

6. The use of sample values of one variable to improve the estimation of another variable.

7. The estimation of a distribution of values over a large area.

8. The estimation of a distribution of values over small areas.

9. The estimation of a distribution of block averages.

10. The assessment of the uncertainty of our various estimates.

The first question, despite being largely qualitative, is very important. Organization and presentation is a vital step in communicating the essential features of a large data set. In the first part of this book we will look at descriptive tools. Univariate and bivariate description are covered in Chapters 2 and 3. In Chapter 4 we will look at various ways of describing the spatial features of a data set. We will then take all of the descriptive tools from these first chapters and apply them to the Walker Lake data sets. The exhaustive data set is analyzed in Chapter 5 and the sample data set is examined in Chapters 6 and 7.

The remaining questions all deal with estimation, which is the topic of the second part of the book. Using the information in the sample data set we will estimate various unknown quantities and see how well we have done by using the exhaustive data set to check our estimates. Our approach to estimation, as discussed in Chapter 8, is first to consider what it is we are trying to estimate and then to adopt a method that is suited to that particular problem. Three important considerations form the framework for our presentation of estimation in this book. First, do we want an estimate over a large area or estimates for specific local areas? Second, are we interested only in some average value or in the complete distribution of values? Third, do we want our estimates to refer to a volume of the same size as our sample data or do we prefer to have our estimates refer to a different volume?

In Chapter 9 we will discuss why models are necessary and introduce the probabilistic models common to geostatistics. In Chapter 10 we will present two methods for estimating an average value over a large area. We then turn to the problem of local estimation. In Chapter 11 we will look at some nongeostatistical methods that are commonly used for local estimation. This is followed in Chapter 12 by a presentation of the geostatistical method known as *ordinary point kriging*. The adaptation of point estimation methods to handle the problem of local block estimates is discussed in Chapter 13.

Following the discussion in Chapter 14 of the important issue of the search strategy, we will look at cross validation in Chapter 15 and show how this procedure may be used to improve an estimation methodology. In Chapter 16 we will address the practical problem of modeling variograms, an issue that arises in geostatistical approaches to estimation.

In Chapter 17 we will look at how to use related information to improve estimation. This is a complication that commonly arises in practice when one variable is undersampled. When we analyze the sample data set in Chapter 6, we will see that the measurements of the second variable, U, are missing at many sample locations. The method of cokriging presented in Chapter 17 allows us to incorporate the more abundant V sample values in the estimation of U, taking advantage of the relationship between the two to improve our estimation of the more sparsely sampled U variable.

The estimation of a complete distribution is typically of more use in practice than is the estimation of a single average value. In many applications one is interested not in an overall average value but in the average value above some specified threshold. This threshold is often some extreme value and the estimation of the distribution above extreme values calls for different techniques than the estimation of the overall mean. In Chapter 18 we will explore the estimation of local and global distributions. We will present the *indicator approach*, one of several advanced techniques developed specifically for the estimation of local distributions.

A further complication arises if we want our estimates to refer to a volume different from the volume of our samples. This is commonly referred to as the *support* problem and frequently occurs in practical applications. For example, in a model of a petroleum reservoir one does not need estimated permeabilities for core-sized volumes but rather for much larger blocks. In a mine, one will be mining and processing volumes much larger than the volume of the samples that are typically available for a feasibility study. In Chapter 19 we will show that the distribution of point values is not the same as the distribution of average block values and present two methods for accounting for this discrepancy.

In Chapter 20 we will look at the assessment of uncertainty, an issue that is typically muddied by a lack of a clear objective meaning for the various uncertainty measures that probabilistic models can provide. We will look at several common problems, discuss how our probabilistic model might provide a relevant answer, and use the exhaustive data set to check the performance of various methods.

The final chapter provides a recap of the tools discussed in the book, recalling their strengths and their limitations. Since this book attempts an introduction to basic methods, many advanced methods have not been touched, however, the types of problems that require more advanced methods are discussed and further references are given.

Before we begin exploring some basic geostatistical tools, we would like to emphasize that the case studies used throughout the book are presented for their educational value and not necessarily to provide a definitive case study of the Walker Lake data set. It is our hope that this book will enable a reader to explore new and creative combinations of the many available tools and to improve on the rather simple studies we have presented here.

2

UNIVARIATE DESCRIPTION

Data speak most clearly when they are organized. Much of statistics, therefore, deals with the organization, presentation, and summary of data. It is hoped that much of the material in these chapters will already be familiar to the reader. Though some notions peculiar to geostatistics will be introduced, the presentation in the following chapters is intended primarily as review.

In this chapter we will deal with univariate description. In the following chapter we will look at ways of describing the relationships between pairs of variables. In Chapter 4 we incorporate the location of the data and consider ways of describing the spatial features of the data set.

To make it easy to follow and check the various calculations in the next three chapters we will use a small 10 x 10 m² patch of the exhaustive data set in all of our examples [1]. In these examples, all of the U and V values have been rounded off to the nearest integer. The V values for these 100 points are shown in Figure 2.1. The goal of this chapter will be to describe the distribution of these 100 values.

Frequency Tables and Histograms

One of the most common and useful presentations of data sets is the frequency table and its corresponding graph, the histogram. A frequency table records how often observed values fall within certain intervals or

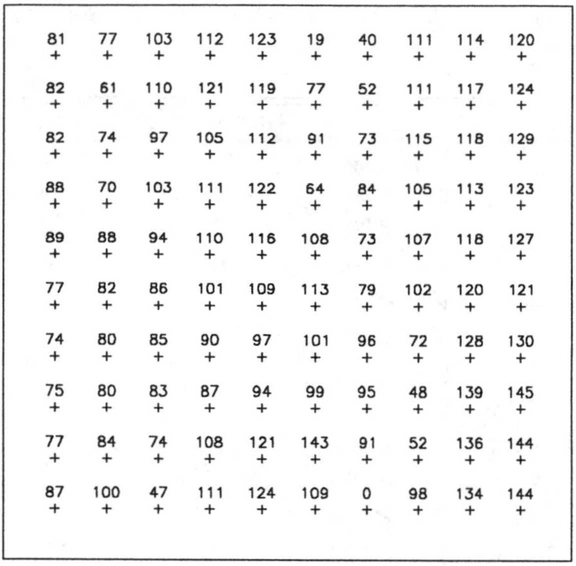

81 77 103 112 123 19 40 111 114 120
82 61 110 121 119 77 52 111 117 124
82 74 97 105 112 91 73 115 118 129
88 70 103 111 122 64 84 105 113 123
89 88 94 110 116 108 73 107 118 127
77 82 86 101 109 113 79 102 120 121
74 80 85 90 97 101 96 72 128 130
75 80 83 87 94 99 95 48 139 145
77 84 74 108 121 143 91 52 136 144
87 100 47 111 124 109 0 98 134 144

Figure 2.1 Relative location map of 100 selected *V* data.

Frequency (%)
20
15
10
5
0
0
50
100
150
200
V (ppm)

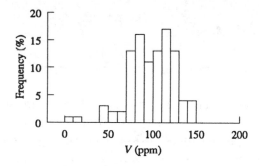

Figure 2.2 Histogram of the 100 selected *V* data.

classes. Table 2.1 shows a frequency table that summarizes the 100 *V* values shown in Figure 2.1.

The information presented in Table 2.1 can also be presented graphically in a histogram, as in Figure 2.2. It is common to use a constant class width for the histogram so that the height of each bar is proportional to the number of values within that class [2].

Table 2.1　Frequency table of the 100 selected *V* values with a class width of 10 ppm.

Class			Number	Percentage
$0 \leq$	V	<10	1	1
$10 \leq$	V	<20	1	1
$20 \leq$	V	<30	0	0
$30 \leq$	V	<40	0	0
$40 \leq$	V	<50	3	3
$50 \leq$	V	<60	2	2
$60 \leq$	V	<70	2	2
$70 \leq$	V	<80	13	13
$80 \leq$	V	<90	16	16
$90 \leq$	V	<100	11	11
$100 \leq$	V	<110	13	13
$110 \leq$	V	<120	17	17
$120 \leq$	V	<130	13	13
$130 \leq$	V	<140	4	4
$140 \leq$	V	$< \infty$	4	4

Cumulative Frequency Tables and Histograms

Most statistical texts use the convention that data are ranked in ascending order to produce cumulative frequency tables and descriptions of cumulative frequency distributions. For many earth science applications, such as ore reserves and pollution studies, the cumulative frequency above a lower limit is of more interest. For such studies, cumulative frequency tables and histograms may be prepared after ranking the data in descending order.

In Table 2.2 we have taken the information from Table 2.1 and presented it in cumulative form. Rather than record the number of values within certain classes, we record the total number of values below certain cutoffs [3]. The corresponding cumulative histogram, shown in Figure 2.3, is a nondecreasing function between 0 and 100%. The percent frequency and cumulative percent frequency forms are used interchangeably, since one can be obtained from the other.

Table 2.2 Cumulative frequency table of the 100 selected V values using a class width of 10 ppm.

Class		Number	Percentage
V	$<$ 10	1	1
V	$<$ 20	2	2
V	$<$ 30	2	2
V	$<$ 40	2	2
V	$<$ 50	5	5
V	$<$ 60	7	7
V	$<$ 70	9	9
V	$<$ 80	22	22
V	$<$ 90	38	38
V	$<$ 100	49	49
V	$<$ 110	62	62
V	$<$ 120	79	79
V	$<$ 130	92	92
V	$<$ 140	96	96
V	$< \infty$	100	100

Normal and Lognormal Probability Plots

Some of the estimation tools presented in part two of the book work better if the distribution of data values is close to a Gaussian or *normal* distribution. The Gaussian distribution is one of many distributions for which a concise mathematical description exists [4]; also, it has properties that favor its use in theoretical approaches to estimation. It is interesting, therefore, to know how close the distribution of one's data values comes to being Gaussian. A normal probability plot is a type of cumulative frequency plot that helps decide this question.

On a normal probability plot the y-axis is scaled in such a way that the cumulative frequencies will plot as a straight line if the distribution is Gaussian. Such graph paper is readily available at most engineering supply outlets. Figure 2.4 shows a normal probability plot of the 100 V values using the cumulative frequencies given in Table 2.2. Note

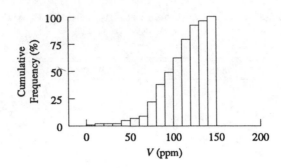

Figure 2.3 Cumulative histogram of the 100 selected V data.

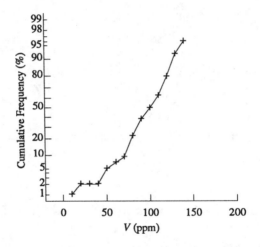

Figure 2.4 A normal probability plot of the 100 selected V data. The y-axis has been scaled in such a way that the cumulative frequencies will plot as a straight line if the distribution of V is Gaussian.

that although most of the cumulative frequencies plot in a relatively straight line, the smaller values of V depart from this trend.

Many variables in earth science data sets have distributions that are not even close to normal. It is common to have many quite small values and a few very large ones. In Chapter 5 we will see several examples of this type from the exhaustive Walker Lake data set. Though the normal distribution is often inappropriate as a model for this type of asymmetric distribution, a closely related distribution, the lognor-

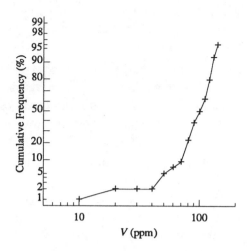

Figure 2.5 A lognormal probability plot of the 100 selected V data. The y-axis is scaled so that the cumulative frequencies will plot as a straight line if the distribution of the logarithm of V is Gaussian.

mal distribution, can sometimes be a good alternative. A variable is distributed lognormally if the distribution of the logarithm of the variable is normal.

By using a logarithmic scale on the x-axis of a normal probability plot, one can check for lognormality. As in the normal probability plot, the cumulative frequencies will plot as a straight line if the data values are lognormally distributed. Figure 2.5 shows a lognormal probability plot of the 100 V values using the same information that was used to plot Figure 2.4. The concave shape of the plot clearly indicates that the values are not distributed lognormally.

Assumptions about the distribution of data values often have their greatest impact when one is estimating extreme values. If one intends to use a methodology that depends on assumptions about the distribution, one should be wary of casually disregarding deviations of a probability plot at the extremes. for example, it is tempting to take the normal probability plot shown in Figure 2.4 as evidence of normality, disregarding the departure from a relatively straight line for the smaller values of V. Departures of a probability plot from approximate linearity at the extreme values are often deceptively small and easy to overlook when the rest of the plot looks relatively straight. However, the estimates derived using such a "close fitted" distribution model may be vastly different from reality.

Probability plots are very useful for checking for the presence of multiple populations. Although kinks in the plots do not necessarily indicate multiple populations, they represent changes in the characteristics of the cumulative frequencies over different intervals and the reasons for this should be explored.

Choosing a theoretical model for the distribution of data values is not always a necessary step prior to estimation, so one should not read too much into a probability plot. The straightness of a line on a probability plot is no guarantee of a good estimate and the crookedness of a line should not condemn distribution-based approaches to estimation. Certain methods lean more heavily on assumptions about the distribution than do others. Some estimation tools built on an assumption of normality may still be useful even when the data are not normally distributed.

Summary Statistics

The important features of most histograms can be captured by a few summary statistics. The summary statistics we use here fall into three categories: measures of location, measures of spread and measures of shape.

The statistics in the first group give us information about where various parts of the distribution lie. The mean, the median, and the mode can give us some idea where the *center* of the distribution lies. The location of other parts of the distribution are given by various quantiles. The second group includes the variance, the standard deviation, and the interquartile range. These are used to describe the variability of the data values. The shape of the distribution is described by the coefficient of skewness and the coefficient of variation; the coefficient of skewness provides information on the symmetry while the coefficient of variation provides information on the length of the tail for certain types of distributions. Taken together, these statistics provide a valuable summary of the information contained in the histogram.

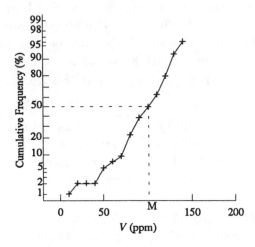

Figure 2.6 Reading the median from a probability plot.

Measures of Location

Mean. The mean, m, is the arithmetic average of the data values [5]:

$$m = \frac{1}{n} \sum_{i=1}^{n} x_i \qquad (2.1)$$

The number of data is n and $x_1, \ldots, x_n$ are the data values. The mean of our 100 V values is 97.55 ppm.

Median. The median, M, is the midpoint of the observed values if they are arranged in increasing order. Half of the values are below the median and half of the values are above the median. Once the data are ordered so that $x_1 \le x_2 \le \ldots \le x_n$, the median can be calculated from one of the following equations:

$$M = \begin{cases} x_{\frac{n+1}{2}} & \text{if } n \text{ is odd} \\ \left(x_{\frac{n}{2}} + x_{\frac{n}{2}+1}\right) \div 2 & \text{if } n \text{ is even} \end{cases} \qquad (2.2)$$

The median can easily be read from a probability plot. Since the y-axis records the cumulative frequency, the median is the value on the x-axis that corresponds to 50% on the y-axis (Figure 2.6).

Both the mean and the median are measures of the location of the center of the distribution. The mean is quite sensitive to erratic high values. If the 145 ppm value in our data set had been 1450 ppm, the mean would change to 110.60 ppm. The median, however, would be unaffected by this change because it depends only on how many values are above or below it; how much above or below is not considered.

For the 100 V values that appear in Figure 2.1 the median is 100.50 ppm.

Mode. The mode is the value that occurs most frequently. The class with the tallest bar on the histogram gives a quick idea where the mode is. From the histogram in Figure 2.2 we see that the 110-120 ppm class has the most values. Within this class, the value 111 ppm occurs more times than any other.

One of the drawbacks of the mode is that it changes with the precision of the data values. In Figure 2.1 we rounded all of the V values to the nearest integer. Had we kept two decimal places on all our measurements, no two would have been exactly the same and the mode could then be any one of 100 equally common values. For this reason, the mode is not particularly useful for data sets in which the measurements have several significant digits. In such cases, when we speak of the mode we usually mean some approximate value chosen by finding the tallest bar on a histogram. Some practitioners interpret the mode to be the tallest bar itself.

Minimum. The smallest value in the data set is the minimum. In many practical situations the smallest values are recorded simply as being below some detection limit. In such situations, it matters little for descriptive purposes whether the minimum is given as 0 or as some arbitrarily small value. In some estimation methods, as we will discuss in later chapters, it is convenient to use a nonzero value (e.g., half the detection limit) or to assign slightly different values to those data that were below the detection limit. For our 100 V values, the minimum value is 0 ppm.

Maximum. The largest value in the data set is the maximum. The maximum of our 100 V values is 145 ppm.

Lower and Upper Quartile. In the same way that the median splits the data into halves, the quartiles split the data into quarters. If the data values are arranged in increasing order, then a quarter of the data falls below the lower or first quartile, Q_1, and a quarter of the data falls above the upper or third quartile, Q_3.

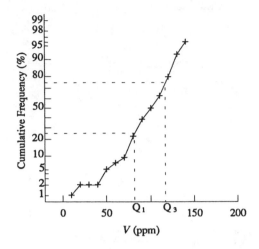

Figure 2.7 The quartiles of a normal probability plot.

As with the median, quartiles can easily be read from a probability plot. The value on the x-axis, which corresponds to 25% on the y-axis, is the lower quartile and the value that corresponds to 75% is the upper quartile (Figure 2.7). The lower quartile of our 100 V values is 81.25 ppm and the upper quartile is 116.25 ppm.

Deciles, Percentiles, and Quantiles. The idea of splitting the data into halves with the median or into quarters with the quartiles can be extended to any other fraction. Deciles split the data into tenths. One tenth of the data fall below the first or lowest decile; two tenths fall below the second decile. The fifth decile corresponds to the median. In a similar way, percentiles split the data into hundredths. The twenty-fifth percentile is the same as the first quartile, the fiftieth percentile is the same as the median and the seventy-fifth percentile is the same as the third quartile.

Quantiles are a generalization of this idea to any fraction. For example, if we wanted to talk about the value below which one twentieth of the data fall, we call it $q_{.05}$ rather than come up with a new -ile name for twentieths. Just as certain deciles and percentiles are equivalent to the median and the quartiles, so too can certain quantiles be written as one of these statistics. For example $q_{.25}$ is the lower quartile, $q_{.5}$ is the median, and $q_{.75}$ is the upper quartile. In this book we will usually use quantiles rather than deciles and percentiles, keeping only the median and the two quartiles as special measures of location.

Measures of Spread

Variance. The variance, σ^2, is given by [6]:

$$\sigma^2 = \frac{1}{n}\sum_{i=1}^{n}(x_i - m)^2 \qquad (2.3)$$

It is the average squared difference of the observed values from their mean. Since it involves squared differences, the variance is sensitive to erratic high values. The variance of the 100 V values is 688 ppm^2.

Standard Deviation. The standard deviation, σ, is simply the square root of the variance. It is often used instead of the variance since its units are the same as the units of the variable being described. For the 100 V values the standard deviation is 26.23 ppm.

Interquartile Range. Another useful measure of the spread of the observed values is the interquartile range. The interquartile range or IQR, is the difference between the upper and lower quartiles and is given by

$$IQR = Q_3 - Q_1 \qquad (2.4)$$

Unlike the variance and the standard deviation, the interquartile range does not use the mean as the center of the distribution, and is therefore often preferred if a few erratically high values strongly influence the mean. The interquartile range of our 100 V values is 35.50 ppm.

Measures of Shape

Coefficient of Skewness. One feature of the histogram that the previous statistics do not capture is its symmetry. The most commonly used statistic for summarizing the symmetry is a quantity called the *coefficient of skewness*, which is defined as

$$coefficient\ of\ skewness = \frac{\frac{1}{n}\sum_{i=1}^{n}(x_i - m)^3}{\sigma^3} \qquad (2.5)$$

The numerator is the average cubed difference between the data values and their mean, and the denominator is the cube of the standard deviation.

The coefficient of skewness suffers even more than the mean and variance from a sensitivity to erratic high values. A single large value can heavily influence the coefficient of skewness since the difference between each data value and the mean is cubed.

Quite often one does not use the magnitude of the coefficient of skewness but rather only its sign to describe the symmetry. A positively skewed histogram has a long tail of high values to the right, making the median less than the mean. In geochemical data sets, positive skewness is typical when the variable being described is the concentration of a minor element. If there is a long tail of small values to the left and the median is greater than the mean, as is typical for major element concentrations, the histogram is negatively skewed. If the skewness is close to zero, the histogram is approximately symmetric and the median is close to the mean.

For the 100 V values we are describing in this chapter the coefficient of skewness is close to zero (-0.779), indicating a distribution that is only slightly asymmetric.

Coefficient of Variation. The coefficient of variation, CV, is a statistic that is often used as an alternative to skewness to describe the shape of the distribution. It is used primarily for distributions whose values are all positive and whose skewness is also positive; though it can be calculated for other types of distributions, its usefulness as an index of shape becomes questionable. It is defined as the ratio of the standard deviation to the mean [7]:

$$CV = \frac{\sigma}{m} \qquad (2.6)$$

If estimation is the final goal of a study, the coefficient of variation can provide some warning of upcoming problems. A coefficient of variation greater than one indicates the presence of some erratic high sample values that may have a significant impact on the final estimates.

The coefficient of variation for our 100 V values is 0.269, which reflects the fact that the histogram does not have a long tail of high values.

Notes

[1] The coordinates of the corners of the 10 x 10 m^2 patch used to illustrate the various descriptive tools are (11,241), (20,241), (20,250), and (11,250).

[2] If the class widths are variable it is important to remember that on a histogram it is the area (not the height) of the bar that is proportional to the frequency.

[3] The example in the text is designed to make it easy to follow how Table 2.2 relates to Table 2.1. Though the choice of classes is necessary for a frequency table and a histogram, it is not required for cumulative frequency tables or cumulative histograms. Indeed, in practice one typically chooses cutoffs for the cumulative frequencies that correspond to the actual data values.

[4] For a description of the normal distribution and its properties see: Johnson, R. A. and Wichern, D. W. , *Applied Multivariate Statistical Analysis.* Englewood Cliffs, New Jersey: Prentice-Hall, 1982.

[5] Though the arithmetic average is appropriate for a wide variety of applications, there are important cases in which the averaging process is not arithmetic. For example, in fluid flow studies the effective permeability of a stratified sequence is the arithmetic mean of the permeabilities within the various strata if the flow is parallel to the strata. If the flow is perpendicular to the strata, however, the harmonic mean, m_H, is more appropriate:

$$\frac{1}{k_{eff}} = \frac{1}{m_H} = \frac{1}{n} \sum_{i=1}^{n} \frac{1}{k_i}$$

where the k_i are the permeabilities of the n strata. For the case where the flow is neither strictly parallel nor strictly perpendicular to the stratification, or where the different facies are not clearly stratified, some studies suggest that the effective permeability is close to the geometric mean, m_G:

$$\log k_{eff} = \log m_G = \frac{1}{n} \sum_{i=1}^{n} \log k_i$$

[6] Some readers will recall a formula for σ^2 from classical statistics that uses $\frac{1}{n-1}$ instead of $\frac{1}{n}$. This classical formula is designed to give an unbiased estimate of the population variance if the data are uncorrelated. The formula given here is intended only to give the sample variance. In later chapters we will look at the problem of inferring population parameters from sample statistics.

[7] The coefficient of variation is occasionally given as a percentage rather than a ratio.

Further Reading

Davis, J. C. , *Statistical and Data Analysis in Geology.* New York: Wiley, 1973.

Koch, G. and Link, R. , *Statistical Analysis of Geological Data.* New York: Wiley, 2 ed., 1986.

Mosteller, F. and Tukey, J. W. , *Data Analysis and Regression.* Reading, Mass.: Addison-Wesley, 1977.

Ripley, B. D. , *Spatial Statistics.* New York: Wiley, 1981.

Tukey, J. , *Exploratory Data Analysis.* Reading, Mass.: Addison-Wesley, 1977.

3

BIVARIATE DESCRIPTION

The univariate tools discussed in the last chapter can be used to describe the distributions of individual variables. We get a very limited view, however, if we analyze a multivariate data set one variable at a time. Some of the most important and interesting features of earth science data sets are the relationships and dependencies between variables.

The Walker Lake data set contains two continuous variables. Figure 3.1 shows the 100 V values we saw in Figure 2.1 along with the U values at the same 100 locations. In this chapter we look at ways of describing the relationship between these two variables.

Comparing Two Distributions

In the analysis of earth science data sets we will often want to compare two distributions. A presentation of their histograms along with some summary statistics will reveal gross differences. Unfortunately, if the two distributions are very similar, this method of comparison will not be helpful in uncovering the interesting subtle differences.

The histograms of the V and U values shown in Figure 3.1 are given in Figure 3.2, and their statistics are presented in Table 3.1 There are some rather major differences between the distributions of the two variables. The U distribution is positively skewed; the V distribution, on the other hand, is negatively skewed. Also, the V values are generally higher than the U values, with a mean value more

81 + 15	77 + 12	103 + 24	112 + 27	123 + 30	19 + 0	40 + 2	111 + 18	114 + 18	120 + 18
82 + 16	61 + 7	110 + 34	121 + 36	119 + 29	77 + 7	52 + 4	111 + 18	117 + 18	124 + 20
82 + 16	74 + 9	97 + 22	105 + 24	112 + 25	91 + 10	73 + 7	115 + 19	118 + 19	129 + 22
88 + 21	70 + 8	103 + 27	111 + 27	122 + 32	64 + 4	84 + 10	105 + 15	113 + 17	123 + 19
89 + 21	88 + 18	94 + 20	110 + 27	116 + 29	108 + 19	73 + 7	107 + 16	118 + 19	127 + 22
77 + 15	82 + 16	86 + 16	101 + 23	109 + 24	113 + 25	79 + 7	102 + 15	120 + 21	121 + 20
74 + 14	80 + 15	85 + 15	90 + 16	97 + 17	101 + 18	96 + 14	72 + 6	128 + 28	130 + 25
75 + 14	80 + 15	83 + 15	87 + 15	94 + 16	99 + 17	95 + 13	48 + 2	139 + 40	145 + 38
77 + 16	84 + 17	74 + 11	108 + 29	121 + 37	143 + 55	91 + 11	52 + 3	136 + 34	144 + 35
87 + 22	100 + 28	47 + 4	111 + 32	124 + 38	109 + 20	0 + 0	98 + 14	134 + 31	144 + 34

Figure 3.1 Relative location map of the 100 selected V and U data. V values are plotted above the "+" symbol and U are below.

than five times that of U. The V median and standard deviation are also greater than their U counterparts.

The statistical summary provided in Table 3.1 allows us to compare, among other things, the medians and the quartiles of the two distributions. A more complete comparison of the various quantiles is given in Table 3.2, which shows the V and U quantiles for several cumulative frequencies. The For example, the first entry tells us that that 5% of the V values are below 48.1 ppm while 5% of the U values fall below 3.1 ppm. The medians and quartiles we saw earlier in Table 3.1 are also included in Table 3.2. The first quartile, 81.3 ppm for V and 14.0 ppm for U, corresponds to the 0.25 quantile; the median, 100.5 ppm for V and 18.0 ppm for U, corresponds to $q_{.5}$; and the upper quartile, 116.8 ppm for V and 25.0 ppm for U, corresponds to $q_{.75}$.

For a good visual comparison of two distributions we can use a graph called a *q-q plot*. This is commonly used when there is some reason to expect that the distributions are similar. A q-q plot is a graph on which the quantiles from two distributions are plotted versus

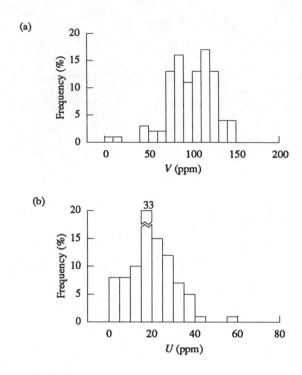

Figure 3.2 The histogram of the 100 V values in (a) and of the corresponding 100 U values in (b).

one another. The information contained in Table 3.2 is presented as a q-q plot in Figure 3.3. The quantiles of the V distribution serve as the x-coordinates while those of the U distribution serve as the y-coordinates. If the two distributions being compared have the same number of data, then the calculation of the quantiles of each distribution is not a necessary step in making a q-q plot. Instead, one can sort the data values from each distribution in ascending order and plot the corresponding pairs of values.

A q-q plot of two identical distributions will plot as the straight line $x = y$. For distributions that are very similar, the small departures of the q-q plot from the line $x = y$ will reveal where they differ. As we have already noted, the distributions of the V and U values within our selected area are very different; therefore, their q-q plot does not come close to the straight line $U = V$.

Table 3.1 Statistical summary of the *V* and *U* values shown in Figure 3.1.

	V	*U*
n	100	100
m	97.6	19.1
σ	26.2	9.81
CV	0.27	0.51
min	0.0	0.0
Q_1	81.3	14.0
M	100.5	18.0
Q_3	116.8	25.0
max	145.0	55.0

Table 3.2 Comparison of the *V* and *U* quantiles.

Cumulative Frequency	Quantile *V*	*U*	Cumulative Frequency	Quantile *V*	*U*
0.05	48.1	3.1	0.55	104.1	19.0
0.10	70.2	7.0	0.60	108.6	20.0
0.15	74.0	8.1	0.65	111.0	21.0
0.20	77.0	11.2	0.70	112.7	22.7
0.25	81.3	14.0	0.75	116.8	25.0
0.30	84.0	15.0	0.80	120.0	27.0
0.35	87.4	15.4	0.85	122.9	29.0
0.40	91.0	16.0	0.90	127.9	33.8
0.45	96.5	17.0	0.95	138.9	37.0
0.50	100.5	18.0			

If a q-q plot of two distributions is some straight line other than *x* = *y*, then the two distributions have the same shape but their location and spread may differ. We have already taken advantage of this property when we constructed the normal probability plots in Figure 2.4. In fact, this is a q-q plot on which we compare the quantiles of the *V* distribution to the quantiles of a standard normal distribution. Likewise,

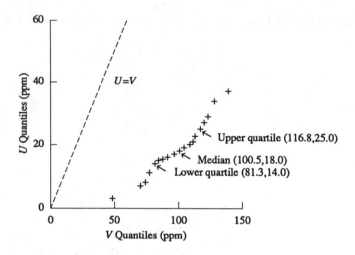

Figure 3.3 A q-q plot of the distribution of the 100 special U values versus the 100 V values. Note the different scales on the axes.

the lognormal probability plot we drew in Figure 2.5 is a comparison of the V quantiles to those of a standard lognormal distribution. The similarity of an observed distribution to any theoretical distribution model can be checked by the straightness of their q-q plot.

Scatterplots

The most common display of bivariate data is the *scatterplot*, which is an x-y graph of the data on which the x-coordinate corresponds to the value of one variable and the y-coordinate to the value of the other variable.

The 100 pairs of V-U values in Figure 3.1 are shown on a scatterplot in Figure 3.4a. Though there is some scatter in the cloud of points, the larger values of V tend to be associated with the larger values of U and the smaller values of V tend to be associated with the smaller values of U.

In addition to providing a good qualitative feel for how two variables are related, a scatterplot is also useful for drawing our attention to aberrant data. In the early stages of the study of a spatially continuous data set it is necessary to check and clean the data; the success of any estimation method depends on reliable data. Even after the data

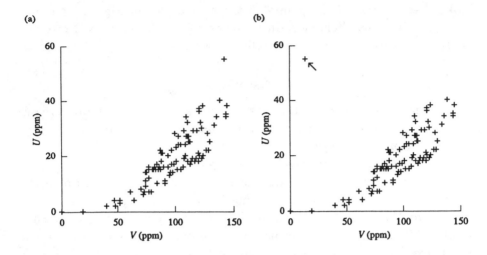

Figure 3.4 Scatterplot of 100 *U* versus *V* values. The actual 100 data pairs are plotted in (a). In (b) the *V* value indicated by the arrow has been "accidentally" plotted as 14 ppm rather than 143 ppm to illustrate the usefulness of the scatterplot in detecting errors in the data.

have been cleaned, a few erratic values may have a major impact on estimation. The scatterplot can be used to help both in the validation of the initial data and in the understanding of later results.

The scatterplot shown in Figure 3.4a does not reveal any obvious errors in the *V* and *U* values. There is one point that plots in the upper right corner of Figure 3.4a with a *U* value of 55 ppm and a *V* value of 143 ppm. Had the *V* value accidentally been recorded as 14 ppm, this pair of values would plot in the upper left corner all by itself, as in Figure 3.4b, and one's suspicion would be aroused by such an unusual pair. Often, further investigations of such unusual pairs will reveal errors that were most likely made when the data were collected or recorded.

A powerful principle underlies this simple concept of using a scatterplot for error checking. We are relying on the general relationship between the two variables to tell us if a particular pair of values is unusual. In the example given in the last paragraph, we expected the *V* value associated with a *U* value of 55 ppm to be quite high, somewhere between 100 and 150 ppm. This reasonable expectation comes from looking at the rest of the points on the scatterplot in Figure 3.4b and extrapolating their behavior. In part two of this book we will present an approach to estimation that relies on this same idea.

Correlation

In the very broadest sense, there are three patterns one can observe on a scatterplot: the variables are either *positively correlated, negatively correlated*, or *uncorrelated*.

Two variables are positively correlated if the larger values of one variable tend to be associated with larger values of the other variable, and similarly with the smaller values of each variable. In porous rocks, porosity and permeability are typically positively correlated. If we drew a scatterplot of porosity versus permeability, we would expect to see the larger porosity values associated with the larger permeability values.

Two variables are negatively correlated if the larger values of one variable tend to be associated with the smaller values of the other. In geological data sets, the concentrations of two major elements are often negatively correlated; in a dolomitic limestone, for example, an increase in the amount of calcium usually results in a decrease in the amount magnesium.

The final possibility is that the two variables are not related. An increase in one variable has no apparent effect on the other. In this case, the variables are said to be uncorrelated.

Correlation Coefficient. The correlation coefficient, ρ, is the statistic that is most commonly used to summarize the relationship between two variables. It can be calculated from:

$$\rho = \frac{\frac{1}{n}\sum_{i=1}^{n}(x_i - m_x)(y_i - m_y)}{\sigma_x \sigma_y} \tag{3.1}$$

The number of data is n; $x_1, \ldots, x_n$ are the data values for the first variable, m_x is their mean, and σ_x is their standard deviation; $y_1, \ldots, y_n$ are the data values for the second variable, m_y is their mean, and σ_y is their standard deviation.

The numerator in Equation 3.1 is called the *covariance*,

$$C_{XY} = \frac{1}{n}\sum_{i=1}^{n}(x_i - m_x)(y_i - m_y) \tag{3.2}$$

and is often used itself as a summary statistic of a scatterplot. The covariance between two variables depends on the magnitude of the data values. If we took all of our V-U pairs from Figure 3.1 and multiplied their values by 10, our scatterplot would still look the same, with the axes relabeled accordingly. The covariance, however, would be 100 times larger. Dividing the covariance by the standard deviations of the two variables guarantees that the correlation coefficient will always be between -1 and $+1$, and provides an index that is independent of the magnitude of the data values.

The covariance of our 100 V-U pairs is 216.1 ppm^2, the standard deviation of V is 26.2 ppm and of U is 9.81 ppm. The correlation coefficient between V and U therefore, is 0.84.

The correlation coefficient and the covariance may be affected by a few aberrant pairs. A good alignment of a few extreme pairs can dramatically improve an otherwise poor correlation coefficient. Conversely, an otherwise good correlation could be ruined by the poor alignment of a few extreme pairs. Earlier, in Figure 3.4, we showed two scatterplots that were identical except for one pair whose V value had been erroneously recorded as 14 ppm rather than 143 ppm. The correlation coefficient of the scatterplot shown in Figure 3.4a is the value we calculated in the previous paragraph, 0.84. With the change of only one pair, the scatterplot shown in Figure 3.4b has a correlation coefficient of only 0.64.

The correlation coefficient is actually a measure of how close the observed values come to falling on a straight line. If $\rho = +1$, then the scatterplot will be a straight line with a positive slope; if $\rho = -1$, then the scatterplot will be a straight line with a negative slope. For $|\rho| < 1$ the scatterplot appears as a cloud of points that becomes fatter and more diffuse as $|\rho|$ decreases from 1 to 0.

It is important to note that ρ provides a measure of the *linear* relationship between two variables. If the relationship between two variables is not linear, the correlation coefficient may be a very poor summary statistic. It is often useful to supplement the linear correlation coefficient with another measure of the strength of the relationship, the rank correlation coefficient [1]. To calculate the rank correlation coefficient, one applies Equation 3.1 to the ranks of the data values rather than to the original sample values:

$$\rho_{rank} = \frac{\frac{1}{n}\sum_{i=1}^{n}(Rx_i - m_{Rx})(Ry_i - m_{Ry})}{\sigma_{Rx}\sigma_{Ry}} \qquad (3.3)$$

Rx_i is the rank of x_i among all the other x values and is usually calculated by sorting the x values in ascending order and seeing where each value falls. The lowest of the x values would appear first on a sorted list and would therefore receive a rank of 1; the highest x value would appear last on the list and would receive a rank of n. Ry_i is the rank of y_i among all the other y values. m_{Rx} is the mean of all of the ranks $Rx_1, \ldots, Rx_n$ and σ_{Rx} is their standard deviation. m_{Ry} is the mean of all of the ranks $Ry_1, \ldots, Ry_n$ and σ_{Ry} is their standard deviation [2].

Large differences between ρ_{rank} and ρ are often quite revealing about the location of extreme pairs on the scatterplot. Unlike the traditional correlation coefficient, the rank correlation coefficient is not strongly influenced by extreme pairs. Large differences between the two may be due to the location of extreme pairs on the scatterplot. A high value of ρ_{rank} and a low value of ρ may be due to the fact that a few erratic pairs have adversely affected an otherwise good correlation. If, on the other hand, it is ρ that is quite high while ρ_{rank} is quite low, then it is likely that the high value of ρ is due largely to the influence of a few extreme pairs.

For the scatterplot shown in Figure 3.4b, the the coefficient of linear correlation is 0.64, while the rank correlation coefficient is 0.80. The single aberrant pair in the upper left corner has less of an influence on the rank correlation than it does on the traditional correlation coefficient.

Differences between ρ and ρ_{rank} may also reveal important features of the relationship between two variables. If the rank correlation coefficient is +1, then the ranks of the two variables are identical: the largest value of x corresponds to the largest value of y, and the smallest value of x corresponds to the smallest value of y. If the rank correlation coefficient is +1, then the relationship between x and y need not be linear. It is, however, monotonic; if the value of x increases, then the value of y also increases. Two variables whose rank correlation coefficient is noticeably higher than their traditional linear correlation coefficient may exhibit a nonlinear relationship. For example, two variables, X and Y, which are related by the equation $Y = X^2$ will have a value of ρ near 0 but a value of ρ_{rank} of 1.

The value of ρ is often a good indicator of how successful we might be in trying to predict the value of one variable from the other with a linear equation. If $|\rho|$ is large, then for a given value of one variable, the other variable is restricted to only a small range of possible values. On the other hand, if $|\rho|$ is small, then knowing the value of one variable does not help us very much in predicting the value of the other.

Linear Regression

As we noted earlier, a strong relationship between two variables can help us predict one variable if the other is known. The simplest recipe for this type of prediction is linear regression, in which we assume that the dependence of one variable on the other can be described by the equation of a straight line:

$$y = ax + b \qquad (3.4)$$

The slope, a, and the constant, b, are given by:

$$a = \rho \, \frac{\sigma_y}{\sigma_x} \qquad b = m_y - a \cdot m_x \qquad (3.5)$$

The slope, a, is the correlation coefficient multiplied by the ratio of the standard deviations, with σ_y being the standard deviation of the variable we are trying to predict and σ_x the standard deviation of the variable we know. Once the slope is known, the constant, b, can be calculated using the means of the two variables, m_x and m_y.

If we use our 100 V-U pairs to calculate a linear regression equation for predicting V from U, we get

$$a = 0.84 \, \frac{26.2}{9.81} = 2.24 \qquad b = 97.6 - 2.24 \cdot 19.1 = 54.7 \qquad (3.6)$$

Our equation to predict V from a known U value is then

$$V = 2.24 \, U + 54.7 \qquad (3.7)$$

In Figure 3.5b this line is superimposed on the scatterplot. Although it looks reasonable through the middle of the cloud, this regression line does not look very good at the extremes. It would definitely overestimate very low values of V. The problem is our assumption that the dependence of V on U is linear. No other straight line would do better than the one we calculated earlier[3].

Equation 3.7 gives us a prediction for V if U is known. We might also be interested in predicting U if V is the variable that is known.

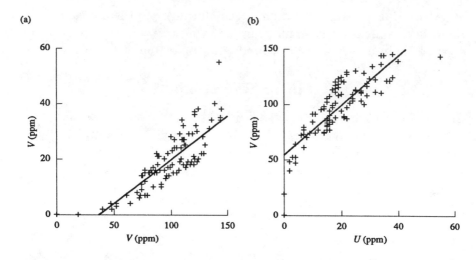

Figure 3.5 Linear regression lines superimposed on the scatterplot. The regression line of U given V is shown in (a), and of V given U in (b).

In Equation 3.5, y is the unknown variable and x is known, so the calculation of a linear regression equation that predicts U from V is:

$$a = 0.84 \, \frac{9.81}{26.2} = 0.314 \qquad b = 19.1 - 0.314 \cdot 97.6 = -11.5 \quad (3.8)$$

The linear regression equation for predicting U from a known V value is then

$$U = 0.314 \, V - 11.5 \qquad\qquad (3.9)$$

This regression line is shown in Figure 3.5a. In this figure we have plotted U on the y-axis and V on the x-axis to emphasize the fact that it is U that is the unknown in this case. We will continue with this convention throughout the book; for scatterplots on which there is a known variable and an unknown variable, we will plot the unknown variable on the y-axis.

A close look at Figure 3.5a and Figure 3.5b reveals that the two regression lines are not the same; indeed Equation 3.9 is not simply a rearrangement of Equation 3.7.

The regression line shown in Figure 3.5a raises an issue that we will confront when we look at estimation in part two. Noticing that the regression line hits the x-axis near a V value of 35 ppm, one might

Table 3.3 Mean values of V within classes defined on the U value.

Class	Number of Pairs	Mean of V
$0 \leq U < 5$	8	40.3
$5 \leq U < 10$	8	72.4
$10 \leq U < 15$	10	85.5
$15 \leq U < 20$	33	97.5
$20 \leq U < 25$	15	106.9
$25 \leq U < 30$	12	113.5
$30 \leq U < 35$	7	125.7
$35 \leq U < \infty$	7	133.9

wonder what the predicted value of U is for a V value of about 5 ppm. Of course, the regression line continues into negative values for U and if we substitute a value of 5 ppm for V into Equation 3.9 we get a predicted value of -6.2 ppm for U. This is clearly a silly prediction; U values are never negative. Simple linear regression does not guarantee positive estimates, so where common sense dictates that the data values are always positive, it is appropriate to set negative predictions to 0, or to consider other forms of regression which that respect this constraint.

Conditional Expectation

The formulas for calculating a linear regression equation are very simple but the assumption of a straight line relationship may not be good. For example, in Figure 3.5a the regression line seems inadequate because the cloud of points has a clear bend in it.

An alternative to linear regression is to calculate the mean value of y for different ranges of x. In Table 3.3 we have calculated the mean value of V for different ranges of U. Each of our 100 $U - V$ pairs has been assigned to a certain class based on its U value, and the mean value of V has been calculated separately for each class.

If we wanted to predict an unknown V value from its corresponding U value, we could assign the new pair to the proper class based on its known U value then use the mean of all the other V values from that

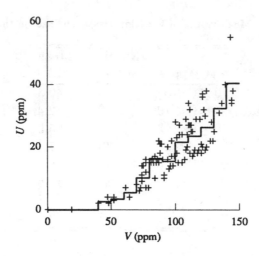

Figure 3.6 A graph of the mean values of V within classes defined on U values.

same class as our predicted value. This results in a prediction curve that looks like the one shown in Figure 3.6. The curve is discontinuous because the predicted value of V jumps to a new value whenever we cross a U class boundary.

This is a type of conditional expectation curve. Within certain classes of U values we have calculated an expected value for V. Though "expected value" has a precise probabilistic meaning, it is adequate for our purposes here to allow it to keep its colloquial meaning, "the value one expects to get." Our expected values are called *conditional* because they are good only for a certain range of U values; if we move to a different class, we expect a different value. The stair step curve shown in Figure 3.6 is obtained by moving through all the possible classes of U and calculating an expected value of V for each class.

Ideally, with a huge number of data, one would like to make a conditional expectation curve with as many classes as possible. As the number of classes increases, the width of each particular class would get narrower and the discontinuities in our conditional expectation curve would get smaller. In the limit, when we have a huge number of very narrow classes, our conditional expectation curve would be a smooth curve that would give us an expected value of V conditional to known U values. When we speak of the conditional expectation curve we are

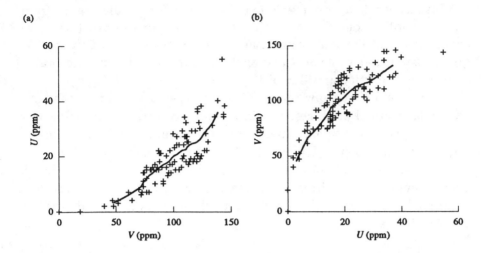

Figure 3.7 Conditional expectation curves superimposed on the scatterplot. The expected value of U given V is given in (a) and the expected value of V given U is shown in (b).

usually referring to this ideal limit. This ideal limit would serve very well as a prediction curve, being preferable to the linear regression line since it is not constrained to any assumed shape.

Regrettably, there are many practical problems with calculating such an ideal limit. From Table 3.3 we can see that if the class width was made any narrower, we would start to run out of pairs in the highest and lowest classes. As the number of pairs within each class decreases, the mean value of V from one class to the next becomes more erratic. This erraticness also increases as the correlation between the two variables gets poorer.

There are many methods for dealing with these practical complications. We have adopted one particular method for use throughout this book [4]. Whenever we present a conditional expectation curve, it will have been calculated using the method that, for the curious, is referenced in the notes at the end of this chapter.

We will not be relying on these conditional expectation curves for prediction but will be using them only as graphical summaries of the scatterplot. It will often be more informative to look at the conditional expectation curve than at the whole scatterplot.

Just as we had two regression lines, one for predicting V from U, and another one for predicting U from V, so too are there two conditional expectation curves, one that gives the expected value of V given a particular value of U and another that gives the expected value of U given a particular value of V.

In Figure 3.7 we show the conditional expectation curves that our particular method produces. It is interesting to note that for predicting V from U, the conditional expectation curve is quite different from the regression line shown in Figure 3.5b, but for the prediction of U from V, the regression line is quite close to the conditional expectation curve. Even though the conditional expectation curve is, in some sense, the ideal prediction curve, linear regression offers a very simple alternative that is often adequate.

Notes

[1] The linear coefficient of correlation given in Equation 3.1 is often referred to in the statistical literature as the Pearson correlation coefficient while the correlation coefficient of the ranks given in Equation 3.3 is often referred to as the Spearman rank correlation coefficient.

[2] All of the numbers from 1 to n appear somewhere in the set of x ranks, $Rx_1, \ldots, Rx_n$, and also in the set of y ranks, $Ry_1, \ldots, Ry_n$. For this reason, the univariate statistics of the two sets are identical. In particular, for large values of n, the values of m_{Rx} and m_{Ry} are both close to $n/2$, and the values of σ^2_{Rx} and σ^2_{Ry} are both close to $n/12$.

[3] There are many assumptions built into the theory that views this particular line as the best. Since at this point we are proposing this only as a tool for summarizing a scatterplot, we defer the discussion of these important assumptions until the second part of the book where we deal specifically with methods that aim at minimizing the variance of the estimation errors.

[4] Summarizing a scatterplot with a conditional expectation curve is often a useful way of defining a nonlinear relationship between two variables. Often the overall shape of the point cloud clearly reveals a relationship between two variables that can be more accurately described by a smooth curve drawn through the cloud

than it can by a straight line. For example, a scatterplot of $\gamma(h)$ and h (commonly called a variogram cloud), most often reveals a nonlinear relationship between $\gamma(h)$ *and h* that is best described by a smooth curve. There are a number of methods one can use for estimating the conditional expectation curves of a scatterplot; the algorithms are known generally as *smoothers*. The particular smoother we have chosen for use throughout this book is based on linear regression within a local sliding neighborhood. The algorithm provides an "optimal" neighborhood size as well as an option for curve estimation using methods resistant to extreme values. A complete description of the smoother with Fortran code is provided in: Friedman, J. H. and Stuetzle, W. , "Smoothing of Scatterplots," Tech. Rep. Project Orion 003, Department of Statistics, Stanford University, July 1982.

Further Reading

Chatterjee, S. and Price, B. , *Regression Analysis by Example*. New York: Wiley, 1977.

Cleveland, W. , "Robust locally weighted regression and smoothing scatterplots," *J. American Statistical Association*, vol. 74, pp. 828–836, 1979.

Mosteller, F. and Tukey, J. W. , *Data Analysis and Regression*. Reading, Mass.: Addison-Wesley, 1977.

Ripley, B. D. , *Spatial Statistics*. New York: Wiley, 1981.

Silverman, B. , "Some aspects of the spline smoothing approach to nonparametric regression curve fitting (with discussion)," *J. Royal Statistical Society, Series B*, vol. 47, pp. 1–52, 1985.

Tukey, J. , *Exploratory Data Analysis*. Reading, Mass.: Addison-Wesley, 1977.

4

SPATIAL DESCRIPTION

One of the things that distinguishes earth science data sets from most others is that the data belong to some location in space. Spatial features of the data set, such as the location of extreme values, the overall trend, or the degree of continuity, are often of considerable interest. None of the univariate and bivariate descriptive tools presented in the last two chapters capture these spatial features. In this chapter we will look at the spatial aspects of our 100 selected data and incorporate their location into our description.

Data Postings

As with the histogram from Chapter 2 and the scatterplot from Chapter 3, our most effective tools for spatial description are visual ones. The simplest display of spatial data is a data posting, a map on which each data location is plotted along with its corresponding data value. Figure 2.1 was a data posting of the V values; Figure 3.1 added the U values.

Postings of the data are an important initial step in analyzing spatial data sets. Not only do they reveal obvious errors in the data locations, but they often also draw attention to data values that may be erroneous. Lone high values surrounded by low values and vice versa are worth rechecking. With irregularly gridded data, a data posting often gives clues to how the data were collected. Blank areas on the

(a)   (b)

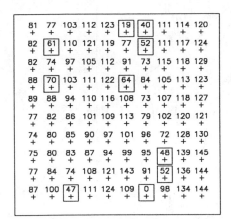

Figure 4.1   Location of the lowest V values in (a), and the highest values in (b).

map may have been inaccessible; heavily sampled areas indicate some initial interest.

Locating the highest and lowest values on a posting of the data may reveal some trends in the data. On Figure 4.1a we have highlighted the 10 lowest values from Figure 2.1. Seven values, 19, 40, 52, 64, 48, 52, and 0 are located in a north-south trough that runs through the area. On Figure 4.1b, a similar display for the 10 highest values, no obvious trend is apparent, with the highest values appearing in the southeast corner.

Contour Maps

The overall trends in the data values can be revealed by a contour map. Contouring by hand is an excellent way to become familiar with a data set. Unfortunately, the size of many data sets makes automatic contouring with a computer an attractive alternative. The contour map of the V values shown in Figure 4.2 was generated by computer.

At this preliminary descriptive stage the details of the contouring algorithm need not concern us as long as the contour map provides a helpful visual display. There are many algorithms that provide an adequate interpolation of the data values. A good automatic contouring

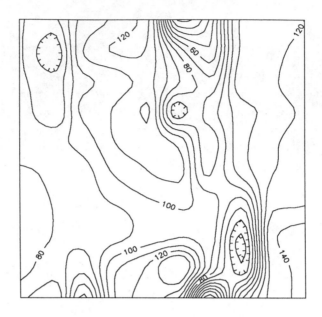

Figure 4.2 Computer generated contour map of 100 selected V data. The contour lines are at intervals of 10 ppm and range from 0 to 140 ppm.

program will also pay attention to aesthetic details such as the use of downhill tick marks to show depressions.

Some of the features we noticed earlier on the data posting become clearer when contoured. The north-south trough is readily apparent, as are the local maximums. Also, some features that were not obvious from the data posting alone are now more prominent. The closeness of the contour lines in the southeastern corner indicates a steep gradient and draws our attention to the fact that the highest data value (145 ppm) is very close to the lowest data value (0 ppm).

Automatic contouring of irregularly gridded data usually requires the data values to be interpolated to a regular grid. Interpolated values are usually less variable than the original data values and make the contoured surface appear smoother. This is an aesthetic asset, but a smoother surface understates the variability and may be misleading from a quantitative point of view. In this book we treat our contour maps as we treat our conditional expectation curves: as helpful qualitative displays with questionable quantitative significance.

5	5	6	7	8	1	2	7	7	8		0 = 0-14 ppm
5	4	7	8	7	5	3	7	7	8		1 = 15-29
5	4	6	7	7	6	4	7	7	8		2 = 33-44
5	4	6	7	8	4	5	7	7	8		3 = 45-59
5	5	6	7	7	7	4	7	7	8		4 = 60-74
5	5	5	6	7	7	5	6	8	8		5 = 75-89
4	5	5	6	6	6	6	4	8	8		6 = 90-104
5	5	5	5	6	6	6	3	9	9		7 = 105-119
5	5	4	7	8	9	6	3	9	9		8 = 120-134
5	6	3	7	8	7	0	6	8	9		9 = 135-149

Figure 4.3 Symbol map of 100 selected V data. Each symbol represents a class of data values as indicated by the legend on the right-hand side of the figure.

Symbol Maps

For many very large regularly gridded data sets, a posting of all the data values may not be feasible, and a contour map may mask many of the interesting local details. An alternative that is often used in such situations is a symbol map. This is similar to a data posting with each location replaced by a symbol that denotes the class to which the data value belongs. These symbols are usually chosen so that they convey the relative ordering of the classes by their visual density. This type of display is especially convenient if one has access to a line printer but not to a plotting device. Unfortunately, the scale on symbol maps is usually distorted since most line printers do not print the same number of characters per inch horizontally as they do vertically.

For a data set as small as our 10 x 10 m^2 grid, a symbol map is probably not necessary since the actual data values are easy to post. In order to show a simple example, however, we present a symbol map in Figure 4.3 that corresponds to the posting from Figure 2.1. In this display we have used the digits 0 through 9 to denote which of the ten classes the V value at each location belongs. An alternative to the symbol map is a grayscale map. In such a map the symbols have been replaced with a suitable shade of grey as shown in Figure 4.4. These maps are much more pleasing to the eye and provide an excellent visual summary of the data [1].

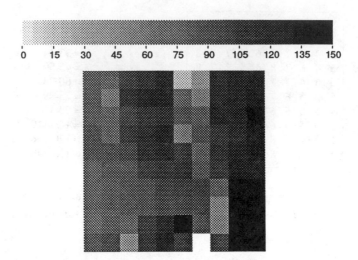

Figure 4.4 Grayscale map of 100 selected V data. The value of each V datum is indicated by its shade of grey as shown by the scale at the top of the figure.

Indicator Maps

An indicator map is a symbol map on which there are only two symbols; in our examples here we use a black box and a white box. With only two symbols one can assign each data point to one of only two classes, so an indicator map simply records where the data values are above a certain threshold and where they are below. Though this may seem at first to be rather restrictive, a series of indicator maps is often very informative. They share the advantage with all symbol maps that they show more detail than a contour map and that they avoid the difficulty of distinguishing between symbols that exist with conventional symbol maps.

In Figures 4.5a-i we show a series of nine indicator maps corresponding to the nine class boundaries from our symbol map in Figure 4.3. Each map shows in white the data locations at which the V value is less than the given threshold and in black the locations at which V is greater than or equal to the threshold. This series of indicator maps records the transition from low values that tend to be aligned in a north-south direction to high values that tend to be grouped in the southeast corner. Different indicator maps give good views of dif-

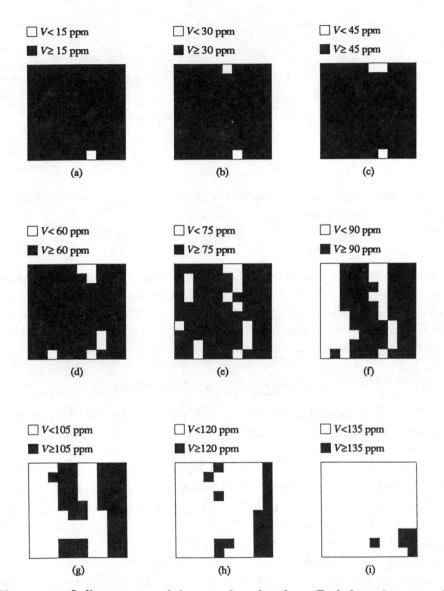

Figure 4.5 Indicator maps of the 100 selected V data. Each figure is a map of indicators defined using the indicated cutoff values. The pattern of indicators in the sequence of maps provides a detailed spatial description of the data. For example, the indicator map defined at 75 ppm reveals the trough of low values seen earlier in the contour and symbol maps of Figures 4.2 and 4.3.

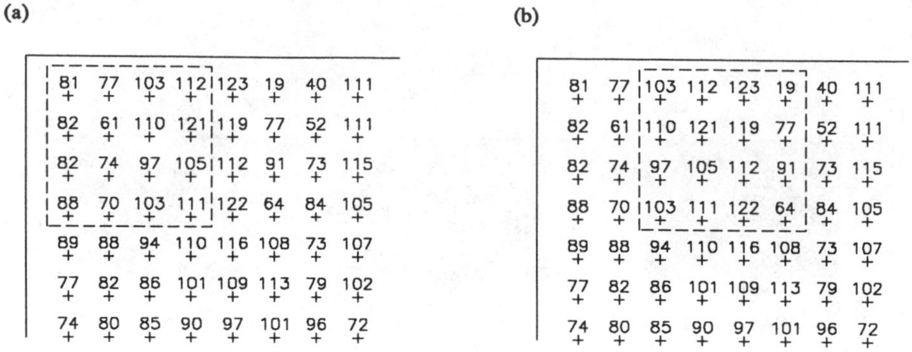

(a)

```
┌─────────────┬──────┐
│ 81  77 103 112│123  19│ 40 111
│ +   +   +   + │ +   + │ +   +
│ 82  61 110 121│119  77│ 52 111
│ +   +   +   + │ +   + │ +   +
│ 82  74  97 105│112  91│ 73 115
│ +   +   +   + │ +   + │ +   +
│ 88  70 103 111│122  64│ 84 105
└──┬──────────┬─┴──────┘ +   +
   89  88  94 110 116 108  73 107
   +   +   +   +   +   +   +   +

   77  82  86 101 109 113  79 102
   +   +   +   +   +   +   +   +

   74  80  85  90  97 101  96  72
   +   +   +   +   +   +   +   +
```

(b)

```
   81  77 ┌─────────────┐ 40 111
   +   +  │103 112 123  19│ +   +
   82  61 │110 121 119  77│ 52 111
   +   +  │ +   +   +   + │ +   +
   82  74 │ 97 105 112  91│ 73 115
   +   +  │ +   +   +   + │ +   +
   88  70 │103 111 122  64│ 84 105
   +   +  └─────────────┘ +   +
   89  88  94 110 116 108  73 107
   +   +   +   +   +   +   +   +

   77  82  86 101 109 113  79 102
   +   +   +   +   +   +   +   +

   74  80  85  90  97 101  96  72
   +   +   +   +   +   +   +   +
```

Figure 4.6 Example of overlapping moving windows for purposes of calculating moving average statistics.

ferent spatial features. Figure 4.5f, for example, gives the best image of the north-south trough we noticed earlier, while Figure 4.5g gives the best image of the local maximums. The usefulness of these maps will become more apparent in Chapter 5 where a series of indicator maps are used to explore several large data sets.

Moving Window Statistics

In the analysis of earth science data sets one is often most interested in the anomalies (i.e., the high grade veins in a gold deposit) or the impermeable layers that condition flow in a petroleum reservoir. A contour map will help locate areas in which the average value is anomalous, but anomalies in the average value are not the only interesting ones.

It is quite common to find that the data values in some regions are more variable than in others. The statistical jargon for such anomalies in the variability is *heteroscedasticity*. Such anomalies may have serious practical implications. In a mine, very erratic ore grades often cause problems at the mill because most metallurgical processes benefit from low variability in the ore grade. In a petroleum reservoir, large fluctuations in the permeability can hamper the effectiveness of many secondary recovery processes.

The calculation of a few summary statistics within moving windows is frequently used to investigate anomalies both in the average value and in the variability. The area is divided into several local neighbor-

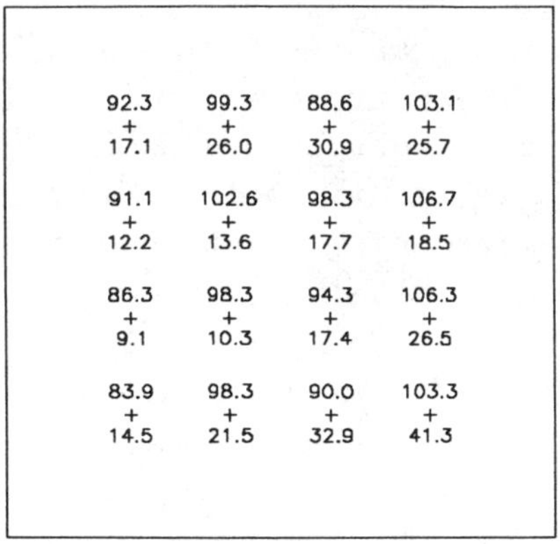

```
92.3    99.3    88.6    103.1
 +       +       +        +
17.1    26.0    30.9     25.7

91.1   102.6    98.3    106.7
 +       +       +        +
12.2    13.6    17.7     18.5

86.3    98.3    94.3    106.3
 +       +       +        +
 9.1    10.3    17.4     26.5

83.9    98.3    90.0    103.3
 +       +       +        +
14.5    21.5    32.9     41.3
```

Figure 4.7 Posting of statistics obtained from moving windows on the 100 *V* data. The mean of each moving window is plotted above the "+", and the standard deviation below.

hoods of equal size and within each local neighborhood, or window, summary statistics are calculated.

Rectangular windows are commonly used, largely for reasons of computational efficiency. The size of the window depends on the average spacing between data locations and on the overall dimensions of the area being studied [2]. We want to have enough data within each neighborhood to calculate reliable summary statistics. If we make our windows too large, however, we will not have enough of them to identify anomalous localities.

Needing large windows for reliable statistics and wanting small windows for local detail may leave little middle ground. A good compromise is often found in overlapping the windows, with two adjacent neighborhoods having some data in common.

In Figure 4.6 we show an example of an overlapping moving window calculation. We have chosen to use a 4 x 4 m^2 window so that we will have 16 data in each local neighborhood. By moving the window only 2 m each time so that it overlaps half of the previous window, we can fit 16 such windows into our 10 x 10 m^2 area. Had we not allowed the windows to overlap, we would have had only four separate local neighborhoods.

For large regularly gridded data sets, overlapping windows are usu-

ally not necessary. For smaller data sets or for ones in which the data
are irregularly spaced, overlapping becomes a useful trick. With irregularly spaced data it is also important to decide how many data will
be required within each window for a reliable calculation of the summary statistics. If there are too few data within a particular window,
it is often better to ignore that window in subsequent analysis than to
incorporate an unreliable statistic.

With enough data in any window, one can calculate any of the
summary statistics we have previously discussed. The mean and the
standard deviation are commonly used, with one providing a measure
of the average value and the other a measure of the variability. If the
local means are heavily influenced by a few erratic high values, one
could use the median and interquartile range instead.

The means and standard deviations within 4 x 4 m^2 windows are
shown in Figure 4.7. As shown in Figure 4.6 and described earlier,
the windows overlap each other by 2 m, giving us a total of 16 local
neighborhood means and standard deviations. We have posted these
values in Figure 4.7, where the center of each window is marked with a
plus sign. The mean of each window is plotted above the + sign while
the standard deviation is plotted below. If we had a larger area and
many more local neighborhoods, a more informative display would be
two contour maps, with one showing the means and the other showing
the standard deviations.

From this posting of moving window means and standard deviations we can see that both the average value and the variability change
locally across the area. The windows with a high average value correspond to the highs we can see on the contour map (Figure 4.2). The
local changes in variability, however, have not been captured by any of
our previous tools. In the southeastern corner we see the highest local
standard deviations, a result of the very low values in the trough being
adjacent to some of the highest values in the entire area. The very
low standard deviation on the western edge reflects the very uniform
V values in that region.

It is interesting to note in this example that the standard deviations
vary much more across the area than the means. It is often tempting
to conclude that uniformity in the local means indicates generally well
behaved data values. Here we see that even though the mean values are
quite similar, ranging from 83.9 to 106.7 ppm, the standard deviations
can be quite different, ranging from 9.1 to 41.3 ppm.

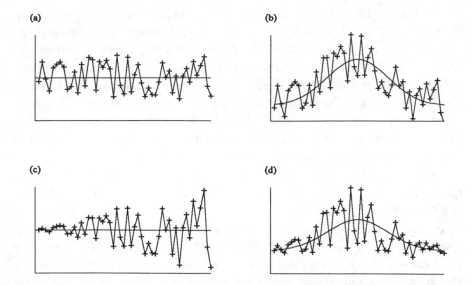

Figure 4.8 Hypothetical profiles of data values illustrating common relationships between the local mean and local variability. In (a) the local mean, represented by the straight line, and the variability are both constant. Case (b) shows a trend in the local mean while the variability remains constant. Case (c) exhibits a constant local mean while the variability contains a trend and case (d) illustrates a trend in both the local mean and the variability.

Proportional Effect

When we look at estimation in later chapters, anomalies in the local variability will have an impact on the accuracy of our estimates. If we are in an area with very uniform values, the prospects for accurate estimates are quite good. On the other hand, if the data values fluctuate wildly our chances for accurate local estimates are poor. This has nothing to do with the estimation method we choose to use; the estimates from *any* reasonable method will benefit from low variability and suffer from high variability.

In a broad sense there are four relationships one can observe between the local average and the local variability. These are shown in Figures 4.8a-d, which represent hypothetical profiles of the data values. On each profile the line that connects the plus signs represents the actual data values; the smoother line represents the local average.

In Figure 4.8a the average and the variability are both constant. The data values fluctuate about the local average, but there is no obvious change in the variability. In Figure 4.8b, there is a trend to the local average; it rises gradually then falls. The variability, however, is still roughly constant. In Figure 4.8c, we see the reverse case, where the local average is constant while the variability changes. The most common case for earth science data is shown in Figure 4.8d, where the local average and variability both change together. As the local average increases, so, too, does the local variability.

For estimation, the first two cases are the most favorable. If the local variability is roughly constant, then estimates in any particular area will be as good as estimates elsewhere; no area will suffer more than others from highly variable data values. It is more likely, however, that the variability does change noticeably. In such a case, it is preferable to be in a situation like the one shown in Figure 4.8d, where the local variability is related to the local average and is, therefore, somewhat predictable. It is useful, therefore, to establish in the initial data analysis if such a predictable relationship does exist.

A scatterplot of the local means and the local standard deviations from our moving window calculations is a good way to check for a relationship between the two. If it exists, such a relationship is generally referred to as a *proportional effect*. One of the characteristics of normally distributed values is that there is usually no proportional effect; in fact, the local standard deviations are roughly constant [3]. For lognormally distributed values, a scatterplot of local means versus local standard deviations will show a linear relationship between the two.

Figure 4.9 shows a scatterplot of the local means versus the local standard deviations from the 16 local neighborhoods shown in Figure 4.7. The correlation coefficient is only 0.27, which is quite low. There is no apparent relationship between the mean and the standard deviation for our 100 selected values. In the next chapter, where we analyze the complete exhaustive data set, we will see that this is generally not the case.

Spatial Continuity

Spatial continuity exists in most earth science data sets. Two data close to each other are more likely to have similar values than two

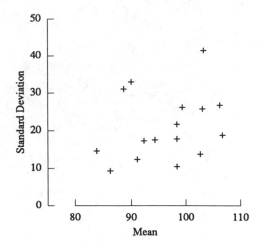

Figure 4.9 Plot of the standard deviation versus the mean obtained from moving windows on the 100 V data.

data that are far apart. When we look at a data posting or a contour map, the values do not appear to be randomly located, but rather, low values tend to be near other low values and high values tend to be near other high values. Often we are interested in zones of anomalously high values where the tendency of high values to be near other high values is very obvious. Indeed, as we remarked earlier, a single very low value surrounded by high ones usually raises our suspicion.

In the last chapter we gave an example in Figure 3.4 where we relied on the strong observed relationship between two variables to detect errors in the data. When we see a solitary extreme value on a map, our intuition warns us that it may be in error because it shows an unusual relationship with the other values. This same intuition drew our attention to the strange point on Figure 3.4.

The tools we used to describe the relationship between two variables can also be used to describe the relationship between the value of one variable and the value of the same variable at nearby locations. We will use the scatterplot to display spatial continuity and, as we shall see, the same statistics we used in the last chapter to summarize the relationship between two variables can also be used to summarize spatial continuity. We will also introduce several new tools that describe spatial continuity.

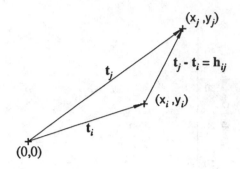

Figure 4.10   An illustration of the vector notation used in the text.

h-Scatterplots

An h-scatterplot shows all possible pairs of data values whose locations
are separated by a certain distance in a particular direction. The
notation we use to describe such pairs requires some explanation. This
is frequently a stumbling block for newcomers to geostatistics so we
have tried to limit the confusion between similar symbols for different
concepts by taking advantage of different fonts or letter styles.

The letters x and y are used to refer to coordinates on a graphical
display. For example, in the last chapter we wrote the equation of a
regression line as $y = ax + b$, and we understood that y meant the
value of the unknown variable and x meant the value of the known
variable. The letters x and y will be used to refer to coordinates that
have spatial significance. For example, the 10 x 10 m^2 grid of data
values that we have been using in these chapters to present descriptive
tools includes all values from x = 11 E to x = 20 E and from y = 241 N
to y = 250 N; here, the x and y are actual eastings and northings.

The location of any point can be described by a vector, as can the
separation between any two points. When describing pairs of values
separated by a certain distance in a particular direction, it is convenient
to use vector notation. In the diagram in Figure 4.10, the location of
the point at (x_i, y_i) can be written as t_i, with the bold font of the t
reminding us that it is a vector. Similarly, the location of the point
at (x_j, y_j) can be written as t_j. The separation between point i and
point j is $t_j - t_i$, which can also be expressed as the coordinate pair
$(x_j - x_i, y_j - y_i)$. Sometimes it will be important to distinguish between

(a)   (b)

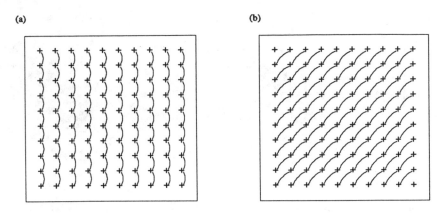

Figure 4.11   Examples of how data can be paired to obtain an h-scatterplot.

the vector going from point i to j and the vector from j to i. We will use the symbol $\mathbf{h}_{ij}$ to refer to the vector going from point i to point j and $\mathbf{h}_{ji}$ for the vector from point j to i.

On our h-scatterplots, the x-axis is labeled $V(\mathbf{t})$ and the y-axis is labeled $V(\mathbf{t} + \mathbf{h})$. The x-coordinate of a point corresponds to the V value at a particular location and the y-coordinate to the V value a distance and direction $\mathbf{h}$ away. If, for example, $\mathbf{h}=(0,1)$ it means that we have taken each data location and paired it with the data location whose easting is the same and whose northing is 1 m larger (i.e., the data location 1 m to the north). Figure 4.11a shows the 90 pairs of data locations separated by exactly 1 m in a northerly direction for our 10 x 10 m^2 area. To take another example, if $\mathbf{h}=(1,1)$ then each data location has been paired with the data location whose easting is 1 m larger and whose northing is also 1 m larger (i.e., the data location at a distance of $\sqrt{2}$ m in a northeasterly direction). Figure 4.11b shows the 81 pairs of data locations separated by $\sqrt{2}$ m in a northeasterly direction for our 10 x 10 m^2 area.

The shape of the cloud of points on an h-scatterplot tells us how continuous the data values are over a certain distance in a particular direction. If the data values at locations separated by $\mathbf{h}$ are very similar then the pairs will plot close to the line $x = y$, a 45-degree line passing through the origin. As the data values become less similar, the cloud of points on the h-scatterplot becomes fatter and more diffuse.

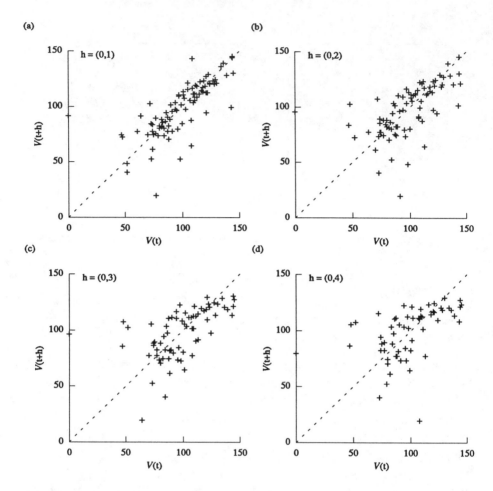

Figure 4.12 h-Scatterplots for four separation distances in a northerly direction between pairs of the 100 *V* values. As the separation distance increases, the similarity between pairs of values decreases and the points on the h-scatterplot spread out further from the diagonal line.

For our 100 selected *V* values the change in continuity in a northerly direction is captured in Figures 4.12a–d, which show the four h-scatterplots for data values located 1 m apart to 4 m apart in a northerly direction. Looking at these four h-scatterplots we see that from **h**=(0,1) to **h**=(0,4) the cloud gets progressively fatter as the points spread out, away from the 45-degree line. Although the 45-

degree line passes through the cloud of points on each h-scatterplot, the cloud is not symmetric about this line. Direction is important in constructing an h-scatterplot; a pair of data values, v_i and v_j, appears on the h-scatterplot only once as the point (v_i, v_j) and not again as (v_j, v_i) [4].

On our four h-scatterplots in Figure 4.12 we notice that there are some points that do not plot close to the rest of the cloud. For example, in each of the scatterplots there is one point that plots near the bottom of the scatterplot. In each case this point involves the 19 ppm V value located on the northern edge of our 10 x 10 m^2 area. In Figure 4.12a, this value is paired with the 77 ppm value immediately to the south accounting for the unusual point which plots far away from the 45-degree line. On the other h-scatterplots in Figure 4.12, the 19 ppm value is paired with the 91 ppm value in (b), the 64 ppm value in (c), and the 108 ppm value in (d).

Whenever we summarize h-scatterplots we should be aware that our summary statistics may be influenced considerably by a few aberrant data values. It is often worth checking how the summary statistics change if certain data are removed.

Correlation Functions, Covariance Functions, and Variograms

As with our other graphical displays, we often need some quantitative summary of the information contained on an h-scatterplot. As we pointed out in the last section, one of the essential features of an h-scatterplot is the fatness of the cloud of points. We already have a way of summarizing this feature—the *correlation coefficient*. As the cloud of points gets fatter, we expect the correlation coefficient to decrease. The correlation coefficients for each of the four scatterplots in Figures 4.12a-d are given in Table 4.1. As we expected, the correlation coefficient steadily decreases and is therefore a useful index for our earlier impression that the cloud of points was getting fatter.

The relationship between the correlation coefficient of an h-scatterplot and **h** is called the *correlation function* or *correlogram*. The correlation coefficient depends on **h** which, being a vector, has both a magnitude and a direction. To graphically display the correlation function, one could use a contour map showing the correlation coefficient of the h-scatterplot as a function of both magnitude and direction.

Table 4.1 Statistics summarizing the fatness of the four h-scatterplots shown in Figures 4.12a-d.

h	Correlation Coefficient	Covariance (ppm^2)	Moment of Inertia (ppm^2)
(0,1)	0.742	448.8	312.8
(0,2)	0.590	341.0	479.2
(0,3)	0.560	323.8	521.4
(0,4)	0.478	291.5	652.9

Though this provides a complete and effective display of $\rho(\mathbf{h})$, it is not a traditional format. Instead, we usually plot separate graphs of the correlation function versus the magnitude of $\mathbf{h}$ for various directions.

In Figure 4.13a we use the data from Table 4.1 to show how the correlation coefficient decreases with increasing distance in a northerly direction. Similar plots for other directions would give us a good impression of how the correlation coefficient varies as a function of both separation distance and direction.

An alternative index for spatial continuity is the covariance. The covariance for each of our h-scatterplots is also given in Table 4.1. We can see that these also steadily decrease in a manner very similar to the correlation coefficient [5]. The relationship between the covariance of an h-scatterplot and $\mathbf{h}$ is called the *covariance function* [6]. Figure 4.13b shows the covariance function in a northerly direction.

Another plausible index for the fatness of the cloud is the moment of inertia about the line $x = y$, which can be calculated from the following:

$$moment\ of\ inertia = \frac{1}{2n} \sum_{i=1}^{n} (x_i - y_i)^2 \qquad (4.1)$$

It is half of the average squared difference between the x and y coordinates of each pair of points on the h-scatterplot, the factor $\frac{1}{2}$ being a consequence of the fact that we are interested in the perpendicular distance of the points from the 45-degree line.

This is a summary statistic we did not consider when we first looked

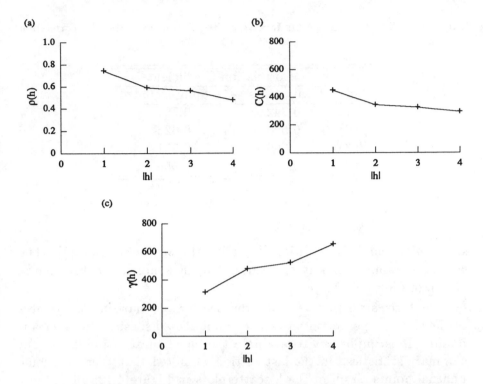

Figure 4.13 Plots of (a) the correlation coefficient, (b) the covariance, and (c) the moment of inertia given in Table 4.1.

at scatterplots in Chapter 3. Though the moment of inertia about the line $x = y$ can be calculated for any scatterplot, it usually has no particular relevance since there is usually no special significance to the 45-degree line. On an h-scatterplot, however, this line does have special significance because we are pairing values of the same variable with each other. All of the points on the h-scatterplot for $h = (0,0)$ will fall exactly on the line $x = y$ since each value will be paired with itself. As $|h|$ increases, the points will drift away from this line and the moment of inertia about the 45-degree line is therefore a natural measure of the fatness of the cloud.

Unlike the other two indices of spatial continuity, the moment of inertia increases as the cloud gets fatter. In Table 4.1 we see that the moment of inertia increases as the correlation coefficient and covariance decrease. The relationship between the moment of inertia of an h-

Table 4.2   Effect of the points involving the 19 ppm value on the correlation coefficient.

	Correlation Coefficient	
h	All points	19 ppm excluded
(0,1)	0.742	0.761
(0,2)	0.590	0.625
(0,3)	0.560	0.551
(0,4)	0.478	0.559

scatterplot and **h** is traditionally called the *semivariogram* [7]. The semivariogram, or simply the variogram, in a northerly direction is shown in Figure 4.13c.

The three statistics we have proposed for summarizing the fatness of the cloud of points on an h-scatterplot are all sensitive to aberrant points. It is important to evaluate the effect of such points on our summary statistics. In the last section we noted that there are some unusual points on all of the h-scatterplots in Figure 4.12; all of these points involve the 19 ppm value on the northern edge of the area. In Table 4.2 we compare the correlation coefficients from Table 4.1 to the correlation coefficients calculated with the 19 ppm value removed.

Table 4.2 shows that a single erratic value can have a significant impact on the correlation coefficient of an h-scatterplot. The covariance and the moment of inertia are similarly affected. The sensitivity of our summary statistics to aberrant points requires that we pay careful attention to the effect of any erratic values. In practice, the correlation function, the covariance function, and the variogram often do not clearly describe the spatial continuity because of a few unusual values. If the shape of any of these functions is not well defined it is worth examining the appropriate h-scatterplots to determine if a few points are having an undue effect.

Although the h-scatterplots contain much more information than any of the three summary statistics we have described, it is quite common to bypass the actual h-scatterplots and go directly to either $\rho(\mathbf{h})$, $C(\mathbf{h})$ or $\gamma(\mathbf{h})$ to describe spatial continuity. It is convenient, therefore, to have the formulas for these functions expressed directly in terms of the data values rather than in terms of x and y coordinates on an h-scatterplot.

The formulas we present below are the same as the ones we presented in Equation 3.1 and Equation 4.1. They have been rewritten directly in terms of the data values and rearranged to reflect the form in which they are usually computed.

The covariance function, $C(\mathbf{h})$, can be calculated from the following [8]:

$$C(\mathbf{h}) = \frac{1}{N(\mathbf{h})} \sum_{(i,j)|\mathbf{h}_{ij}=\mathbf{h}} v_i \cdot v_j \; - \; m_{-\mathbf{h}} \cdot m_{+\mathbf{h}} \qquad (4.2)$$

The data values are $v_1, \ldots, v_n$; the summation is over only the $N(\mathbf{h})$ pairs of data whose locations are separated by $\mathbf{h}$. $m_{-\mathbf{h}}$ is the mean of all the data values whose locations are $-\mathbf{h}$ away from some other data location:

$$m_{-\mathbf{h}} = \frac{1}{N(\mathbf{h})} \sum_{i|\mathbf{h}_{ij}=\mathbf{h}} v_i \qquad (4.3)$$

$m_{+\mathbf{h}}$ is the mean of all the data values whose locations are $+\mathbf{h}$ away from some other data location:

$$m_{+\mathbf{h}} = \frac{1}{N(\mathbf{h})} \sum_{j|\mathbf{h}_{ij}=\mathbf{h}} v_j \qquad (4.4)$$

The values of $m_{-\mathbf{h}}$ and $m_{+\mathbf{h}}$ are generally not equal in practice.

The correlation function, $\rho(\mathbf{h})$, is the covariance function standardized by the appropriate standard deviations:

$$\rho(\mathbf{h}) = \frac{C(\mathbf{h})}{\sigma_{-\mathbf{h}} \cdot \sigma_{+\mathbf{h}}} \qquad (4.5)$$

$\sigma_{-\mathbf{h}}$ is the standard deviation of all the data values whose locations are $-\mathbf{h}$ away from some other data location:

$$\sigma_{-\mathbf{h}}^2 = \frac{1}{N(\mathbf{h})} \sum_{i|\mathbf{h}_{ij}=\mathbf{h}} v_i^2 \; - \; m_{-\mathbf{h}}^2 \qquad (4.6)$$

$\sigma_{+\mathbf{h}}$ is the standard deviation of all the data values whose locations are $+\mathbf{h}$ away from some other data location:

$$\sigma_{+\mathbf{h}}^2 = \frac{1}{N(\mathbf{h})} \sum_{j|\mathbf{h}_{ij}=\mathbf{h}} v_j^2 \; - \; m_{+\mathbf{h}}^2 \qquad (4.7)$$

Like the means, the standard deviations, σ_{-h} and σ_{+h}, are usually not equal in practice.

The variogram, $\gamma(\mathbf{h})$, is half the average squared difference between the paired data values:

$$\gamma(\mathbf{h}) = \frac{1}{2N(\mathbf{h})} \sum_{(i,j)|\mathbf{h}_{i,j}=\mathbf{h}} (v_i - v_j)^2 \qquad (4.8)$$

The values of $\rho(\mathbf{h})$, $C(\mathbf{h})$ and $\gamma(\mathbf{h})$ are unaffected if we switch all of the i and j subscripts in the preceding equations. For example, Equation 4.8 would become

$$\gamma(\mathbf{h}) = \frac{1}{2N(\mathbf{h})} \sum_{(j,i)|\mathbf{h}_{ji}=\mathbf{h}} (v_j - v_i)^2 \qquad (4.9)$$

Instead of summing over all (j, i) pairs that are separated by $\mathbf{h}$, we could sum over all (i, j) pairs that are separated by $-\mathbf{h}$ and Equation 4.9 would become

$$\gamma(\mathbf{h}) = \frac{1}{2N(\mathbf{h})} \sum_{(i,j)|\mathbf{h}_{ij}=-\mathbf{h}} (v_i - v_j)^2 \qquad (4.10)$$

The right-hand side is equal to $\gamma(-\mathbf{h})$, so we have the result that

$$\gamma(\mathbf{h}) = \gamma(-\mathbf{h}) \qquad (4.11)$$

This result entails that the variogram calculated for any particular direction will be identical to the variogram calculated in the opposite direction. The correlation function and the covariance function share this property. For this reason we commonly combine opposite directions when we are describing spatial continuity. For example, rather than speak of the spatial continuity in the northerly direction, as we have been doing so far, it is more common to speak of spatial continuity in the north-south direction, since any of our summary statistics will have the same values in the northerly direction as in the southerly.

Cross h-Scatterplots

We can extend the idea of an h-scatterplot to that of a cross h-scatterplot. Instead of pairing the value of one variable with the value of the same variable at another location, we can pair values of different variables at different locations.

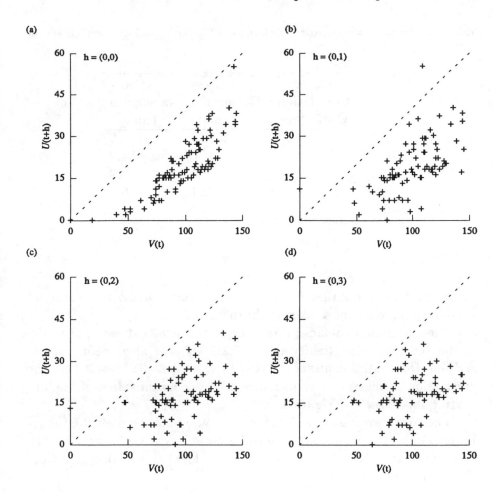

Figure 4.14 Cross h-scatterplots of the 100 V and U values at various separation distances in the north-south direction.

In Figure 4.14 we show the cross h-scatterplot between V and U for various values of $\mathbf{h}$. The x-coordinate of each point is the V value at a particular data location and the y-coordinate is the U value at a separation distance $|\mathbf{h}|$ to the north. The scatterplot of V values versus the U values that we saw in Figure 3.4a can be thought of as a cross h-scatterplot for $\mathbf{h} = (0,0)$. The x-coordinate of each point corresponded to the V value at a particular location and the y-coordinate to the U value at the same location. A comparison between the four scatterplots

Table 4.3 Statistics summarizing the fatness of the four cross h-scatterplots shown in Figures 4.14a-d.

h	Cross Correlation Coefficient	Cross Covariance (ppm^2)	Cross variogram (ppm^2)
(0,0)	0.84	218.3	0.0
(0,1)	0.60	144.0	54.2
(0,2)	0.45	94.2	80.7
(0,3)	0.36	73.1	89.5
(0,4)	0.28	60.1	111.0

in Figure 4.14 shows that the relationship between the two variables becomes progressively weaker as |h| increases.

The correlation coefficient and the covariance that we used to describe the spatial continuity of one variable, are also useful for describing the spatial continuity between variables. Table 4.3 lists the correlation coefficient, the covariance, and variogram values of the four scatterplots shown in Figure 4.14.

The cross-covariance and cross-correlation function describe the relationship between these statistics of a cross h-scatterplot and h. The cross-covariance function between two variables can be calculated from the following equation:

$$C_{uv}(\mathbf{h}) = \frac{1}{N(\mathbf{h})} \sum_{(i,j)|\mathbf{h}_{ij}=\mathbf{h}} u_i \cdot v_j \;-\; m_{u_{-\mathbf{h}}} \cdot m_{v_{+\mathbf{h}}} \qquad (4.12)$$

The data values of the first variable are $u_1, \ldots, u_n$ and the data values of the second variable are $v_1, \ldots, v_n$. As in Equation 4.2, the summation is over only those pairs of data whose locations are separated by h. $m_{u_{-\mathbf{h}}}$ is the mean value of the first variable over those data locations which are $-\mathbf{h}$ away from some other v - type data location:

$$m_{u_{-\mathbf{h}}} = \frac{1}{N(\mathbf{h})} \sum_{i|\mathbf{h}_{ij}=\mathbf{h}} u_i \qquad (4.13)$$

$m_{v_{+\mathbf{h}}}$ is the mean value of the second variable over those locations that

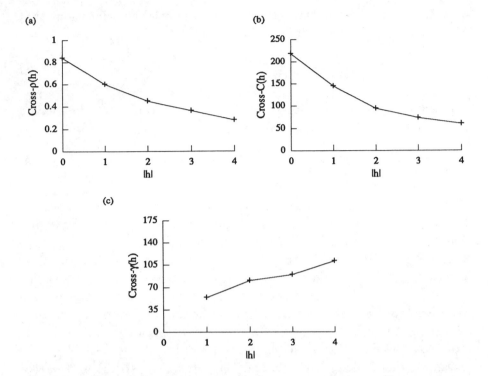

Figure 4.15 The cross-correlation, cross-covariance, and cross-variogram functions for the 100 selected V and U values are shown in (a), (b), and (c), respectively.

are $+\mathbf{h}$ away from some other u - type data location:

$$m_{v_{+\mathbf{h}}} = \frac{1}{N(\mathbf{h})} \sum_{j|\mathbf{h}_{ij}=\mathbf{h}} v_j \qquad (4.14)$$

The cross-correlation function is given by the equation

$$\rho_{uv}(\mathbf{h}) = \frac{C_{uv}(\mathbf{h})}{\sigma_{u_{-\mathbf{h}}} \cdot \sigma_{v_{+\mathbf{h}}}} \qquad (4.15)$$

$\sigma_{u_{-\mathbf{h}}}$ is the standard deviation of the first variable at locations that are $-\mathbf{h}$ away from some other data location:

$$\sigma^2_{u_{-\mathbf{h}}} = \frac{1}{N(\mathbf{h})} \sum_{i|\mathbf{h}_{ij}=\mathbf{h}} u_i^2 \; - \; m^2_{u_{-\mathbf{h}}} \qquad (4.16)$$

$\sigma_{v_{+\mathbf{h}}}$ is the standard deviation of the second variable at locations that are $+\mathbf{h}$ away from some other data location:

$$\sigma_{v_{+\mathbf{h}}}^2 = \frac{1}{N(\mathbf{h})} \sum_{j|\mathbf{h}_{ij}=\mathbf{h}} v_j^2 \; - \; m_{v_{+\mathbf{h}}}^2 \qquad (4.17)$$

Equation 4.8, which we used to define the variogram, can be extended to a cross-variogram equation:

$$\gamma_{uv}(\mathbf{h}) = \frac{1}{2N(\mathbf{h})} \sum_{(i,j)|\mathbf{h}_{ij}=\mathbf{h}} (u_i - u_j) \cdot (v_i - v_j) \qquad (4.18)$$

This is no longer the moment of inertia about the line $x = y$, since the line $x = y$ now has no particular significance. On a cross h-scatterplot there is no reason for pairs of values to plot on this line.

In Figure 4.15a-c, we show the cross correlation function, the cross covariance function, and the cross variogram between our 100 selected V and U values. The cross correlation and the cross covariance functions will not be the same if we calculate them in the opposite direction. Reversing the x and y values on a cross h-scatterplot entails switching not only the direction of $\mathbf{h}$ but also the order of the variables. For example, this means that although $C_{uv}(\mathbf{h}) \neq C_{uv}(-\mathbf{h})$, it is true that $C_{uv}(\mathbf{h}) = C_{vu}(-\mathbf{h})$. The cross variogram, however, is the same if we reverse the direction; that is $\gamma_{uv}(\mathbf{h}) = \gamma_{uv}(-\mathbf{h})$.

Notes

[1] Even without a printer capable of printing grayscale maps, it is possible to accomplish a similar effect with a symbol map. For large maps, symbols whose visual densities are noticeably different, such as ., !, +, and #, can provide a better impression of the class ordering than can numerical symbols.

[2] The size of the moving window used to compute moving average statistics will also depend on the coefficient of variation of the data. If the coefficient of variation is very large, more sample values will be required to obtain reliable statistics. If the coefficient of variation is greater than 1, for example, perhaps as many as 20 to 50 values per window may be required.

[3] Univariate normally distributed data can be spatially arranged so that a proportional effect does exist. If the data are multivariate

normal, however, then a proportional effect cannot exist since the local standard deviation is independent of the local mean.

[4] Often, h-scatterplots are plotted in their symmetric form where the pair (v_i, v_j) is plotted twice, once as the point (v_i, v_j), and once as the point (v_j, v_i). This double plotting of one pair symmetrizes the h-scatterplot and makes it more difficult to identify aberrant pairs.

[5] The shape of the correlation function is not strictly identical to that of the covariance function since the standard deviations of $V(\mathbf{t})$ and $V(\mathbf{t+h})$ change from one h-scatterplot to the next.

[6] The functions we defined in Equations 4.2 and 4.5 are often referred to more strictly as autocovariance and autocorrelation functions, signifying that the values of one variable have been compared to the values of the same variable. We will continue to refer to these simply as correlation and covariance functions unless there is some risk of ambiguity or if we wish to emphasize the difference between an autocorrelation and a cross-correlation.

[7] The prefix *semi* comes from the $\frac{1}{2}$ in Equation 4.1. It has become common, however, to refer to the semivariogram simply as the variogram. Throughout the text, we will continue to use this conventional, though theoretically sloppy, jargon.

[8] An alternative equation, perhaps more familiar to geostatisticians, for calculating the covariance is:

$$C(h) = \frac{1}{N(h)} \sum_{(i,j)|h_{ij}=h} v_i v_j \quad - \quad \left(\frac{1}{n}\sum_{k=1}^{n} v_k\right)^2$$

where the total number of data is n. In Equation 4.2 the centering term is the product of two different lag means; in this equation the centering term is the square of the mean of all the data. The two formulas are not equivalent; the one given in Equation 4.2 is preferred since the alternative given above is known to be slightly biased as an estimator for the underlying covariance function.

Further Reading

David, M. , *Geostatistical Ore Reserve Estimation*. Amsterdam: El-sevier, 1977.

Isaaks, E. H. and Srivastava, R. M. , "Spatial continuity measures for probabilistic and deterministic geostatistics," *Mathematical Geology*, vol. 20, no. 4, pp. 313–341, 1988.

Journel, A. G. and Huijbregts, C. J. , *Mining Geostatistics*. London: Academic Press, 1978.

Matern, B. , *Spatial variation*. Meddelanden Fran Statens Skogs-forskningsinstitut, Stockholm, vol. 49, no. 5, 144 pp, 1960.

Matheron, G. F. , "La théorie des variables régionalisées et ses applications," *Les Cahiers du Center de Morphologie Mathématique de Fontainebleau*, Fascicule 5, École Supérieure des Mines de Paris, 1970.

5

THE EXHAUSTIVE DATA SET

Using the descriptive tools presented in the last three chapters we will now try to describe our data sets. This chapter deals with the description of the 78,000 data in the exhaustive data set, and the next two chapters will deal with the 470 data in the sample data set.

This chapter serves two purposes. As we try to describe the exhaustive data set we will encounter certain difficulties peculiar to the description of large, very dense data sets. For example, contour maps become difficult to construct. One of the aims of this chapter will be to look at some ways of overcoming the practical difficulties that such data sets pose. The second purpose of this chapter is the presentation of the correct answers to many of the estimation problems we will consider later. This is one of the privileges we have with the Walker Lake data sets. While we will be using only the data contained in the sample data set for our various estimation studies, we are able at any time to peek behind the curtain and see how well we are doing.

The Distribution of V

The distribution of the 78,000 V values is summarized in Table 5.1. The data values span several orders of magnitude, from 0 to 1,631 ppm. They are also strongly positively skewed; this makes it difficult to construct a single informative frequency table and its corresponding histogram. The frequency table for the 50 ppm class interval given in Table 5.1 manages to cover most of the distribution. The corresponding

Table 5.1 Frequency table of all 78,000 V values in the exhaustive data set using a class width of 50 ppm to show most of the distribution and a class width of 10 ppm to show detail from 0 to 250 ppm.

50 ppm Class Width			10 ppm Class Width		
Class	Number	Percentage	Class	Number	Percentage
0-50	16,802	21.5	0	5,942	7.6
50-100	7,466	9.6	0-10*	3,223	4.1
100-150	6,346	8.1	10-20	2,386	3.1
150-200	5,953	7.6	20-30	2,078	2.7
200-250	5,662	7.3	30-40	1,681	2.2
250-300	5,131	6.6	40-50	1,492	1.9
300-350	4,666	6.0	50-60	1,548	2.0
350-400	4,219	5.4	60-70	1,478	1.9
400-450	3,759	4.8	70-80	1,558	2.0
450-500	3,331	4.3	80-90	1,526	2.0
500-550	2,876	3.7	90-100	1,356	1.7
550-600	2,533	3.2	100-110	1,321	1.7
600-650	2,089	2.7	110-120	1,323	1.7
650-700	1,713	2.2	120-130	1,256	1.6
700-750	1,347	1.7	130-140	1,199	1.5
750-800	1,050	1.3	140-150	1,247	1.6
800-850	810	1.0	150-160	1,234	1.6
850-900	601	0.8	160-170	1,202	1.5
900-950	482	0.6	170-180	1,179	1.5
950-1,000	323	0.4	180-190	1,158	1.5
1,000-1,050	252	0.3	190-200	1,180	1.5
1,050-1,100	172	0.2	200-210	1,195	1.5
1,100-1,150	128	0.2	210-220	1,112	1.4
1,150-1,200	90	0.1	220-230	1,069	1.4
1,200-1,250	62	0.1	230-240	1,145	1.5
1,250-1,300	53	0.1	240-250	1,141	1.5
1,300-1,350	28	<0.1	250+	35,771	45.9
1,350-1,400	22	<0.1			
1,400-1,450	13	<0.1	* excludes 0 ppm values		
1,450-1,500	6	<0.1			
1,500+	14	<0.1			

(a)

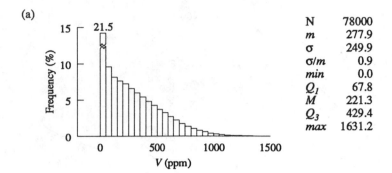

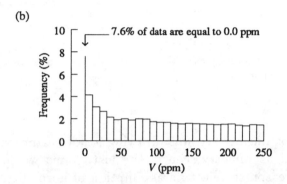

N	78000
m	277.9
σ	249.9
σ/m	0.9
min	0.0
Q_1	67.8
M	221.3
Q_3	429.4
max	1631.2

(b)

Figure 5.1 Histogram and univariate statistics of the 78,000 *V* values using a class width of 50 ppm in (a) and a class width of 10 ppm in (b).

histogram, shown in Figure 5.1a, gives a good impression of the overall shape of the distribution. Unfortunately, nearly a quarter of the values fall within the 0 - 50 ppm class and our frequency table contains no information on how these low values are distributed.

To shed more light on one part of a distribution it is sometimes necessary to use smaller class widths. In Table 5.1 we have provided a second frequency table using a class width of 10 ppm to show more of the detail for the low *V* values. The number of 0 ppm values is also tabulated and plotted as a spike on the histogram of Figure 5.1b [1]. The proportion of zeros has practical implications for some estimation techniques, particularly those that involve some transformation of the original data values.

The normal probability plot of the *V* values shown in Figure 5.2a

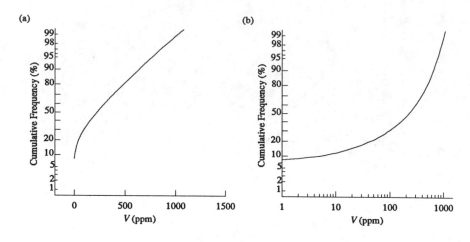

Figure 5.2 The normal probability plot of the 78,000 V values is given in (a), and the lognormal probability plot in (b).

is quite definitely not linear. Without drawing a normal probability plot, the pronounced asymmetry seen in the histograms would have served as sufficient reason to reject an assumption of normality. This asymmetry may suggest that a lognormal distribution is a more likely model. The lognormal probability plot in Figure 5.2b, however, shows that an assumption of lognormality is not appropriate either.

The table that accompanies Figure 5.1a presents a statistical summary of the distribution of V values. The positive skewness evident on the histogram is also reflected in the difference between the median (221.3 ppm) and the mean (277.9 ppm). Though the coefficient of variation (0.9) is moderate, there are still some very high values that may prove to be problematic for estimation.

The Distribution of U

Frequency tables for the U values are given in Table 5.2; the corresponding histograms are shown in Figures 5.3a and 5.3b. We have again chosen to use two different class widths, one that allows us to capture most of the distribution and another that gives us more detailed information on the low U values. We have also separated the

Table 5.2 Frequency table of all 78,000 *U* values in the exhaustive data set using a class width of 50 ppm to show most of the distribution and a class width of 10 ppm to show details of the distribution between 0 and 250 ppm.

50 ppm Class Width			10 ppm Class Width		
Class	Number	Percentage	Class	Number	Percentage
0-50	37,626	48.2	0	4,551	5.8
50-100	7,550	9.7	0-10*	17,638	22.6
100-150	4,714	6.0	10-20	5,988	7.7
150-200	3,216	4.1	20-30	4,117	5.3
200-250	2,537	3.3	30-40	3,018	3.9
250-300	2,157	2.8	40-50	2,314	3.0
300-350	1,980	2.5	50-60	1,958	2.5
350-400	1,765	2.3	60-70	1,637	2.1
400-450	1,568	2.0	70-80	1,441	1.8
450-500	1,522	2.0	80-90	1,353	1.7
500-550	1,374	1.8	90-100	1,161	1.5
550-600	1,191	1.5	100-110	1,095	1.4
600-650	1,014	1.3	110-120	1,036	1.3
650-700	877	1.1	120-130	903	1.2
700-750	712	0.9	130-140	885	1.1
750-800	581	0.7	140-150	795	1.0
800-850	618	0.8	150-160	755	1.0
850-900	527	0.7	160-170	629	0.8
900-950	467	0.6	170-180	673	0.9
950-1,000	431	0.6	180-190	602	0.8
1,000-1,050	416	0.5	190-200	557	0.7
1,050-1,100	380	0.5	200-210	546	0.7
1,100-1,150	356	0.5	210-220	519	0.7
1,150-1,200	323	0.4	220-230	509	0.7
1,200-1,250	306	0.4	230-240	475	0.6
1,250-1,300	282	0.4	240-250	488	0.6
1,300-1,350	249	0.3	250+	22,357	28.7
1,350-1,400	239	0.3			
1,400-1,450	203	0.3	* excludes 0 ppm values		
1,450-1,500	188	0.2			
1,500+	2,631	3.4			

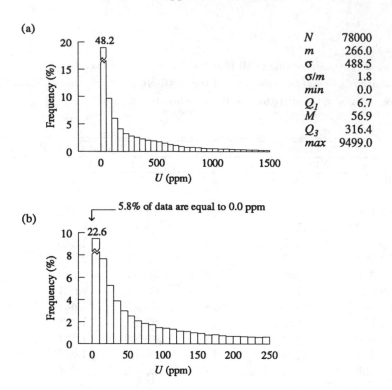

Figure 5.3 Histogram and univariate statistics of the 78,000 U values using a class width of 50 ppm in (a) and a class width of 10 ppm in (b).

values that are exactly 0 ppm, showing them as a spike on the histogram.

Like the V values, the U values also span several orders of magnitude, ranging from 0 ppm to almost 10,000 ppm. Nearly 6% of the values are exactly 0 ppm, a slightly lower fraction than for V. A strong positive skewness is evident from the histograms and supported by the fact that the mean is almost five times larger than the median.

The cumulative probability plots shown in Figures 5.4a and 5.4b again show a curved line revealing that neither normality nor lognormality are appropriate assumptions for the distributions of U values.

The statistical summary given along with Figure 5.3a provides further evidence that the distribution of U values is more skewed than the distribution of V values. The difference between the median (56.9

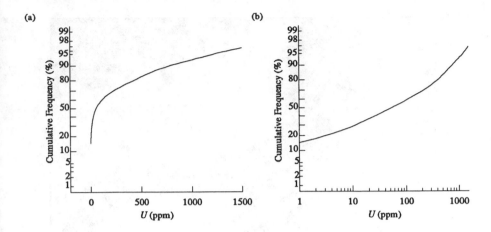

Figure 5.4 The normal probability plot of the 78,000 U values is given in (a), and the lognormal probability plot in (b).

ppm) and the mean (266 ppm) is much larger than for V and the coefficient of variation is also higher. The coefficient of variation (1.84) is high, a warning that erratic values may pose problems in estimation.

The Distribution of T

The third variable, T, is discrete, having only two possible values. We could compile a frequency table, draw a histogram and calculate summary statistics for T; however, this is not necessary since the univariate distribution of T can be completely described by noting that 16% of the T data are type 1 and 84% are type 2. Additional summary statistics of a dichotomous variable provide no further information.

Just as important as knowing the fraction of each type is knowing their location. Figure 5.5 is a map of the two types. The type 1 data, shown as the cross-hatched areas, cover all of Walker Lake itself and extend along the Soda Spring Valley to the southeast (see Figure 1.1). There are also four isolated patches of type 1 within the map area, with the largest of these being located in the northeast corner.

From the map in Figure 5.5 it appears that the two types can easily be separated by a simple boundary. Such a spatial arrangement, however, is not always the case with discrete variables. The different

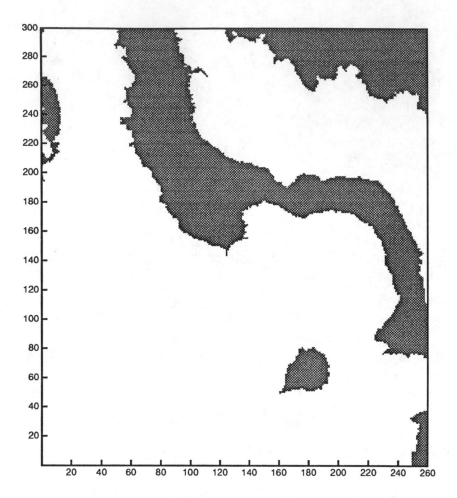

Figure 5.5 Location map of the third variable, T. The cross hatched area shows the location of the type 1 T values. The remaining area corresponds to type 2 values of T.

types can be intermingled at a very small scale, making discrimination by a simple boundary a practical impossibility. If we wish to distinguish between such spatially mixed types we will have to resort to some method other than one which establishes a simple boundary. When we look at indicator maps of the exhaustive data set later in this chapter we will see this mixed type of spatial arrangement and still later we will present methods for handling small scale spatial mixtures.

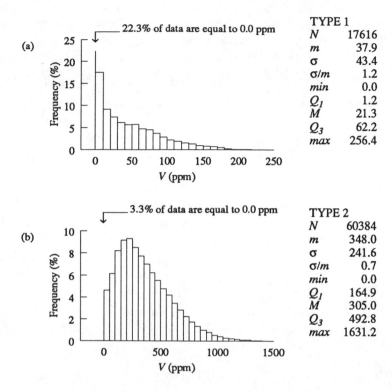

Figure 5.6 Histogram and univariate statistics of, (a) V given that T is type 1 and (b) V given that T is type 2.

Recognition of Two Populations

At the beginning of this chapter, when we presented the univariate descriptions of V and U, we lumped all $78,000$ data locations together and presented each variable as one single group or population. The natural separation of the T values into distinct regions raises the possibility that we should perhaps recognize two different populations.

The definition of meaningful populations is fundamental to any statistical approach. There are no rules that dictate when to divide a larger group into smaller subgroups, but the following questions may serve as useful guidelines:

- Is the distinction meaningful? There should be some good reason, often a purely subjective one, for the distinction. For example, in the Walker Lake data set one could use the east-west line at 200N to separate the data into two populations. Such a distinction, however, is quite arbitrary. There is nothing in the data to suggest that this particular line has any special significance.

- Are there enough data within each population? Each population has to contain enough data to allow reliable calculation of statistical parameters. As we shall see later, covariance functions and variograms calculated on only a few data are often meaningless.

- What is the final goal of the study? In many situations, the recognition of separate populations contributes little to the goal of the study. For example in a study of fluid flow in a petroleum reservoir core plug measurements may be accompanied by some description of the rock. While this may allow the definition of different populations based on the color of the rock, such a distinction does not contribute to the study of the flow properties. If the final goal is accurate local estimation, however, then a division of the data into separate populations, if possible, will likely improve the accuracy of the estimates.

Histograms of the V and U values for each T type are shown in Figures 5.6 and 5.7 along with a statistical summary. The type 1 data generally have much lower V and U values than the type 2 data, though it is interesting to note that the maximum U value, 9,499 ppm, is a type 1 datum. The standard deviation of the type 1 data is also considerably lower for both V and U. The coefficient of variation, however, is larger for the type 1 data, indicating that although the type 1 values are generally lower, they also tend to be more erratic with respect to the mean.

In the exhaustive data set there are certainly enough data of each type to make the differences recorded in Figures 5.6 and 5.7 significant [2]. As we proceed with the analysis of the exhaustive data set we will continue to examine the impact of treating the two types as separate populations.

The V-U Relationship

A scatterplot of the 78,000 V-U pairs at each data location is shown in Figure 5.8a. (The curious bands visible in the cloud of points are an

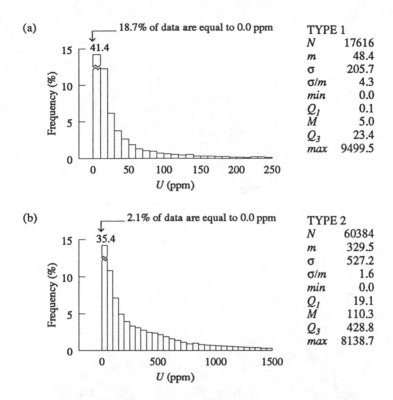

Figure 5.7 Histogram and univariate statistics of, (a) U given that T is type 1 and (b) U given that T is type 2.

artifact of the original elevation data and are explained in Appendix A.) Though the large number of points somewhat obscures the relationship, there is a fair positive linear correlation ($\rho = 0.65$) between the two variables. The relationship between V and U is not linear, and the strength of their relationship may not be well described by the linear correlation coefficient.

Having noted earlier that the T type might be useful for defining two populations, we can check the V-U relationship separately for each type. The 78,000 points plotted in Figure 5.8a have been separated into two groups according to their T type. The scatterplot for the type 1 data is shown in Figure 5.8c and for the type 2 data in Figure 5.8b. These two scatterplots show that the T type plays an important role in the V-U relationship. If we wanted to predict V values from U values,

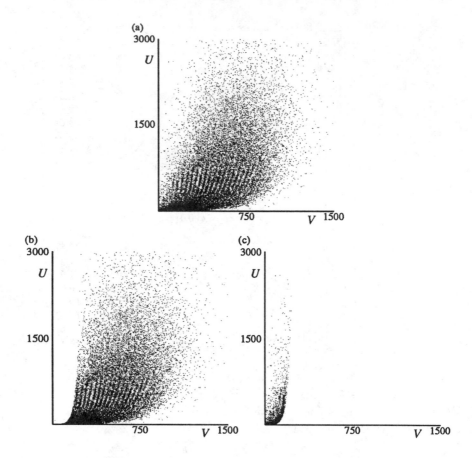

Figure 5.8 (a) Scatterplot of all 78,000 values of V and U (b) 60,384 values of V and U for $T = 2$, and (c) 17,616 values of V and U for $T = 1$.

knowing the T type would significantly improve the accuracy of our predictions.

Spatial Description of V

The huge number of data in the exhaustive data set causes problems in displaying the spatial features of the data values. A posting of the 78,000 values would require a very large map and would not be very informative. Many automated contouring programs could produce a contour map of the 78,000 values. Unfortunately, the skewness and the short scale variability of the data values make the resulting display a confusing and often illegible jumble of contour lines. An alternative that we will consider later is to do some local averaging before trying to construct a contour map.

A symbol map would be an effective display of the exhaustive data set, particularly if the symbols were chosen so that their visual density corresponds to the magnitude of the values. Even this format, however, requires more space than we have here, so we will use indicator maps instead.

Figures 5.9a-i show the nine indicator maps for the V values at cutoffs equal to the nine deciles from $q_{.1}$ to $q_{.9}$. Each map shows in black the data locations at which the V value is above the cutoff and in white those locations at which the value is below the cutoff.

These nine indicator maps give us a good image of how the V values are spatially distributed. We can see that most of the very lowest values (the white patches on Figure 5.9a) occur near Walker Lake itself and along the Soda Spring Valley. Most of the highest values (the black patches on Figure 5.9i) are located west of Walker Lake along the Wassuk Range.

These regions of generally low or high values are different from the distinct regions we saw when we mapped the two T types in Figure 5.5. The western edge of Walker Lake is the only region in which we can find a simple boundary between the values above cutoff and those below cutoff; elsewhere, the black and white are intermingled. Even in areas that are predominantly black there are occasional patches of white. This shows that although most of the values in some region may be above a specified threshold, there is a small but not negligible chance of finding values below the threshold. Similarly, even in areas where the V values are generally below some threshold, erratic high values can often be found.

These indicator maps also reveal certain peculiar features of the exhaustive data set that deserve some explanation. There are several north-south stripes that appear unnatural. The most prominent of these are the two that hang like icicles slightly northwest of the center of Figure 5.10b. There are also some sets of roughly parallel bands that appear in certain regions. The best example of these bands is seen near the top of the western edge of Figure 5.10c.

Both of these features are artifacts of the procedure used to create the digital elevation model from which our variables are derived.

Appendix A contains a more detailed explanation of the origin of the artifacts. Though these features are artifacts whose origins we know, we have chosen not to correct them. Such corrections would make all of the estimation methods that we discuss later appear to work better since none of the methods will have much success with these peculiarities. It is not our intention, however, to mask the difficulty of estimation problems.

Attempts to correct these artifacts would also be arbitrary and would rob the data set of its value as a reproducible standard. Interested readers can obtain the digital elevation model that was used to generate the elevation data set. Were we to make several ad hoc alterations to this data, the studies we present could not be reproduced or extended by others.

Spatial Description of U

Indicator maps for the U values at cutoffs equal to the nine deciles of the U distribution are shown in Figures 5.10a-i. The fair correlation between the U and V values that we noted earlier is evident from the similarity between the indicator maps for U and those for V. The same major features can be found on the maps of both variables.

There is a slight but noticeable difference, however, between the spatial arrangements of the two variables. A comparison of the indicator maps for the highest decile (Figures 5.9i and 5.10i) shows that the very high U values are more widely scattered than the very high V values. The black dots, which represent the highest 10% of U values can be found over much of Figure 5.10i; on Figure 5.9i, the highest 10% of the V values are consolidated in fewer areas. The same is true of the lowest values; the white dots on Figure 5.10a appear to be scattered over a larger area than are those on Figure 5.9a. This tendency of the dots on the V indicator maps to be more clustered suggests that the V values are slightly more continuous over short distances. Later, we will see that the covariance functions quantitatively verify this observation.

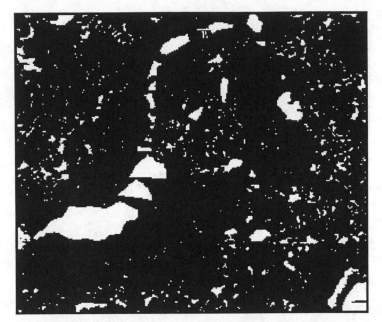

Figure 5.10a Indicator map of *U* for the first decile, 0.12 ppm.

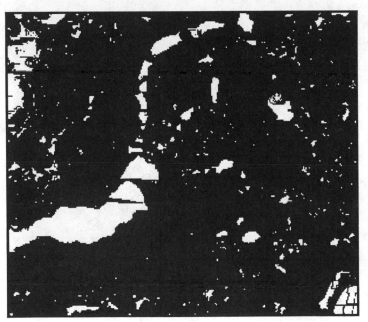

Figure 5.9a Indicator map of *V* for the first decile, 4.79 ppm.

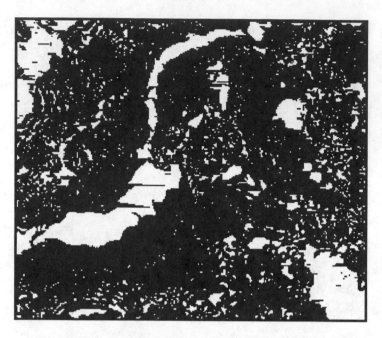

Figure 5.10b Indicator map of *U* for the second decile, 3.10 ppm.

Figure 5.9b Indicator map of *V* for the second decile, 42.06 ppm.

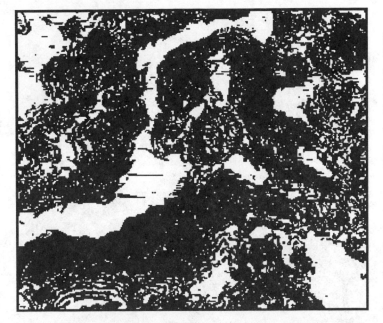

Figure 5.10c   Indicator map of *U* for the third decile, 11.79 ppm.

Figure 5.9c   Indicator map of *V* for the third decile, 93.82 ppm.

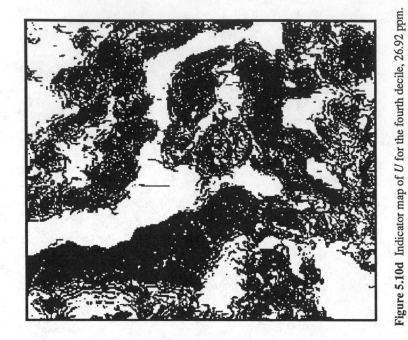

Figure 5.10d Indicator map of *U* for the fourth decile, 26.92 ppm.

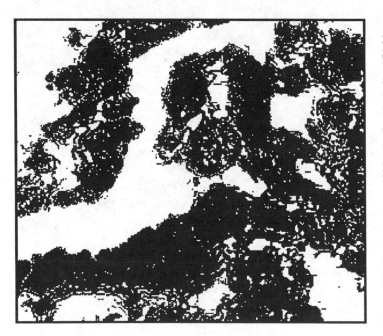

Figure 5.9d Indicator map of *V* for the fourth decile, 154.82 ppm.

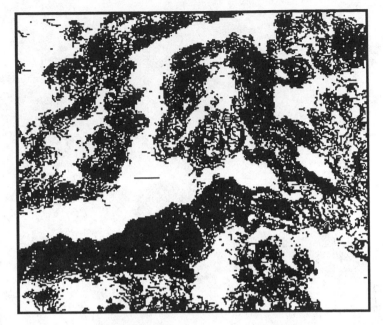

Figure 5.10e Indicator map of *U* for the fifth decile, 56.90 ppm.

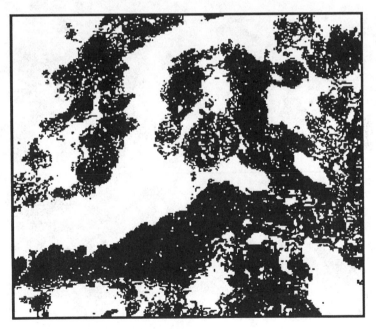

Figure 5.9e Indicator map of *V* for the fifth decile, 221.25 ppm.

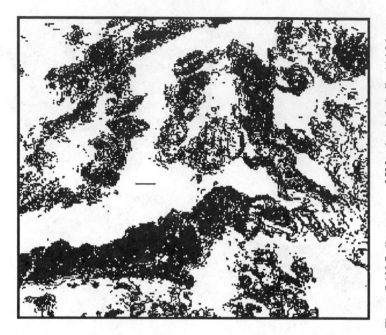

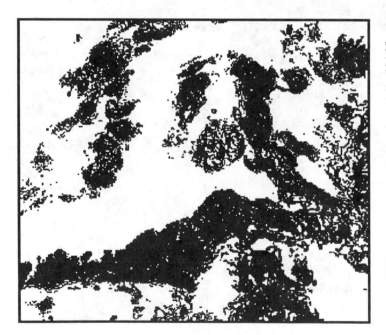

Figure 5.10f Indicator map of *U* for the sixth decile, 114.96 ppm.

Figure 5.9f Indicator map of *V* for the sixth decile, 294.05 ppm.

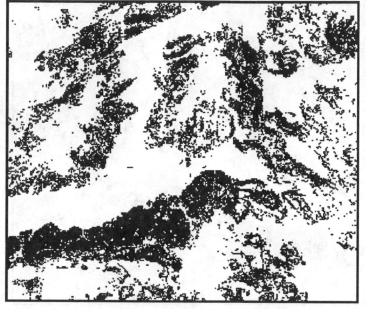

Figure 5.10g Indicator map of U for the seventh decile, 228.09 ppm.

Figure 5.9g Indicator map of V for the seventh decile, 380.04 ppm.

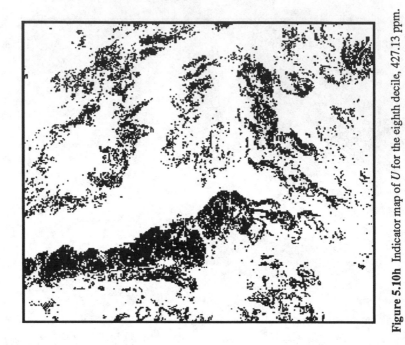

Figure 5.10h   Indicator map of *U* for the eighth decile, 427.13 ppm.

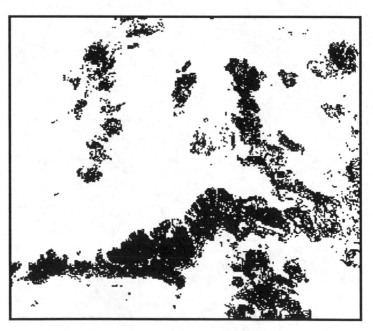

Figure 5.9h   Indicator map of *V* for the eighth decile, 485.19 ppm.

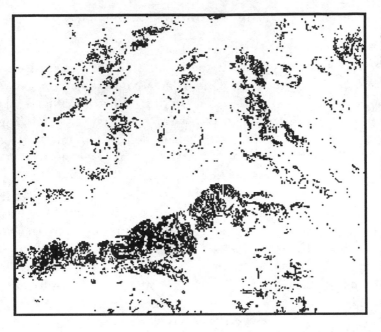

Figure 5.10i Indicator map of *U* for the ninth decile, 784.18 ppm.

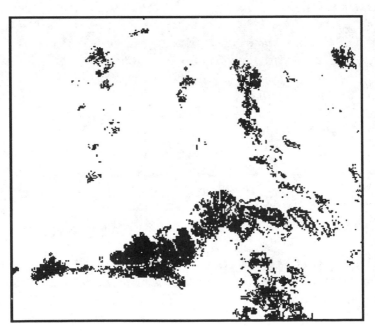

Figure 5.9i Indicator map of *V* for the ninth decile, 634.01 ppm.

Moving Window Statistics

To better display the anomalies in both the average value and in the variability, we have calculated the mean and the standard deviation within 10 x 10 m^2 windows. Our exhaustive data set consists of 780 such windows, each containing 100 data. This is sufficient for our purposes so there is no reason to overlap the windows as we discussed in the previous chapter. The only reason for overlapping the windows is to provide a larger number of moving averages when the number of data is small.

The 780 moving window means for the V values have been contoured in Figure 5.11. Though this contour map shows the main features we saw earlier on the indicator maps, it presents a much smoother version. Less of the intricate detail is visible on the contour map than on the indicator maps; for example, the artifacts discussed earlier cannot be seen on the contour map. This smoothing is due to the size of the moving window and to the automated contouring procedure, which strives to produce an aesthetically pleasing map.

The standard deviations within our 780 10 x 10 m^2 moving windows are contoured in Figure 5.12. Comparing this map to the corresponding map of the moving window means (Figure 5.11) reveals that the mean and the standard deviation do not strongly resemble one another, which suggests that there is not a strong proportional effect. A scatterplot of the mean versus the standard deviation for the 780 windows (Figure 5.13a) confirms that the relationship between the local mean and the local standard deviation is not very strong.

The 780 moving window means for the U values have been contoured in Figure 5.14; the corresponding standard deviations are contoured in Figure 5.16. The map of U means shows the same features that we saw on the V map and on the indicator maps. As with the V values, the price of an aesthetic overall display is the loss of local detail.

If we repeat the calculation of the moving window statistics for 20 x 20 m^2 windows and contour the resulting 195 window means, we get the map shown in Figure 5.15. The same broad features are still evident but their outlines are even smoother than they were for the 10 x 10 m^2 windows. By experimenting with the size of the moving window, one can find a contour map that balances local detail and overall appearance. Supplemented by a display that provides more

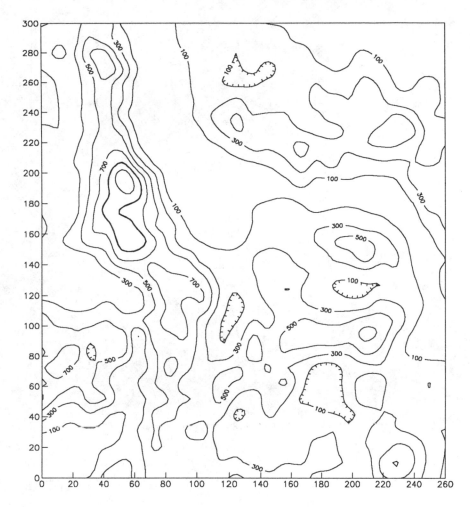

Figure 5.11 Contour map of exhaustive averages of V for 10 x 10 m^2 blocks. The contour interval is 200 ppm with the first contour at 100 ppm.

local detail, such as our indicator maps, a contour map of moving window means serves as a good display of the important spatial features of a very dense data set.

The scatterplot of the 780 local U means and standard deviations (Figure 5.13b) shows that the mean and standard deviation are roughly proportional to one another. A comparison of the two scatterplots in this figure shows that the relationship between the moving average

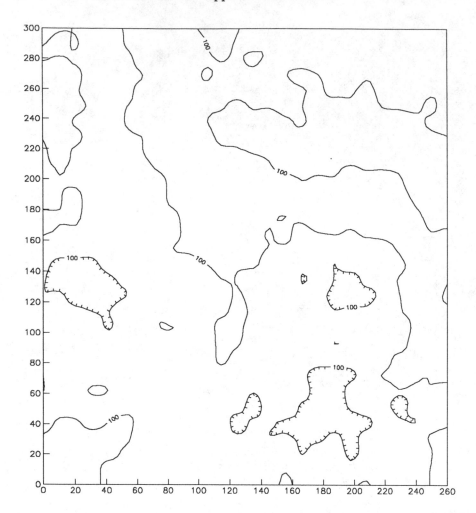

Figure 5.12 Contour map of exhaustive standard deviations of V for 10 x 10 m^2 blocks. The contour interval is 200 ppm with the first contour at 100 ppm.

means and standard deviations is weaker for V than for U. The correlation coefficient for the 780 U m-σ pairs is 0.921 whereas for the 780 V m-σ pairs it is 0.798. On the U scatterplot the strong linear relationship between small values of m and σ can still be observed, though somewhat weaker, even for the very largest values. On the V scatterplot the relationship observed near the origin rapidly becomes weaker and above 400 ppm the m and σ pairs appear to be uncorre-

(a)   (b)

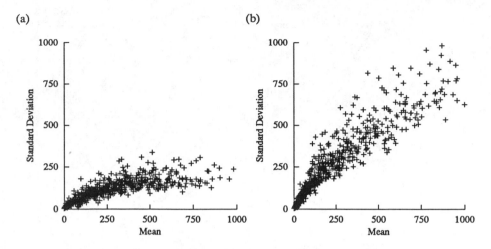

Figure 5.13 Scatterplot of standard deviations versus means. In (a), the standard deviations of V within 10 x 10 m^2 blocks are plotted versus the mean of V within the same blocks. The corresponding plot for the U values is shown in (b).

lated. Also, the relationship between the mean and standard deviation for V does not appear to be linear [3].

Spatial Continuity

In the last chapter we suggested three methods for describing spatial continuity: the correlation function, the covariance function, and the variogram. As descriptive tools, any one of these three serves as well as the other. For the purpose of estimation, however, these three functions are not equivalent. In the second part of this book we will see that the classical theory of estimation places most relevance on the covariance function. For this reason we will use the covariance function to describe our exhaustive data set.

As we observed when we first looked at the covariance function, it depends on both the magnitude of **h** and also on its direction. In Figure 5.17 we have contoured 40,000 $C(\mathbf{h})$ values on a 1 x 1 meter grid, showing the covariance of all h-scatterplots in every direction to a distance of at least 100 m. The values of h_x (i.e., the magnitude of **h** in the east-west direction) are recorded along the bottom edge of the diagram and range from -100 to $+100$ m. The values of h_y (i.e.,

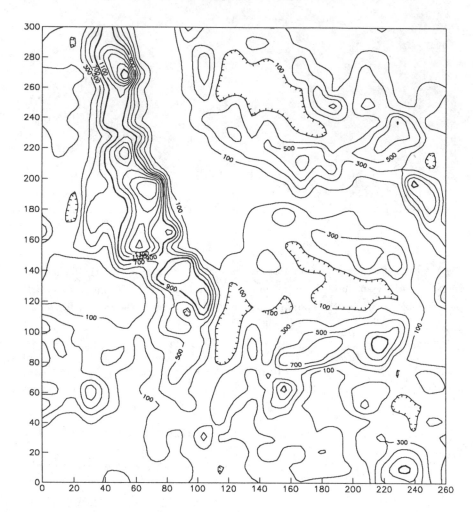

Figure 5.14 Contour map of exhaustive average of U within 10 x 10 m^2 blocks. The contour interval is 200 ppm with the first contour at 100 ppm.

the magnitude of **h** in the north-south direction) are recorded along the left edge of the diagram, also ranging from -100 to $+100$ m. The value of $C(\mathbf{h})$ for $\mathbf{h} = (0,0)$ plots in the center of the diagram.

As we saw in the previous chapter, the relationship between the paired data values on an h-scatterplot becomes stronger as the distance between the paired data locations decreases. If we let this distance shrink to 0, then each of the data values will be paired with itself

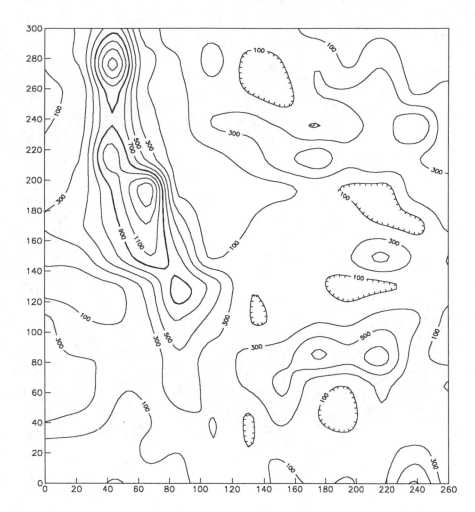

Figure 5.15 Contour map of exhaustive average of U within 20 x 20 m^2 blocks. The contour interval is 200 ppm with the first contour at 100 ppm.

and the resulting h-scatterplot would appear as a straight line. The covariance of this unique h-scatterplot is larger than for any other one we could construct. We can see on Figure 5.17 that $C(\mathbf{h})$ does indeed reach its maximum value at the center of the diagram where $\mathbf{h} = (0,0)$.

In the last chapter we also noted that $C(\mathbf{h}) = C(-\mathbf{h})$. This is confirmed by the symmetry of our contour map about $\mathbf{h} = (0,0)$.

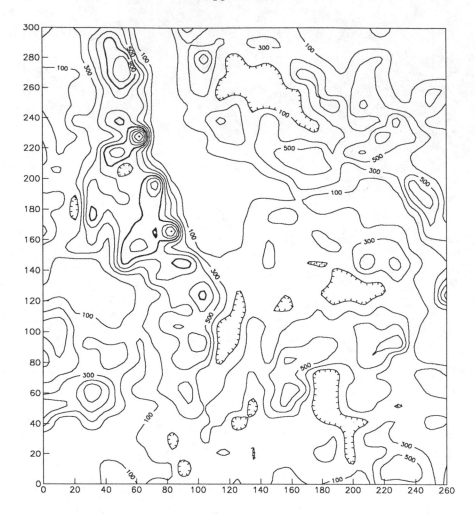

Figure 5.16 Contour map of 10x10 exhaustive U block standard deviations. The contour interval is 200 ppm with the first contour at 100 ppm.

For example, the local minimum near $\mathbf{h} = (30, 50)$ can also be found reflected through the origin near $\mathbf{h} = (-30, -50)$.

The anisotropy is the most striking feature of the contoured $C(\mathbf{h})$ values. The decline in continuity in the north-south direction is not the same as that in the east-west direction. There is an axis of maximum continuity, roughly N14°W, along which the covariance function decreases very slowly. Roughly perpendicular to this is an axis of min-

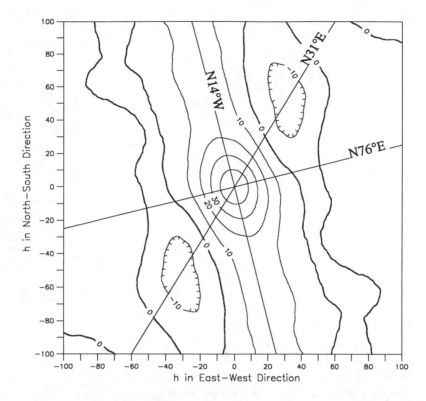

Figure 5.17 Contour map of the exhaustive covariance function for V. The values contoured are the covariance values of all possible h-scatterplots in every direction to a distance of at least 100 meters. The contour interval is 10,000 ppm^2. The covariance value for $\mathbf{h} = (0,0)$ is located at the center of the map. The two lines N14°W and N76°E are the directions of maximum and minimum continuity; the line N31°E lies midway between these two axes. Profiles of the covariance function along these directions are shown in Figure 5.18.

imum continuity along which the covariance function decreases very rapidly.

Though the type of display we have used in Figure 5.17 provides an effective summary of the spatial continuity, it is not commonly used in practice. It is very time consuming to compute and requires a large number of regularly gridded data. More commonly the covariance function is simply plotted in one direction. This type of display can be seen as a cross-sectional slice of the $C(\mathbf{h})$ surface shown in Figure 5.17.

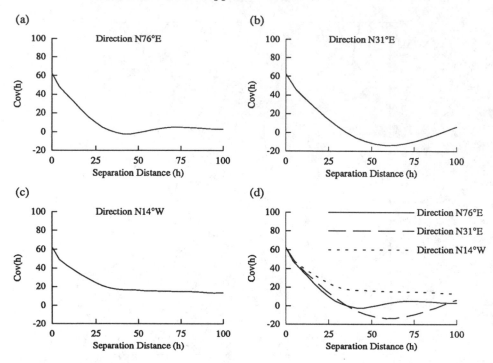

Figure 5.18 (a), (b), and (c) are cross sectional profiles of the exhaustive covariance function for V in three directions. All three profiles are plotted in (d). The vertical axis in all plots is labeled in units of thousands of ppm^2.

On the contour map of the $C(\mathbf{h})$ surface in Figure 5.17 we have marked three directions of interest: the axes of maximum and minimum continuity and one intermediate direction. If we could slice the surface along these lines and view it from the side, we would see the three covariance functions plotted in Figures 5.18a-c. Because of the symmetry of $C(\mathbf{h})$ we have plotted only half of each sectional view. The three covariance functions are superimposed on Figure 5.18d to make their comparison easier in the different directions. In all three directions $C(\mathbf{h})$ drops rather steadily over 25 meters from its maximum value of 62,450 ppm^2. In the N14°W direction, the value of $C(\mathbf{h})$ does not continue to drop as quickly as in the other two directions. For pairs of data separated by 50 m in this direction, the covariance between their V values is approximately 20,000 ppm^2; in the other directions the covariance has already decreased to 0 ppm^2.

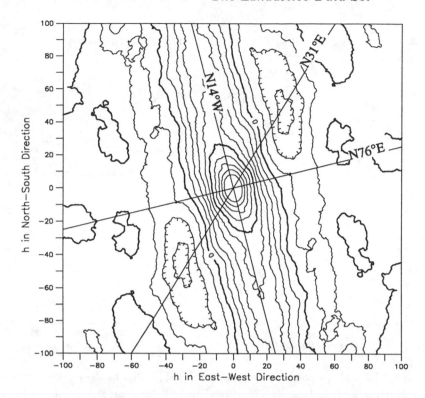

Figure 5.19 Contour map of the exhaustive covariance function for U. The values contoured are the covariance values of all possible h-scatterplots in every direction to a distance of at least 100 meters. The contour interval is 10,000 ppm^2. The lines N14°W and N76°E are the directions of maximum and minimum continuity; the line N31°E lies along a direction midway between these two axes. Profiles of the covariance function along these directions are shown in Figure 5.20.

The covariance function for the U values is contoured in Figure 5.19 using the same contour interval as was used for the V map in Figure 5.17. The greater variability of the U values that we noticed in the summary statistics and in the indicator maps is again apparent. The magnitude of the covariance function is larger than for V but its overall shape is similar. The nature of the anisotropy appears to be the same, with N14°W remaining the direction of maximum continuity.

The plots of the covariance function along the same three directions we used earlier are shown in Figures 5.20a-c and superimposed

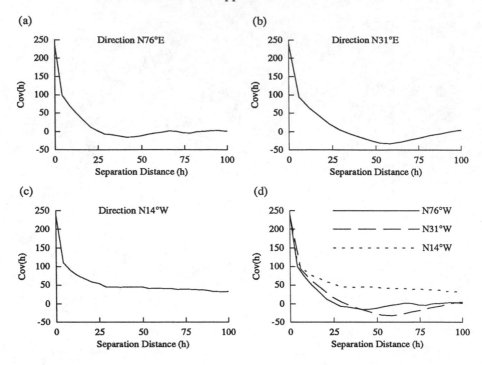

Figure 5.20 (a), (b), and (c) are cross sectional profiles of the exhaustive covariance function for U in three specific directions. All three profiles are plotted in (d). The vertical axis on all the plots is labeled in units of thousands of ppm^2.

on Figure 5.20d. Though their overall appearance is similar to the V covariance functions, there are some important differences. In particular, the rapid decrease in the first few meters is much more pronounced for U than for V.

This difference in the behavior of the covariance function for small values of $|h|$ provides confirmation of our earlier qualitative observation. When we looked at the indicator maps of V and U we noticed that the extreme V values seemed to be less scattered than their U counterparts. As we noted at the time, this suggested that the V values were slightly more continuous. A very gradual change in the value of $C(h)$ near the origin indicates strong spatial continuity while a sudden decrease points to a lot of short-scale variability.

The U-V cross-covariance function is contoured in Figure 5.21. Unlike the autocovariances we contoured earlier, the cross-covariance is

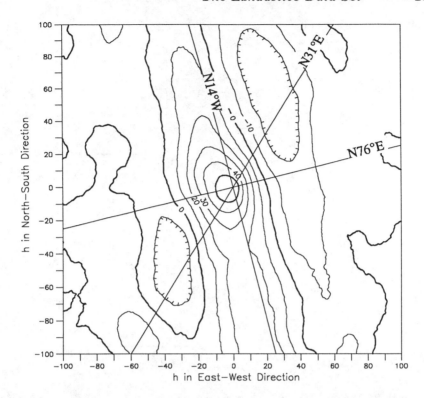

Figure 5.21 Contour map of the exhaustive cross-covariance function between U and V. The values contoured are the cross-covariance values in every direction to a distance of at least 100 meters, with a contour interval of 10,000 ppm^2. Note that the cross-covariance at **h** is not equal to the cross-covariance at -**h** . Though the lines N14°w and N76°E are no longer exactly the directions of maximum and minimum continuity, they have been retained for modeling purposes. The line N31°E lies along a direction midway between these two axes. Profiles of the cross-covariance function along these directions for **h** and -**h** are shown in Figure 5.22.

not symmetric [4]. When we view the cross-covariance function in the three directions, it is no longer adequate to show only the $|\mathbf{h}| > 0$ half of the function. In Figures 5.22a-c we show the cross-covariance for $|\mathbf{h}| > 0$ with the solid line and the cross-covariance for $|\mathbf{h}| < 0$ with the dotted line. Clearly, the cross-covariance for $|\mathbf{h}| > 0$ is quite different from that for $|\mathbf{h}| < 0$.

The general features of the cross-covariance are similar to the au-

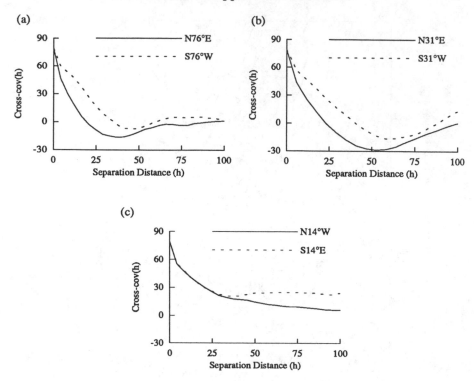

Figure 5.22   (a), (b), and (c) are cross sectional profiles in three specific directions of the exhaustive cross-covariance function between U and V. Since the cross-covariance is not symmetric, each plot shows the profile of the cross-covariance for **h** and for **-h**.

tocovariances of V and U. The anisotropy still exists, though the direction of maximum continuity is rotated slightly from N14°W. The decrease in the first few meters lies somewhere between the rapid decrease we saw on the U autocovariance function and the very slight decrease we saw for V.

The effect of T on the spatial continuity should be considered since we have already seen that the T type may have an important influence on the statistical characteristics of our continuous variables. Earlier, when we were interested in the effect of the T type on the V-U scatterplot, we separated the cloud of points into two groups according to their type. To explore the effect of the T type on spatial continuity we can divide our h-scatterplots into separate groups.

On the previous h-scatterplot, some of the paired values will both be type 1 data, some of the paired values will both be type 2 and the remainder of the paired values will have one of each type. To examine the possible difference in spatial continuity between type 1 data and type 2 data, we can extract from each h-scatterplot those paired values that belong to each type and calculate the covariance separately for the two types.

The contour map of the $C(\mathbf{h})$ surface of V is shown for type 1 and type 2 data separately in Figure 5.23. Since the type 1 data cover a relatively thin strip that meanders southeast from Walker Lake (see Figure 5.5) there is a limit to the distance over which we can compare data values. The two contour maps of the covariance function shown in Figure 5.23 show the $C(\mathbf{h})$ surface for distances of up to 30 m in every direction; beyond this, there are not enough pairs of data for the type 1 h-scatterplots to be meaningful.

When we compared the summary statistics of the V values for the two different types we noticed that the type 1 data generally had much smaller values. Unlike the correlation coefficient, which ranges from -1 to $+1$ regardless of the data values, the covariance is directly influenced by the magnitude of the data values. It is evident from our contour maps of the $C(\mathbf{h})$ surface that the covariance function of the type 1 data has a smaller magnitude than the covariance function of the type 2 data.

More important than the difference in magnitude is the difference in the nature of the anisotropy. From the two maps shown in Figure 5.23, it is apparent that the ratio of the major axis to the minor axis is much more severe for the type 1 data (Figure 5.23a) than for the type 2 data (Figure 5.23b). Furthermore, the orientation of the anisotropy axes changes from one data type to the other. For the type 1 data, the major axis is oriented approximately N14°W while for the type 2 data the axis is almost due north.

The difference between the covariance functions for the two different types is even more dramatic for the U values. In Figure 5.24 the difference in the anisotropy of the covariance function of U for the two types is more severe than it was for V. Using contour lines close to the origin as a guide in Figure 5.24, it appears that the anisotropy ratio is nearly 5:1 for the type 1 data and only 2:1 for the type 2 data. The covariance function for the type 1 data decreases very rapidly, reaching 0 ppm^2 within only 10 m in an east-west direction. The decrease is

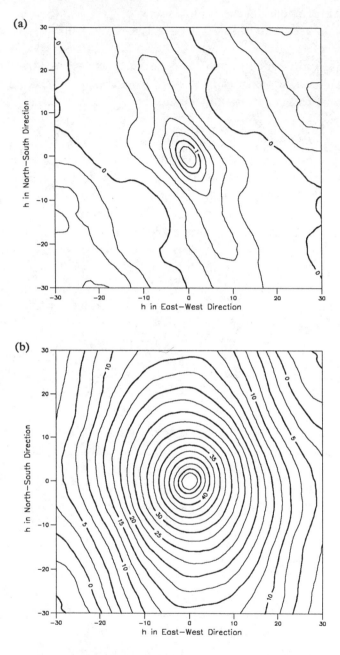

Figure 5.23 Contour maps of the exhaustive covariance functions of V for each T type. (a) shows the contoured covariance function of V for type 1, and (b) for type 2. The contour labels are in thousands of ppm^2.

(a)

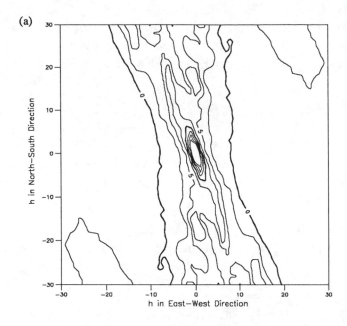

(b)

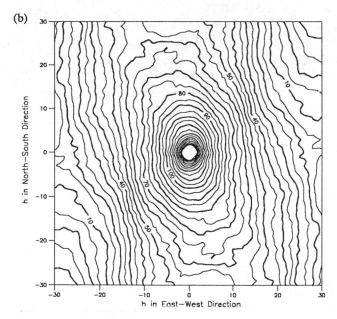

Figure 5.24 Contour maps of the exhaustive covariance functions of U for each type of T. (a) shows the contoured covariance function of U for type 1, and (b) for type 2. The contour labels are in thousands of ppm^2.

not as rapid for the type 2 data; in an east-west direction $C(\mathbf{h})$ has not reached 0 ppm^2 within the 30 m shown on the map. As with the V variable, the orientation of the direction of maximum continuity is rotated more to the west for the type 1 data than for the type 2 data.

Notes

[1] Presenting the number of 0 ppm values as a spike takes a small liberty with the proper histogram format. Since the area of the bars should be proportional to their frequency, a class of zero width should have an infinite height. The spike, however, is conventionally understood to record the frequency by its height.

[2] Statistical tests exist for determining whether or not two sample data sets are significantly different. Such tests usually require statistical independence among the data within each sample data set and thus are not very appropriate for correlated data.

[3] Having noted earlier that the type variable, T, has an important influence on the statistics, it would be good to check the proportional effect separately for each type. The nonlinear relationship we observe in Figure 5.13a may be due to a mixture of moving window statistics for two distinct populations.

[4] The cross covariance between U and V is not symmetric and provides an excellent example of the *lag effect*, which is due to an offset between the locations of extreme values of the two variables. For example in many ore deposits, for example, the direction of flow of hydrothermal fluids, combined with the fact that certain minerals precipitate earlier than others, may cause enrichment of some minerals to lag behind others. In the Walker Lake data sets, the highest V values are located slightly to the west of the highest U values. The result is that the cross-continuity between the two variables is not symmetric. For example, when comparing a U value to nearby V values there is likely to be more similarity in the V values to the west than in those to the east. This is evident in the contour map of $C_{UV}(\mathbf{h})$ shown in Figure 5.21; the spatial continuity decreases more rapidly to the east than it does to the west. While the cross-covariance function and the cross-correlogram can capture such features, the cross-variogram cannot.

6

THE SAMPLE DATA SET

In practice we are never lucky enough to have exhaustive sampling of the area of interest. Instead, we have samples of only a tiny fraction of the total area. Our task is to infer properties of the whole area from this limited sample information. To duplicate this frustrating feature of reality we have selected a small subset of our exhaustive data set to serve as a sample data set. In all of the estimation problems we will tackle later we will pretend that this sample data set is the only information we have. The exhaustive data set will be used only to check our various estimates and to help us understand the strengths and weaknesses of the different methods we consider.

In the next two chapters, we will describe the sample data set using the same tools that helped us analyze the exhaustive data set. This chapter will cover the univariate and bivariate description and will begin the spatial description. The description of the spatial continuity in the sample data set will be dealt with separately in the next chapter. These two chapters provide a preliminary exploratory analysis of the sample data. Their goal is to familiarize us with the sample data set and to uncover its relevant aspects.

Familiarity with the data is an asset in any statistical study. Minor oversights and major conceptual flaws can plague any study. Time spent familiarizing oneself with the data is often rewarded by a quick recognition of when an error has occurred.

The exploratory analysis should also begin the process of understanding what might be important about the data set. When we look at estimation in the second part of this book we will see that there are many ways of handling the data. From our analysis of the exhaustive data set we already know that the T type may play an important role and that the data could, therefore, be treated as two separate groups. Such important considerations are possible only if we have taken the time to explore our data. Many of the features we uncover may later turn out to be useless, but a few dead ends should not discourage us from trying to discover the key ingredients of our data.

Before we begin the analysis of the sample data set it is important to clarify a potential source of confusion. In this chapter we will be summarizing distributions using the same statistical tools that we used in the last chapter. For example, we will soon see that the mean of the V values in the sample data set is 436.5 ppm. In the last chapter we saw that the mean of the V values was 278 ppm. The 436.5 ppm value is a sample statistic; the 278 ppm value is an exhaustive statistic, often referred to as a *population parameter*.

In most studies it is the population parameters that are of greater interest. The sample statistics tell us only about the samples; what we really want to know is what the samples can tell us about the entire population from which they are drawn. In a study of the concentration of some pollutant, for example, we are not really interested in the average concentration of the pollutant in the samples we have collected. What we actually want to know is the concentration of the pollutant over some larger region. Sample statistics are stepping stones to the final goal of understanding more about the entire population.

As we proceed with our analysis of the sample data set we will casually be comparing the sample statistics with the exhaustive statistics we calculated earlier. This is not intended to suggest that the two should be the same or that one is a good estimate of the other. Indeed, they will often be very different. In the second part of this book we will examine the many reasons that the two are not the same and we will look at several alternatives for estimating population parameters from the information contained in the sample data set.

In this chapter, where our goal is purely descriptive, we will continue to refer to the summary statistics by the symbols we introduced earlier; for example, m will still be used to refer to the mean even though it is not the same mean we discussed in the previous chapter. Later, when we start to deal with the topic of estimation, it will be important to maintain a clear distinction between the sample statistics, which we know, and the population parameters, which we do not know. At that time we will alter our notation slightly to help maintain this distinction.

Data Errors

Information extracted from a data set or any inference made about the population from which the data originate can only be as good as the original data. One of the most tedious and time-consuming tasks in a geostatistical study is error checking. Though one would like to weed out all the errors at the outset, it seems that they often manage to remain hidden until the analysis is already started.

Our sample data set happens to be error-free; it has not suffered the bumpy journey from data collection to laboratory analysis to entry on a computer. Beyond offering our sympathy to the readers when they encounter a major data-cleaning exercise, we can offer a few helpful suggestions. These are not guaranteed to produce perfectly clean data but they will catch gross errors:

- Sort the data and examine the extreme values. If they appear excessive, investigate their origin and try to establish their authenticity. Erratic extremes may be the result of a misplaced decimal point. Original sample diaries or sampling logs, if they still exist, are useful sources of information.

- Locate the extreme values on a map. Note their location with respect to anomalous areas. Are they located along trends of similar data values or are they isolated? Be suspicious of isolated extremes.

- Check coordinate errors by sorting and examining coordinate extremes. Are they within expected limits?

- Examine a posting of the data. Do the samples plot where they should?

When trying to sort out inconsistencies in the data, it often helps to understand how the data set came to be in its current form. The evolution of the data set should be probed with questions such as: Were all the samples collected in one program? If not, was the sampling procedure consistent from one program to the next? Were different people involved in the sampling? If so, how did their methodologies differ? Are all the samples the same size? Were sample locations surveyed? If not, how were the sample locations determined? Is the data presently in the form in which it was received from the laboratory? If not, how does it differ? Are there missing samples? If so, how were these data treated?

There are many telltale signs of procedural changes in data collection. The numbering of the samples may reveal different sampling programs; changes in the recorded precision of the data values may be the result of switching to different methods of measuring the values. One should try to recognize those events that could lead to inconsistencies. If they exist, such inconsistencies should be checked and resolved before the statistical analysis proceeds.

The Sampling History

We begin our analysis with a complete listing of the data (Table 6.1) and a posting of the data values (Figures 6.1 and 6.2). Not only are these a necessary first step in error checking, they also act as valuable references throughout the exploratory analysis.

Looking through the listing of the sample data set in Table 6.1 we see that there are 470 sample locations. V, U and T measurements exist at most of these locations. The first 195 samples, however, are missing the U measurement. This is our first hint that the data were not collected in one campaign. We get some confirmation of this when we look at the postings of the data values.

In Figure 6.1, a posting of the V values, we notice that a grid of samples at roughly 20 x 20 m^2 spacing covers the entire area, with additional sampling providing denser coverage of certain areas. On the posting of the U values in Figure 6.2 we discover that the samples that are missing their U measurement are those located on the preliminary 20 x 20 m^2 grid.

Further study of the data postings leads to the observation that the more densely sampled areas generally have high values. This is quite common in practice. The areas of interest are typically those with the highest grades, the strongest concentrations, the greatest porosities, etc., and additional sampling is often aimed at delineating such anomalous zones.

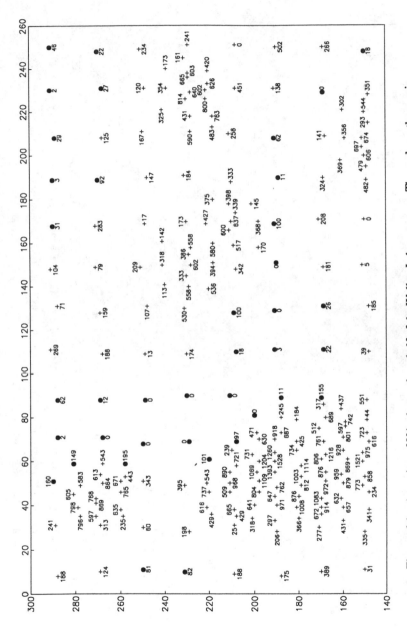

Figure 6.1a A posting of *V* in the northern half of the Walker Lake map area. The scale on the y-axis ranges from 140 to 300 meters north. The symbol ● marks data locations at which T = 1, while the + sign indicates locations at which T = 2.

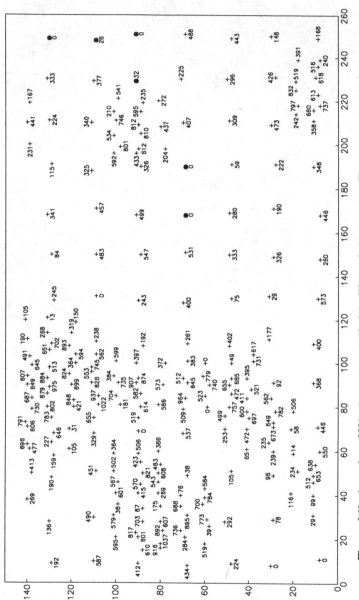

Figure 6.1b A posting of *V* in the southern half of the Walker Lake map area. The scale on the y-axis ranges from 0 to 150 meters north. The symbol ● marks data locations at which T = 1, while the + sign indicates locations at which T = 2.

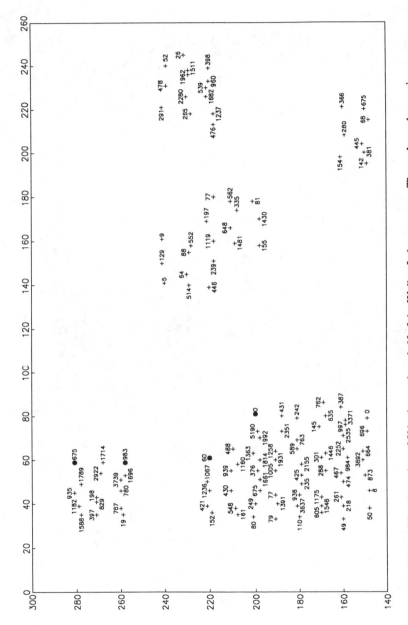

Figure 6.2a A posting of *U* in the northern half of the Walker Lake map area. The scale on the y-axis ranges from 140 to 300 meters north. The symbol ● marks data locations at which T = 1, while the + sign indicates locations at which T = 2.

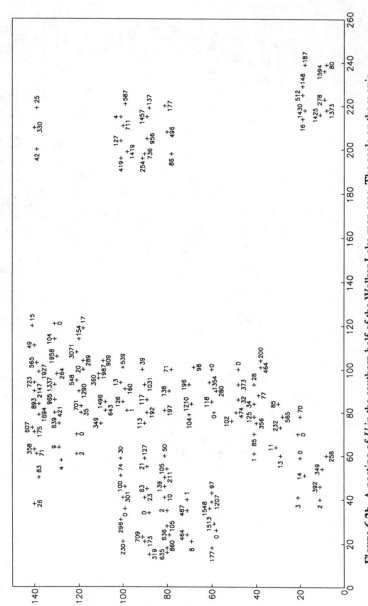

Figure 6.2b A posting of U in the southern half of the Walker Lake map area. The scale on the y-axis ranges from 0 to 150 meters north.

Table 6.1   The Sample Data Set

No.	X	Y	V	U	T	No.	X	Y	V	U	T
1	11	8	0.	N/A	2	50	69	90	0.	N/A	2
2	8	30	0.	N/A	2	51	68	110	329.1	N/A	2
3	9	48	224.4	N/A	2	52	68	128	646.3	N/A	2
4	8	68	434.4	N/A	2	53	69	148	616.2	N/A	2
5	9	90	412.1	N/A	2	54	69	169	761.3	N/A	2
6	10	110	587.2	N/A	2	55	70	191	918.0	N/A	2
7	9	129	192.3	N/A	2	56	69	208	97.4	N/A	1
8	11	150	31.3	N/A	2	57	69	229	0.	N/A	1
9	10	170	388.5	N/A	2	58	68	250	0.	N/A	1
10	8	188	174.6	N/A	2	59	71	268	0.	N/A	1
11	9	209	187.8	N/A	2	60	71	288	2.4	N/A	1
12	10	231	82.1	N/A	1	61	91	11	368.3	N/A	2
13	11	250	81.1	N/A	1	62	91	29	91.6	N/A	2
14	10	269	124.3	N/A	2	63	90	49	654.7	N/A	2
15	8	288	188.0	N/A	2	64	91	68	645.5	N/A	2
16	31	11	28.7	N/A	2	65	91	91	907.2	N/A	2
17	29	29	78.1	N/A	2	66	91	111	826.3	N/A	2
18	28	51	292.1	N/A	2	67	89	130	975.3	N/A	2
19	31	68	895.2	N/A	2	68	88	149	551.1	N/A	2
20	28	88	702.6	N/A	2	69	89	170	155.5	N/A	1
21	30	110	490.3	N/A	2	70	89	188	10.7	N/A	1
22	28	130	136.1	N/A	2	71	90	211	0.	N/A	1
23	28	150	335.0	N/A	2	72	90	230	0.	N/A	1
24	30	171	277.0	N/A	2	73	88	249	0.	N/A	1
25	28	190	206.1	N/A	2	74	88	269	12.1	N/A	1
26	31	209	24.5	N/A	2	75	88	288	62.2	N/A	1
27	28	229	198.1	N/A	2	76	109	11	399.6	N/A	2
28	30	250	60.3	N/A	2	77	111	31	176.6	N/A	2
29	31	269	312.6	N/A	2	78	108	49	402.0	N/A	2
30	31	289	240.9	N/A	2	79	109	68	260.6	N/A	2
31	49	11	653.3	N/A	2	80	108	88	192.0	N/A	2
32	49	29	96.4	N/A	2	81	110	109	237.6	N/A	2
33	51	48	105.0	N/A	2	82	109	129	702.0	N/A	2
34	49	68	37.8	N/A	2	83	110	148	38.5	N/A	2
35	50	88	820.8	N/A	2	84	111	169	22.1	N/A	1
36	51	109	450.7	N/A	2	85	111	191	2.7	N/A	1
37	48	129	190.4	N/A	2	86	110	208	17.9	N/A	1
38	49	151	773.3	N/A	2	87	109	230	174.2	N/A	2
39	51	168	971.9	N/A	2	88	109	249	12.9	N/A	2
40	48	190	762.4	N/A	2	89	109	268	187.8	N/A	2
41	50	211	968.3	N/A	2	90	111	291	268.8	N/A	2
42	49	231	394.7	N/A	2	91	130	9	572.5	N/A	2
43	51	250	343.0	N/A	2	92	131	31	29.1	N/A	2
44	50	268	863.8	N/A	2	93	130	48	75.2	N/A	2
45	51	290	159.6	N/A	1	94	128	70	399.9	N/A	2
46	71	9	445.8	N/A	2	95	129	90	243.1	N/A	2
47	71	29	673.3	N/A	2	96	131	109	0.	N/A	2
48	70	51	252.6	N/A	2	97	129	128	244.7	N/A	2
49	68	70	537.5	N/A	2	98	131	148	185.2	N/A	2

An Introduction to Applied Geostatistics

Table 6.1 The Sample Data Set (Cont.)

No.	X	Y	V	U	T	No.	X	Y	V	U	T
99	131	169	26.0	N/A	1	148	190	248	146.6	N/A	2
100	129	191	0.	N/A	1	149	189	270	92.0	N/A	1
101	128	209	100.3	N/A	1	150	189	290	2.5	N/A	1
102	130	231	530.3	N/A	2	151	211	11	358.1	N/A	2
103	131	248	107.4	N/A	2	152	209	30	473.3	N/A	2
104	128	269	159.3	N/A	2	153	211	49	308.8	N/A	2
105	131	288	70.7	N/A	2	154	210	70	406.8	N/A	2
106	148	8	260.2	N/A	2	155	209	90	812.1	N/A	2
107	149	29	326.0	N/A	2	156	210	111	339.7	N/A	2
108	150	49	332.7	N/A	2	157	211	130	223.9	N/A	2
109	151	69	531.3	N/A	2	158	208	151	673.5	N/A	2
110	150	89	547.2	N/A	2	159	209	168	141.0	N/A	2
111	150	109	482.7	N/A	2	160	208	191	61.8	N/A	1
112	150	129	84.1	N/A	2	161	210	211	258.3	N/A	2
113	150	151	4.7	N/A	2	162	211	228	590.3	N/A	2
114	149	169	180.6	N/A	2	163	211	250	166.9	N/A	2
115	151	190	0.	N/A	1	164	208	268	125.2	N/A	2
116	148	208	342.4	N/A	2	165	208	289	29.3	N/A	1
117	150	228	602.3	N/A	2	166	231	10	617.6	N/A	2
118	149	251	209.1	N/A	2	167	231	28	425.9	N/A	2
119	149	271	79.4	N/A	2	168	230	50	295.7	N/A	2
120	148	291	104.1	N/A	2	169	230	71	224.9	N/A	2
121	168	8	446.0	N/A	2	170	229	91	31.7	N/A	1
122	171	29	189.9	N/A	2	171	229	110	377.4	N/A	2
123	169	49	280.4	N/A	2	172	230	131	333.3	N/A	2
124	168	69	0.	N/A	1	173	228	148	351.0	N/A	2
125	168	91	499.3	N/A	2	174	229	169	0.	N/A	1
126	171	109	457.3	N/A	2	175	231	191	137.6	N/A	2
127	168	131	341.2	N/A	2	176	231	208	451.2	N/A	2
128	171	150	0.	N/A	2	177	229	228	639.5	N/A	2
129	171	171	208.3	N/A	2	178	231	249	119.9	N/A	2
130	169	191	99.7	N/A	1	179	231	268	27.2	N/A	1
131	170	210	636.6	N/A	2	180	230	291	2.1	N/A	1
132	170	230	173.1	N/A	2	181	249	9	167.7	N/A	2
133	169	249	17.0	N/A	2	182	250	30	147.8	N/A	2
134	168	271	283.1	N/A	2	183	249	48	442.7	N/A	2
135	168	290	30.9	N/A	1	184	251	69	487.7	N/A	2
136	190	11	348.5	N/A	2	185	251	91	0.	N/A	1
137	191	28	222.4	N/A	2	186	248	109	28.2	N/A	1
138	191	48	59.1	N/A	2	187	249	130	0.	N/A	1
139	190	69	0.	N/A	1	188	248	150	18.3	N/A	1
140	190	89	326.0	N/A	2	189	250	169	266.3	N/A	2
141	188	111	325.1	N/A	2	190	250	190	502.3	N/A	2
142	191	129	114.7	N/A	2	191	251	208	0.	N/A	2
143	189	149	481.6	N/A	2	192	251	229	240.9	N/A	2
144	190	169	324.1	N/A	2	193	249	251	234.4	N/A	2
145	190	189	10.9	N/A	1	194	248	270	22.4	N/A	1
146	188	210	332.9	N/A	2	195	250	291	45.6	N/A	1
147	191	231	184.4	N/A	2	196	40	71	76.2	1.1	2

Table 6.1 The Sample Data Set (Cont.)

No.	X	Y	V	U	T	No.	X	Y	V	U	T
197	21	69	284.3	7.8	2	246	59	281	148.8	675.0	1
198	28	80	606.8	105.3	2	247	39	279	798.0	1182.1	2
199	29	59	772.7	1512.7	2	248	59	258	194.9	983.3	1
200	41	81	269.5	9.8	2	249	38	260	635.2	766.6	2
201	18	80	1036.7	860.4	2	250	78	28	781.6	565.4	2
202	39	60	783.8	1207.3	2	251	60	29	238.6	12.7	2
203	18	60	519.4	177.1	2	252	70	41	472.0	84.9	2
204	41	90	414.9	23.4	2	253	70	21	58.1	0.3	2
205	21	90	601.4	173.1	2	254	78	41	600.3	124.6	2
206	31	101	579.2	296.5	2	255	61	41	64.9	0.8	2
207	41	100	601.4	300.6	2	256	78	20	505.9	70.0	2
208	21	100	594.6	229.7	2	257	80	131	801.6	421.1	2
209	60	8	550.1	258.3	2	258	58	128	158.8	4.3	2
210	40	11	99.4	2.2	2	259	71	140	606.3	175.1	2
211	51	18	233.6	14.2	2	260	70	121	30.7	0.0	2
212	59	20	14.4	0.1	2	261	79	138	730.1	1694.5	2
213	41	21	115.9	3.1	2	262	80	119	421.2	35.1	2
214	59	90	506.2	126.9	2	263	61	121	104.8	1.8	2
215	51	101	502.4	73.8	2	264	79	149	44.1	0.0	2
216	50	81	608.0	210.7	2	265	71	160	801.1	2535.0	2
217	59	101	363.9	30.4	2	266	78	159	742.0	3371.5	2
218	60	81	385.6	50.3	2	267	80	168	689.1	634.6	2
219	60	151	1521.1	3691.8	2	268	69	181	424.6	762.6	2
220	38	148	340.9	50.0	2	269	79	181	184.3	241.7	2
221	50	160	879.1	474.2	2	270	80	188	245.2	431.1	2
222	50	138	413.4	83.0	2	271	70	198	630.0	1992.1	2
223	61	158	868.9	983.8	2	272	81	200	0.	0.	1
224	39	160	657.4	217.8	2	273	100	48	48.7	0.0	2
225	61	139	477.0	71.5	2	274	80	49	757.4	473.8	2
226	38	140	268.5	26.2	2	275	90	58	739.8	280.2	2
227	61	170	806.4	301.1	2	276	88	39	520.7	76.8	2
228	39	170	914.4	1548.5	2	277	100	60	0.	0.	2
229	49	179	811.5	234.9	2	278	80	59	0.	0.	2
230	58	179	1113.6	2154.7	2	279	101	38	730.5	464.0	2
231	39	181	1008.0	3637.4	2	280	101	68	383.1	97.8	2
232	60	191	1528.1	1930.9	2	281	79	70	508.8	103.9	2
233	40	190	970.9	1391.1	2	282	90	79	573.3	138.3	2
234	51	198	1109.0	1660.8	2	283	100	78	372.4	70.9	2
235	60	198	1203.9	1813.7	2	284	81	81	585.8	197.2	2
236	40	200	641.3	249.1	2	285	100	91	397.2	38.9	2
237	58	208	720.6	1160.1	2	286	80	89	614.5	192.3	2
238	38	209	665.3	547.8	2	287	91	99	734.9	159.6	2
239	50	221	543.3	1066.6	2	288	101	100	599.3	539.3	2
240	61	220	101.1	59.5	1	289	81	98	181.2	1.3	2
241	39	221	615.9	420.9	2	290	98	111	744.8	1987.0	2
242	59	268	543.1	1714.2	2	291	81	108	1022.3	643.0	2
243	41	271	868.8	828.7	2	292	90	120	899.3	1290.3	2
244	49	278	583.0	1788.8	2	293	100	118	363.7	20.5	2
245	51	260	670.7	3738.9	2	294	98	130	513.2	263.9	2

Table 6.1 The Sample Data Set (Cont.)

No.	X	Y	V	U	T	No.	X	Y	V	U	T
295	90	140	648.8	2147.5	2	344	239	220	420.1	398.3	2
296	99	138	645.4	1927.1	2	345	218	218	763.5	1236.7	2
297	121	131	13.0	0.2	2	346	35	71	687.8	486.8	2
298	111	140	190.3	48.8	2	347	24	71	735.8	463.9	2
299	108	121	893.0	3070.9	2	348	34	88	86.9	0.1	2
300	120	141	104.7	14.6	2	349	23	91	817.0	708.8	2
301	119	118	150.4	16.9	2	350	54	10	637.9	349.4	2
302	158	228	558.4	551.6	2	351	46	11	512.3	392.0	2
303	140	229	558.0	513.9	2	352	55	89	423.4	21.2	2
304	150	241	318.5	129.2	2	353	45	89	569.6	62.8	2
305	151	218	394.3	239.2	2	354	53	150	858.0	873.0	2
306	161	241	141.9	8.6	2	355	46	148	234.0	7.5	2
307	141	240	112.5	4.6	2	356	55	168	876.0	288.1	2
308	160	218	580.4	1118.7	2	357	43	170	1082.8	1174.9	2
309	139	220	535.9	445.7	2	358	55	191	1392.6	1004.7	2
310	178	211	398.2	561.8	2	359	44	191	646.6	76.8	2
311	159	209	517.3	1480.8	2	360	55	211	889.7	938.8	2
312	169	221	427.2	197.2	2	361	46	211	509.2	429.7	2
313	170	198	367.6	1429.8	2	362	54	269	613.1	2922.4	2
314	180	218	374.7	77.5	2	363	43	271	767.8	198.1	2
315	178	201	144.8	81.3	2	364	73	29	649.4	231.7	2
316	158	198	169.8	154.5	2	365	64	31	235.4	10.8	2
317	219	88	235.1	136.9	2	366	75	129	782.8	639.3	2
318	198	90	611.7	735.8	2	367	64	129	227.3	8.6	2
319	211	100	746.4	710.6	2	368	73	149	722.9	696.1	2
320	208	80	436.6	495.8	2	369	64	151	974.5	664.1	2
321	221	99	540.9	586.8	2	370	75	171	512.2	144.8	2
322	199	98	801.0	1419.0	2	371	63	168	1215.8	1446.1	2
323	220	81	272.1	177.3	2	372	73	188	687.1	2351.5	2
324	198	78	204.1	86.0	2	373	64	191	1259.9	1257.6	2
325	220	150	543.9	675.0	2	374	93	48	684.5	373.2	2
326	200	150	606.2	381.1	2	375	86	48	471.9	31.8	2
327	208	159	356.0	280.4	2	376	93	70	512.1	196.1	2
328	210	140	440.9	330.3	2	377	84	69	963.9	1210.0	2
329	221	160	301.8	365.6	2	378	93	90	874.0	1031.3	2
330	198	161	369.4	154.5	2	379	86	89	582.4	117.0	2
331	219	139	166.8	24.5	2	380	96	111	553.2	360.5	2
332	200	139	230.9	42.2	2	381	85	108	937.3	1495.5	2
333	239	8	240.3	80.2	2	382	93	131	883.6	1336.8	2
334	218	8	737.1	1373.4	2	383	86	131	879.0	965.3	2
335	229	19	518.6	147.7	2	384	114	131	268.4	104.4	2
336	239	18	390.7	186.7	2	385	106	130	651.5	1957.7	2
337	218	18	797.4	1429.7	2	386	155	229	386.4	88.2	2
338	238	229	602.6	1510.9	2	387	145	230	333.2	63.5	2
339	218	228	430.8	265.2	2	388	174	208	339.2	335.0	2
340	231	239	354.1	478.0	2	389	166	211	600.3	647.6	2
341	230	221	602.4	538.9	2	390	215	89	595.2	1457.0	2
342	240	239	172.6	51.9	2	391	205	89	809.6	955.8	2
343	221	241	324.8	290.9	2	392	215	148	293.3	67.7	2

Table 6.1 The Sample Data Set (Cont.)

No.	X	Y	V	U	T	No.	X	Y	V	U	T
393	204	151	697.3	444.5	2	432	35	259	235.0	18.9	2
394	236	9	515.9	1593.8	2	433	84	30	562.0	85.3	2
395	223	9	613.2	277.6	2	434	84	41	411.4	34.0	2
396	236	229	665.3	1962.0	2	435	75	40	696.7	356.2	2
397	226	230	813.6	2279.8	2	436	73	141	790.9	607.3	2
398	35	80	174.8	2.0	2	437	63	140	696.5	357.8	2
399	24	79	891.8	635.7	2	438	84	138	687.3	893.4	2
400	36	61	699.6	1547.8	2	439	76	159	597.5	997.3	2
401	26	58	39.5	0.3	2	440	84	161	437.4	387.2	2
402	16	80	915.6	634.8	2	441	86	169	317.4	761.7	2
403	43	60	584.0	97.4	2	442	73	199	470.7	5190.1	2
404	15	88	610.0	319.3	2	443	76	51	498.7	101.8	2
405	46	99	566.8	100.2	2	444	94	61	778.7	1354.0	2
406	36	99	38.1	0.0	2	445	85	60	523.3	117.5	2
407	54	80	483.0	105.0	2	446	104	38	617.1	200.2	2
408	46	81	542.6	138.6	2	447	93	41	395.3	28.4	2
409	54	161	959.3	466.7	2	448	75	90	518.9	113.0	2
410	43	161	631.9	261.1	2	449	94	101	383.7	12.9	2
411	65	160	928.3	2252.5	2	450	85	100	704.1	126.0	2
412	33	160	431.0	48.6	2	451	104	109	562.3	908.6	2
413	36	170	672.3	605.5	2	452	75	110	655.3	349.0	2
414	53	179	1003.4	425.4	2	453	95	121	823.6	548.4	2
415	44	180	876.4	937.8	2	454	83	119	847.7	701.4	2
416	65	181	734.1	589.3	2	455	94	140	607.5	723.2	2
417	34	180	366.0	110.2	2	456	103	139	491.2	565.3	2
418	33	191	296.5	79.1	2	457	114	120	319.5	154.2	2
419	55	199	1069.2	376.4	2	458	104	118	594.0	289.2	2
420	46	198	804.3	674.6	2	459	196	91	433.5	254.1	2
421	63	201	731.1	1363.3	2	460	215	101	209.6	4.0	2
422	34	201	318.1	79.6	2	461	204	101	533.8	127.3	2
423	65	210	238.6	488.3	2	462	196	101	592.4	419.4	2
424	35	208	428.9	161.2	2	463	195	149	478.7	141.9	2
425	46	220	737.4	1236.1	2	464	216	11	660.2	1424.8	2
426	36	219	429.1	152.3	2	465	225	19	832.2	512.2	2
427	35	271	597.4	397.0	2	466	214	19	242.5	15.6	2
428	53	258	442.6	1696.4	2	467	245	231	161.2	26.1	2
429	46	260	765.2	779.8	2	468	233	220	626.0	959.7	2
430	45	281	605.5	934.8	2	469	226	221	800.1	1681.5	2
431	35	278	795.9	1588.3	2	470	213	218	482.6	476.2	2

The sampling of the Walker Lake area was conducted in three stages. In the first stage, 195 samples were located on a roughly regular grid at a spacing of 20 x 20 m^2. In the second stage, additional sampling was done near the highest V values from the first stage. Each of the original 195 samples whose V value was greater than 500 ppm was surrounded by eight extra samples located approximately on a 10 x 10 m^2 grid. This second stage added 150 samples to our data set.

From these first two stages, the major anomalies are apparent. The third stage of sampling attempted to delineate these zones better by adding samples along existing east-west section lines. Two extra samples, one roughly 5 m to the east and the other roughly 5 m to the west, were added on either side of any of the previous 345 samples whose V value exceeded 500 ppm. The 125 samples added in this final stage bring the total number of samples in our data set to 470.

V and T measurements exist at the sample locations from all three stages; the U measurements exist only for the second and third stages. Like the preferential sampling in anomalous areas, this undersampling of some variables is common, especially when the sampling has occurred in separate programs. In our Walker Lake example, one can imagine that after the V results from the first stage of sampling became available, new information or changing economic conditions might have created interest in the U values.

Univariate Description of V

A histogram of the 470 V values in the sample data set is given in Figure 6.3 along with its summary statistics. The strong positive skewness evident in the histogram is confirmed by the summary statistics; the coefficient of variation is high and the mean is slightly greater than the median.

There are several extreme values in the data set. The maximum is more than three times greater than the mean. In our estimation studies these extreme values will severely inflate estimated values in their vicinity. For this reason, such extreme values are often dismissed as "outliers" in practice and simply deleted from the data set or arbitrarily adjusted to reduce their influence. Our sample extremes, however, are actual data values originating from the exhaustive data set and we have no justification for applying these methods. We will have to incorporate these extremes in our analysis.

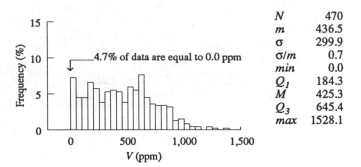

Figure 6.3 Histogram and univariate statistics of the 470 V sample values.

Comparing the sample statistics given in Figure 6.3 with the exhaustive statistics we calculated in the last chapter, we can start to appreciate the difficulty of estimating population parameters. The sample mean of 436 ppm is a very poor estimate of the exhaustive mean of 278 ppm; the median and the standard deviation are also larger than their exhaustive counterparts. The sample coefficient of variation, on the other hand, compares reasonably well with the exhaustive coefficient of variation. Apart from the fact that it is positively skewed with a minimum value of 0 ppm, there is little about the sample distribution that corresponds to the exhaustive distribution.

The differences between our sample and exhaustive distributions are not surprising given that more than half of our samples are intentionally located in areas with anomalously high V values. Our samples do not fairly cover the entire area and their statistics, therefore, are not representative of the entire area.

The effect of the preferential sampling is made clear in Table 6.2, where the summary statistics for the V distribution are calculated separately for each sampling program. The statistics from the first program, which covered the area uniformly with a 20 x 20 m² grid, are quite close to the exhaustive statistics we calculated in the previous chapter. The sample mean is within 10% of the exhaustive mean; the sample standard deviation also compares very well with the exhaustive standard deviation.

The statistics from the second and third programs reflect the sampling strategy. The mean value of the second program is more than twice that of the first, and in the third program the mean has nearly

Table 6.2   Comparison of V statistics by sampling program

	Exhaustive Statistics	Sample Statistics Program 1	Program 2	Program 3
n	78,000	195	150	125
m	278	275	502	610
σ	250	250	295	247
CV	0.90	0.91	0.59	0.41
min	0	0	0	0
Q_1	68	62	269	440
M	221	209	518	608
Q_3	429	426	675	781
max	1,631	975	1,528	1,392

tripled. We notice similar increases in the median and the standard deviation. The coefficient of variation decreases in the second and third campaigns because fewer of the very low values are being encountered. Recall that the coefficient of variation can be viewed as a measure of skewness. The distribution of the V values in the last two campaigns is less skewed than in the first, which is a natural consequence of the focus on high V values.

The effect of the sampling strategy on the distribution of the V values can also be revealed by a series of q-q plots. Figure 6.4 shows the q-q plots of the V sample distribution versus the exhaustive distribution for each of the three sampling campaigns; the dashed line on this figure shows the line that a q-q plot would follow if the two distributions were identical. We can see that the exhaustive V distribution is much closer to the V distribution from the first sampling program than to either of the sample distributions from the second and third programs. In these final two sampling programs, the strategy of locating samples near high V values from the first program makes the sample V distribution a poor approximation of the exhaustive distribution.

The comparisons in Table 6.2 and Figure 6.4 demonstrate that additional sampling does not necessarily make the sample data set more representative of the entire area. As we mentioned earlier, the strategy used in the sampling of the Walker Lake area is quite a common one.

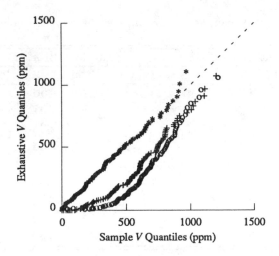

Figure 6.4 A comparison of the distribution of V for the three sampling campaigns to the exhaustive distribution of V The quantiles of V from the first sample program are plotted with the * symbol, the second with the +, and the third with the o.

In practice, we should be aware that the sampling strategy can distort the sample distribution, making it bear only a superficial resemblance to the exhaustive distribution.

The fact that the samples from the second and third campaigns cause the sample statistics to become less representative of the exhaustive statistics does not mean that the sampling strategy was a bad one. We have already seen that the data values become more erratic in areas where the local mean is high. Additional sampling in anomalously high areas makes good sense because it will improve our estimates in exactly those areas where the proportional effect makes them least accurate. Improved accuracy is also usually needed in areas of extreme values. We will see that there are several simple ways of dealing with clustered samples. As long as we are alert to the effects of clustering, the additional sampling in anomalous areas is a definite asset.

Univariate Description of U

Since the 195 samples from the first sampling campaign are missing the U measurement, there are only 275 U measurements in the sample

Table 6.3 Comparison of U statistics by sampling program.

	Exhaustive Statistics	Sample Statistics		
		Program 1	Program 2	Program 3
n	78,000		150	125
m	266		601	628
σ	489		801	724
CV	1.84	Not	1.33	1.15
min	0	Available	0	0
Q_1	7		67	111
M	57		254	397
Q_3	316		782	936
max	9,500		3,739	5,190

data set. The distribution of these 275 values is shown in Figure 6.5. This distribution is more skewed than the sample distribution of V values we saw in Figure 6.3. The sample U mean is higher than the V mean; however, the median is lower for U than for V.

It would be dangerous to infer that these same observations hold for the exhaustive V and U distributions. We have already seen that the strategy used in the second and third sampling campaigns had a large effect on the distribution of the V samples. All of our U data come from these last two campaigns. The differences we observe when we compare Figures 6.3 and 6.5 could therefore be a result of the fact that we are comparing the 275 U values from the last two campaigns with the 470 V values from all three campaigns.

A more legitimate approach to comparing the two distributions would be to compare them by sampling campaigns. Table 6.3 gives a statistical summary of the distributions of the U samples in the last two sampling campaigns. When we compare these statistics to those given in Table 6.2 for the same campaigns we again see the differences we noted earlier in our more naive comparison. The U distribution seems to be more skewed, with a higher mean and a lower median. Though this comparison is certainly more reasonable than our previous one, we should still be cautious about inferring that these differences also hold for the exhaustive distributions.

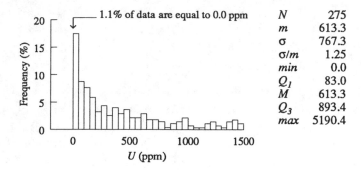

Figure 6.5 Histogram and univariate statistics of the 275 U sample values.

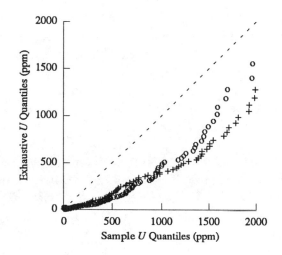

Figure 6.6 A comparison of the distribution of U for the last two sampling campaigns to the exhaustive distribution of U using q-q plots. The quantiles of U from the second sample campaign are plotted with the + symbol, while those from the third are plotted with the o.

The postings of the V and U values in Figure 6.1 and 6.2 make it clear that the first phase of sampling was the only one of the three phases that evenly covered the entire area. The second and third phases, the only two for which we have U measurements, cover only those areas that indicated high V values from the first phase. The differences we observe between our V and U samples may not hold over the entire area. If we had the U measurements at the 195 locations in the first sampling program, we could feel more confident about extending our observations from the sample data set to the exhaustive distributions. Without these 195 U measurements we have to be cautious about such inferences.

Checking the exhaustive V and U statistics, we can see that this caution is warranted. The V distribution is less skewed and does have a higher median. Its mean, however, is not lower than the U mean. The sampling strategy has distorted the V and U sample distributions enough that the V mean appears to be somewhat lower. The reason for this can be better understood if we make q-q plots of our U sample distributions for each of the sampling campaigns.

Figure 6.6 shows the two q-q plots of the sample U distributions versus the exhaustive U distribution for the two sample campaigns in which U was measured. The quantiles of U from the second sampling campaign are plotted using the + symbol while those from the third campaign are plotted using the o. The sampling strategy clearly affects the sample distribution of U in the same way that it affected the V distribution. The positive correlation of the two variables causes sampling aimed at locating high V values also to pick up high U values. This correlation, however, is not so strong that it affects the two sample distributions equally. The sampling strategy, which was based on the V values, produced a larger bias in the U sample distribution than in the V sample distribution. Indeed, we can see that the shift in the q-q plots from the second campaign to the third is larger on the q-q plots for U in Figure 6.6 than it is on the q-q plots for V in Figure 6.4.

The 195 V data from the first sampling phase gave us a reasonable approximation of the exhaustive V distribution. Without the U measurements from this initial phase we do not have a similarly good approximation of the U distribution. The posting of the U samples shows that we simply do not have adequate coverage over the whole area. For the remainder of this chapter we should keep in mind that our description of the U sample data is limited to those areas where we have U measurements and does not apply to those areas where U has not been sampled. Later, when we try to estimate the exhaustive U mean, we should not expect much success. When we tackle this estimation problem, we will look at ways of using the V values to help us in those areas where U information is lacking.

The Effect of the T Type

In the postings of the V and U samples given in Figure 6.1 and 6.2, the samples for which T is type 1 are shown with a • symbol while those for which T is type 2 are shown with a +. Forty-five of the samples in our sample data set, about 10%, are type 1; the remaining 425, about 90%, are type 2.

As with the other variables, the sample distribution of T is affected by the sampling strategy. In the first phase of sampling, 79% of the samples are type 2. This rises to 97% in the second phase and to 100% in the third phase. This tendency of the last two sampling programs to encounter more type 2 samples indicates that high V values are more likely to be associated with type 2 samples. This is easily checked by looking at the posting of the V values in Figure 6.1. Most of the type 1 samples have very low values; for about a third of them, the V value is 0 ppm.

When we look at the U posting in Figure 6.2 we notice a similar tendency. Unfortunately, 41 of the type 1 samples plotted on the V posting are from the first sampling program; the four type 1 samples collected in the final two programs are the only ones for which we have a U measurement. Without the benefit of the U measurements from the first sampling program, the difference in the U values of the two types is not as obvious as it is for V.

As we continue our analysis of the sample data set, we will often be hampered by the lack of U measurements for type 1 samples. Though the T type seems to be an important factor, we are unable to describe its influence on U with only four samples of type 1.

The general tendency of type 1 samples to have lower values is confirmed by the statistical summary given in Table 6.4. With the exception of the coefficient of variation and median, every summary statistic is lower for type 1 samples. The higher coefficient of variation and median indicate that the distribution of U is more strongly skewed for the type 1 samples.

The V-U Relationship

A scatterplot showing the relationship between the V and U values at the same sample locations is given in Figure 6.7. There are 275 pairs of V-U values, one for each of the locations at which a U measurement exists. We have used different symbols for the type 1 and type 2

Table 6.4 Comparison of V and U statistics by sample type.

	V		U	
	Type 1	Type 2	Type 1	Type 2
n	45	425	4	271
m	40	479	429	616
σ	52	284	479	772
CV	1.29	0.59	1.12	1.25
min	0	0	0	0
Q_1	0	241	15	85
M	18	477	367	335
Q_3	72	663	906	893
max	195	1,528	983	5,190

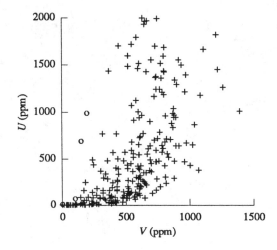

Figure 6.7 A scatterplot of the 275 U and V sample data. The type 1 points are shown with the symbol 1 while the type 2 points are shown with the 2.

samples so that we can also see what influence the T type has on the V-U relationship.

The correlation coefficient of all 275 pairs is 0.55. Though there are only four points of type 1, these all plot to the left of the type 2 points, suggesting that the T type separates the cloud of points into two portions. The T type plays a major role in the V-U relationship. Were we to predict V from the U values, knowledge of the T type may improve our estimates.

If we check our exhaustive statistics from the previous chapter, we see that the preferential sampling, which severely affected

the univariate statistics, has not had such a large effect on the correlation coefficient. For the 78,000 V-U pairs in the exhaustive data set, we calculated a correlation coefficient of 0.65, somewhat higher than our sample value of 0.55. The relationship between V and U is not as strong for high values as it is for low ones. On the exhaustive scatterplot we saw in Figure 5.8, the cloud of points became more diffuse at high values. Our sampling strategy, which gives us a disproportionate number of high values, causes the sample correlation to appear somewhat weaker than the actual exhaustive correlation.

Spatial Description

We began our analysis of the sample data set with a posting (Figure 6.2), that showed the locations of all of the V samples along with their values. Though this is a valuable map for detecting errors and checking various results, it does not effectively convey the spatial features of our samples.

A contour map certainly gives us our best overall view of the important spatial features. Unfortunately, in practice it is often used exclusively with no other displays to provide the finer detail which the contour map masks. It bears repetition that contouring thatrgely an aesthetic exercise and therefore often requires that the actual fluctuations in the data values be smoothed. Many interesting features are not visible or are even camouflaged on a contour map.

Another benefit of exploring other displays to complement the contour map is that we increase our knowledge of the peculiar details of our samples. We start to remember, for example, the locations of extreme values, the areas that are sparsely sampled or the areas where the values are highly variable. We should not allow automated contouring to deprive us of this valuable familiarity with the sample data set.

Figure 6.8 shows the locations of the samples with extremely high or extremely low V values. Like the complete posting of data values, this is often a useful tool for detecting errors. Also, by isolating the

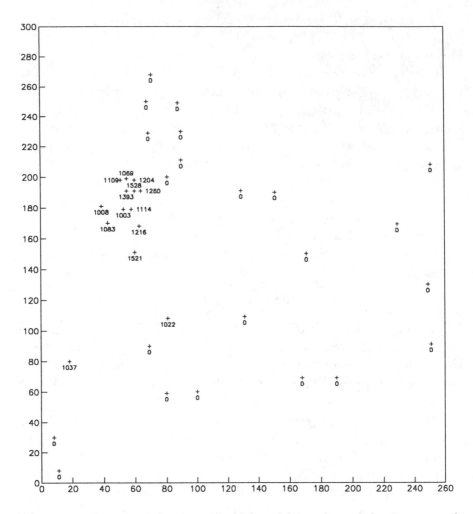

Figure 6.8 Posting of the extremely high and low values of the 470 *V* sample data.

extreme values, we begin to develop an appreciation of the spatial arrangement of the values. We notice that 12 of the highest *V* values all occur in the Wassuk Range area. The 22 samples with a *V* value of 0 ppm are scattered throughout the map area, with a small group located near Walker Lake itself. Later, we will look at how we can summarize this difference between the continuity of the high values and that of the low values.

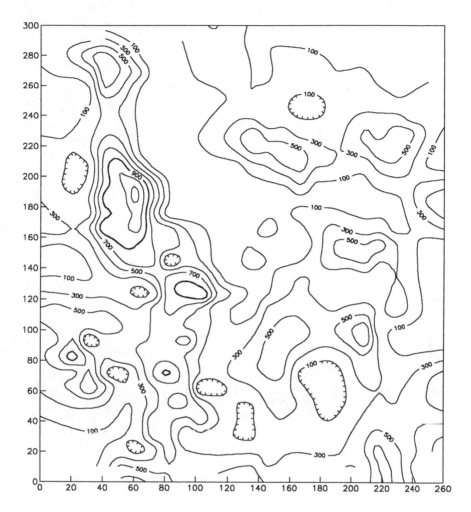

Figure 6.9 Contour map of the 470 V sample data. The contour interval is 200 ppm and begins at 100 ppm.

Figure 6.8 also reveals that extremely high values can be located very close to extreme lows. Several of the 12 highest V values are located only a few meters away from one of the 0 ppm values. Such dramatic short scale variability has a serious impact on the accuracy of estimates. In areas where the data values fluctuate wildly, we should not expect the same accuracy we get in areas where the values are not as erratic.

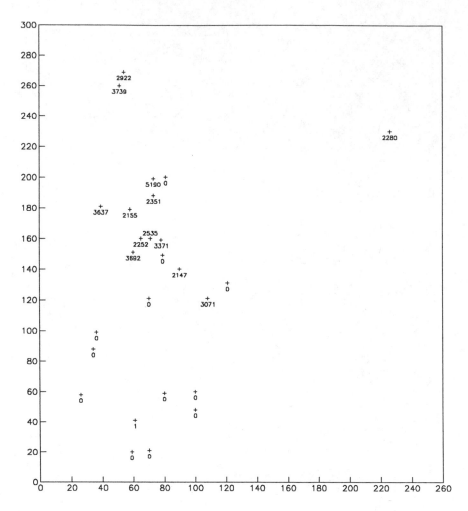

Figure 6.10   Posting of the extremely high and low values of the 275 U sample data.

A contour map of the V values is shown in Figure 6.9. Groups of samples with large V values create several anomalies over the entire Walker Lake map area. The most prominent of these is the large anomaly that covers the Wassuk Range in the northwest portion of the map area. The 12 highest values we plotted on Figure 6.8 all occur within this major anomaly.

The existence of this large area of high V values was apparent

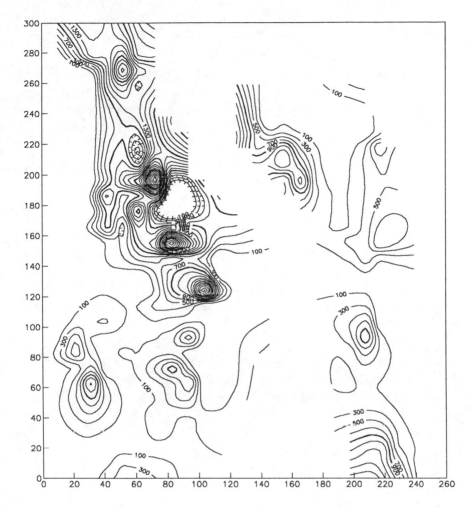

Figure 6.11 Contour map of the 275 *U* sample data. The contour interval is 200 ppm and begins at 100 ppm.

even from our original posting. The dense coverage in the northwestern portion of Figure 6.2 pointed to a major anomaly. Even though the Wassuk Range covers only one tenth of the map area, it contains nearly one third of the samples. The other areas of dense sampling in Figure 6.2 also correspond to anomalies on the contour map.

When we compare the contour map based on our sample data set to our spatial displays of the exhaustive data set (Figures 5.11 and 5.14), we notice a few discrepancies. Our sampling strategy has succeeded in delineating the major high features. Several of the low features, however, are not well defined or are missing completely.

The largest zone of low values is poorly represented on our sample contour map. The exhaustive indicator maps and the exhaustive contour map of the moving window means both show that this zone of low values covers all of Walker Lake itself and extends unbroken along the Soda Spring Valley. On our sample contour map the east-west portion of this zone is not apparent. An examination of the posting of V values shows that the trail of very low values does indeed disappear into some rather high ones near 250E,160N. Ironically, had the 266 ppm value in this area been slightly higher, it would have caused extra samples to be located nearby in the second and third sampling programs; these additional samples likely would have established that the very low values are actually continuous across this area. Regrettably, these additional samples are not available and we have to accept the consequences of the sampling strategy, which did little to help us delineate the lowest values.

When we try to display the spatial features of our U values, we are limited by our sample coverage. As the posting of our U samples (Figure 6.1) shows, much of the map area has no U measurements. A further complication, one which we noticed earlier, is that the U measurements that we do have are generally high ones. The strategy used for the second and third sampling campaigns gave us few additional low V measurements, but we could still rely on those low V measurements we had in the 195 samples from the first campaign. Since none of these initial samples was measured for U the only low U measurements we have are the few we pick up in the second and third campaigns.

A posting of the extreme U values (Figure 6.10) is similar to the corresponding display for V. The highest values again plot in the Wassuk Range area, but are more scattered than the highest V values. The 0 ppm values are also widely scattered. The fact that we see fewer 0 ppm values on the U map than we did on the V map does not mean that the V variable generally reaches its minimum value more often. We should not forget that we are missing the U measurements from the first 195 samples.

The fair correlation between V and U leads us to expect a fair similarity in their major spatial features. The contour map of the U values (Figure 6.11) confirms this similarity, showing many of the same

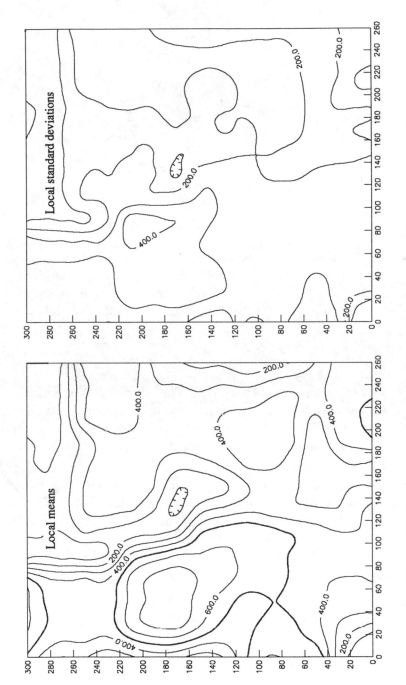

Figure 6.12 Contour maps of moving window statistics for *V*. The local means are contoured in the left map while the local standard deviations are on the right.

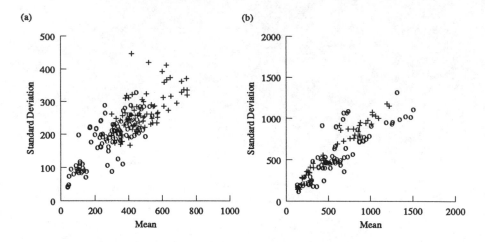

Figure 6.13 Scatterplots of moving window standard deviations versus means for (a) the 470 sample *V* data and (b) the 275 *U* sample data. For moving windows containing less than 20 data values the standard deviation and mean were plotted using the "o" symbol, otherwise a "+" symbol was used. Note that the scale of the two plots differs.

anomalies we saw on the *V* contour map. The Wassuk Range anomaly remains the largest and appears more pronounced further to the south while the location of the other major high anomalies remains roughly the same, with their shapes being somewhat different.

Parts of our *U* contour map have been left blank. This is a good precaution to take when there are large unsampled areas. Automatic contouring packages usually manage to fill these unsampled areas with contour lines. This often involves extrapolating data values over large distances and the results can be quite misleading. It is safer to be humble and admit our ignorance by leaving such areas blank.

Proportional Effect

The posting of the extremely high and low values alerted us to the possibility of very large fluctuations in the data values over short distances. As we noted, it is important to know if the local variability changes across the area. If it does, as is usually the case, it is useful

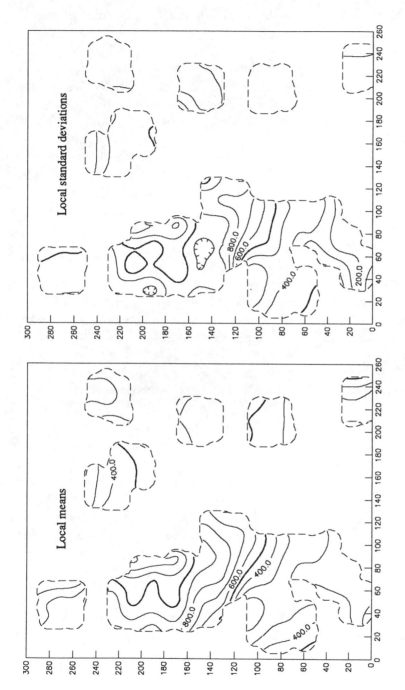

Figure 6.14 Contour maps of moving window statistics for *U*. The local means are contoured in the left map while the local standard deviations are on the right.

to know how the changes in local variability are related to changes in the local mean.

Summary statistics have been calculated within 60 x 60 m^2 windows that overlap and cover the entire area. If we do not overlap these 60 x 60 m^2 windows, we will be able to calculate only 20 local means and standard deviations. By moving the windows 20 m, so that they overlap their neighboring windows by 40 m, we can get 195 local mean and standard deviation calculations. The overlap of our windows causes some of the data to be used in the calculation of several means and standard deviations. This is not a major problem since our intent is simply to establish whether or not a proportional effect exists.

The size of the moving windows guarantees that we get at least nine V samples within each window except along the edges of the map area. If fewer than five samples are found, then their statistics will not be considered since they may be unreliable.

Contour maps of the local means and local standard deviations are shown in Figure 6.12. The large windows cause considerable smoothing, leaving the Wassuk Range anomaly as the only distinct feature. Comparing the local means in Figure 6.12a to the local standard deviations in Figure 6.12b we see they are related. A scatterplot of the m-σ pairs from each moving window (Figure 6.13a) not only confirms this relationship but also reveals that it is approximately linear. The correlation coefficient is 0.81.

The same analysis for the U values produces a similar result. The lack of samples in many areas causes us to reject a total of 58 of the 60 x 60 m^2 moving windows. The results for the remaining 137 windows, all of which contain at least five U samples, are shown in Figure 6.14. Where there are enough data to permit reasonable contouring, the map of the local U means is similar to that of the local standard deviations. The scatterplot of the m-σ pairs from each moving window (Figure 6.13) confirms that a proportional effect exists. The relationship is roughly linear and is well defined ($\rho = 0.88$).

Further Reading

Koch, G. and Link, R. , *Statistical Analysis of Geological Data*. New York: Wiley, 2 ed., 1986.

Castle, B. and Davis, B. , "An overview of data handling and data analysis in geochemical exploration," Tech. Rep. 308, Fluor

Daniel Inc., Engineering Mining and Metals, 10 Twin Dolphin Drive, Redwood City, CA, 94065, March 1984. Association of Exploration Geochemists short course, Reno, Nevada.

7

THE SAMPLE DATA SET:

SPATIAL CONTINUITY

In this chapter we complete the exploratory analysis of the sample data set by describing the spatial continuity of the two continuous variables, V and U, as well as their cross-continuity.

The analysis of spatial continuity in a sample data set is often very frustrating, sometimes seemingly hopeless. There is much that can be learned, however, from failed attempts at describing spatial continuity. If a particular tool, whether a variogram, a covariance function or a correlogram, fails to produce a clear description, exploring the causes of the disappointing results often leads to new insights into the data set. If these insights suggest improvements to the conventional tools, one should not hesitate to adapt the tools presented in Chapter 4 to the peculiarities of the sample data set. Indeed, one of the goals of this chapter is to demonstrate that this process of customizing the available tools in response to disappointing results is often the key to a successful analysis of the spatial continuity.

At this stage, our goal is purely descriptive and we are free to choose whatever tools produce good descriptions. Later, when we try to incorporate the spatial continuity into our estimation procedures, the way we choose to measure spatial continuity will be more restricted. We will defer discussion of those restrictions to later chapters and concentrate in this chapter solely on good description.

Sample h-Scatterplots and Their Summaries

Before we start to look at the spatial continuity in the sample data set, we need to look at how h-scatterplots and their summaries are slightly altered in practice to accommodate sample data sets. The problem we encounter when we try to construct an h-scatterplot from a sample data set is that for any **h** we choose there is enough randomness in our sample locations that very few pairs of samples are separated exactly by **h**.

Despite the best plan, practical sampling campaigns rarely manage to locate samples exactly at the desired locations. One of the awkward realities of earth science studies is that field conditions have a significant impact on the final state of the sample data set. In most other disciplines that use statistical methods, the data are gathered in the safe sterility of a laboratory and are not influenced by such diverse factors as last night's rainfall and this morning's bear sighting. In earth science studies, those responsible for actually collecting the samples often prefer to avoid swamps, dense undergrowth, and steep hills. It may simply be more expedient to sample some location other than the intended one. Surveying errors or the lack of a coherent sampling plan also introduce a certain degree of randomness to sample locations.

In drawing our samples from the exhaustive Walker Lake data set we randomly shifted the intended location of our samples by a small amount in order to mimic the randomness inherent in the locations of real samples. The result, as seen on the posting in Figure 6.2, is that the samples are not perfectly gridded.

To take a specific example, the third sampling campaign, which added 125 samples to our data set, was intended to add samples 5 m on either side of existing samples with high V values. We might expect, therefore, to have 125 pairs of samples located 5 m apart in an east-west direction. Unfortunately, if we construct an h-scatterplot for $\mathbf{h} = (5,0)$ we discover that only 11 pairs of samples have exactly this spacing. The remaining pairs we expected to find are only approximately 5 m apart in an east-west direction; for example, 55 pairs are off by 1 m.

We have to tolerate this inevitable randomness and accept any pair whose separation is close to **h** on our h-scatterplots. In practice, we specify tolerances both on the distance of **h** and on its direction. A typical example is shown in Figure 7.1 in which we use a tolerance of $\pm$ 1 m on the distance and $\pm$ 20 degrees on the direction of **h**. Any

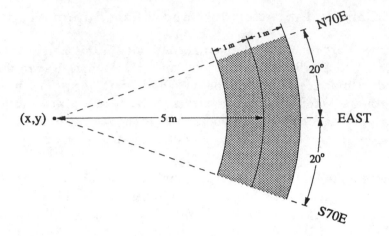

Figure 7.1 An illustration of the tolerances on **h** for the selection of data pairs in an h-scatterplot. A tolerance of ±1 is allowed on the magnitude of **h** and a tolerance of ± 20 degrees is allowed on the direction. Any sample falling within the shaded area would be paired with the sample at (x,y).

point that falls within the shaded area on this diagram is accepted as being 5 m away in an easterly direction [1].

The summaries we proposed earlier for h-scatterplots remain essentially the same, but their equations should be altered to reflect the fact that we no longer require that pairs be separated exactly by **h**. For example, equation 4.9 would now be written

$$\gamma(\mathbf{h}) = \frac{1}{2N(\mathbf{h})} \sum_{(i,j)|\mathbf{h}_{i,j}\approx\mathbf{h}} (v_i - v_j)^2 \qquad (7.1)$$

The only difference between this and our original definition is that we now choose to sum over the $N(\mathbf{h})$ pairs whose separation is approximately **h**. The way we choose to specify "approximately" will depend on the sample data set. The most common approach is the one shown in Figure 7.1, where we allow a certain tolerance on the distance and another tolerance on the direction. As we analyze the spatial continuity of V in the following sections, we will look at how these tolerances are chosen in practice.

An Outline of Spatial Continuity Analysis

Although a series of h-scatterplots provides the most complete description of spatial continuity, it usually contains too much information and requires some type of summary. In Chapter 4 we introduced three functions for summarizing spatial continuity: $\rho(\mathbf{h})$, the correlation function; $C(\mathbf{h})$, the covariance function; and $\gamma(\mathbf{h})$, the variogram. All of these use some summary statistic of the h-scatterplots to describe how spatial continuity changes as a function of distance and direction. Any one of them is adequate for purely descriptive purposes. The variogram, however, is the most traditional choice. Though the covariance and correlation functions are equally useful and, in some other disciplines, more traditional, our approach in this chapter will be to begin with the variogram and resort to other summaries only when there appears to be no way to improve the sample variogram.

It is appropriate to introduce at this point some terminology that is used to describe the important features of the variogram.

Range. As the separation distance between pairs increases, the corresponding variogram value will also generally increase. Eventually, however, an increase in the separation distance no longer causes a corresponding increase in the average squared difference between pairs of values and the variogram reaches a plateau. The distance at which the variogram reaches this plateau is called the *range*.

Sill. The plateau the the variogram reaches at the range is called the *sill*.

Nugget Effect. Though the value of the variogram for $\mathbf{h} = 0$ is strictly 0, several factors, such as sampling error and short scale variability, may cause sample values separated by extremely small distances to be quite dissimilar. This causes a discontinuity at the origin of the variogram. The vertical jump from the value of 0 at the origin to the value of the variogram at extremely small separation distances is called the *nugget effect*. The ratio of the nugget effect to the sill is often referred to as the *relative nugget effect* and is usually quoted in percentages.

One typically begins the analysis of spatial continuity with an omnidirectional variogram for which the directional tolerance is large enough that the direction of any particular separation vector, $\mathbf{h}_{ij}$, becomes unimportant. With all possible directions combined into a single variogram, only the magnitude of $\mathbf{h}_{ij}$ is important. An omnidirectional variogram can be thought of loosely as an average of the various directional variograms. It is not a strict average since the sample locations may cause certain directions to be over represented. For example, in the omnidirectional variograms presented in the this chapter the sampling strategy of the third campaign causes more east-west pairs than north-south pairs.

The calculation of an omnidirectional variogram does not imply a belief that the spatial continuity is the same in all directions; it merely serves as a useful starting point for establishing some of the parameters required for sample variogram calculations. Since direction does not play a role in omnidirectional variogram calculations, one can concentrate on finding the distance parameters that produce the clearest structure. An appropriate increment between successive lags and a distance tolerance can usually be chosen after a few trials.

Another reason for beginning with omnidirectional calculations is that they can serve as an early warning for erratic directional variograms. The omnidirectional variogram contains more sample pairs than any directional variogram and is, therefore, more likely to show a clearly interpretable structure. If the omnidirectional variogram does not produce a clear structure, one should not expect much success with directional variograms. If the omnidirectional variogram is messy, then one should try to discover the reasons for the erraticness. An examination of the h-scatterplots may reveal that a single sample value is having a large influence on the calculations. A map of the locations of erratic pairs can also reveal unforeseen problems. If the reasons for the erraticness can be identified, then one should adapt the variogram calculation to account for the problems. This may involve entirely removing certain samples from the data set, or removing particular pairs of samples only from particular h-scatterplots [2].

If repeated attempts at improving the clarity of the sample variograms prove fruitless, then one should consider a different measure of spatial continuity. In addition to the correlation and covariance functions we used earlier, there are several *relative* variograms that one can consider using. Though these relative variograms lack the theoretical pedigree of the three measures we introduced in Chapter 4, experience has shown that they can be very useful in producing clear descriptions of the spatial continuity.

Once the omnidirectional variograms are well behaved, one can proceed to explore the pattern of anisotropy with various directional variograms. In many practical studies, there is some prior information about the axes of the anisotropy. In a mineral deposit, there may be geologic information about the ore genesis that suggests directions of maximum and minimum continuity. For example, in a sedimentary hosted mineral deposit, the direction of maximum continuity will most likely parallel the stratigraphy. In the study of the concentration of an airborne pollutant, such information might come from knowledge of the prevailing wind direction; for a pollutant transported by groundwater, hydrogeologic information about the contaminated aquifer could be helpful in choosing directions for variogram calculations.

Without such prior information, a contour map of the sample values may offer some clues to the directions of minimum and maximum continuity. One should be careful, however, in relying solely on a contour map even though they usually show the maximum and minimum continuity directions quite well. Automated contouring typically involves a first step of interpolation to a regular grid; this step can produce artifacts. The appearance of elongated anomalies on a contour map may be due to the gridding procedure rather than to an underlying anisotropy. One very good approach to identifying the anisotropy if no prior information exists is to try to produce a picture of the entire variogram surface, similar to the one shown of the exhaustive covariance function in Figure 7.6 With only a few samples, this may not be possible. An alternative, and perhaps the more common approach, is to calculate several directional variograms and plot a rose diagram, which shows the variogram range or slope at the origin as a function of direction.

Once the directions of maximum and minimum continuity have been established, one needs to chose a directional tolerance that is large enough to allow sufficient pairs for a clear variogram, yet small enough that the character of the variograms for separate directions is not blurred beyond recognition. At this stage one may have to repeat the earlier process of examining h-scatterplots, trying to unravel the reasons for erratic behavior and adapting the variogram accordingly.

The analysis of spatial continuity is rarely a straightforward process and one should be prepared for several iterations of the steps outlined earlier. For readers struggling with the variogram analysis of their own uncooperative data sets, it may be encouraging to know that the following analysis, despite its appearance of orderly progress, involved several false starts and retraced steps.

Table 7.1 Omnidirectional sample variogram for V with a 5 m lag*.

No. of Pairs	Lag	$\gamma(h)$	No. of Pairs	Lag	$\gamma(h)$
22	2.1	11,294.1	3,920	55.0	94,415.1
488	5.4	42,671.4	5,324	60.2	88,848.9
1,720	10.4	51,932.4	4,442	64.8	96,309.2
1,856	14.8:	71,141.8	5,478	70.2	96,397.3
3,040	20.3	70,736.9	4,696	74.8	90,704.6
2,412	24.9	86,745.2	5,762	80.2	92,560.6
3,550	30.1	84,077.8	5,084	84.9	88,104.0
2,816	34.8	99,986.6	5,666	90.1	95,530.9
4,092	40.3	89,954.4	4,458	94.8	101,174.8
3,758	44.9	86,155.0	2,890	98.8	94,052.1
4,248	50.2	98,319.3			

*The plot is shown in Figure 7.2.

Choosing the Distance Parameters

There are two distance parameters that need to be chosen. One is the distance between successive h-scatterplots, usually referred to as the *lag spacing* or *lag increment*; the other is the tolerance we will allow on the distance. The sampling pattern may suggest a reasonable lag increment. If the samples are located on a pseudo-regular grid, the grid spacing is also usually a good lag spacing. If the sampling is random, one can use as an initial lag spacing an estimate of the average spacing between neighboring samples.

If the sampling pattern is noticeably anisotropic, with the sample spacing being much smaller in some directions than in others, the distance parameters will depend on the direction. A typical example is a sample data set consisting of assays from drill cores, where the sample spacing in the vertical direction is much smaller than the sample spacing horizontally. In such situations, an omnidirectional variogram is not recommended for establishing the distance parameters. Instead, one should group sample pairs that share similar spacing. For example, with a drill hole data set a variogram that combined all horizontal directions would be appropriate for establishing the horizontal distance

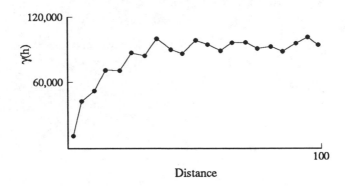

Figure 7.2 Omnidirectional sample variogram for V with a 5 m lag.

Table 7.2 Omnidirectional sample variogram for V at a 10 m lag[*].

No. of Pairs	Lag	$\gamma(h)$	No. of Pairs	Lag	$\gamma(h)$
178	3.6	32,544.3	9,782	60.3	91,285.2
3,044	11.0	55,299.8	10,060	70.3	93,809.2
5,140	20.4	75,224.6	10,628	80.3	92,357.8
6,238	30.2	88,418.6	10,454	90.1	95,010.5
7,388	40.5	90,544.1	4,856	97.8	97,349.3
7,954	50.1	95,689.7			

[*] The plot is shown in Figure 7.3.

parameters, while a separate variogram for the vertical direction would be used for establishing the vertical distance parameters.

The most common choice for the lag tolerance is half the lag spacing. If the samples are located on a regular grid or on a pseudo-regular grid, one may choose a lag tolerance smaller than half the lag spacing. While this can result in some pairs not being used in the variogram calculation, it can also make the structure clearer.

The existence of samples located very close to each other may affect the choice of the distance parameters. One may want to include an additional lag for small separation distances and use a small tolerance for this first lag so that any duplicate or twin samples are grouped together, providing a sample variogram point close to the origin.

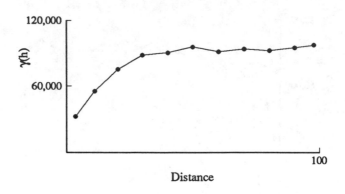

Figure 7.3 Omnidirectional sample variogram for *V* with a 10 m lag.

Figure 7.2 shows the omnidirectional variogram for the 470 *V* values in the sample data set. Successive lags are 5 m apart and each lag has a tolerance of 2.5 m. This initial choice of lag increment and tolerance was based on the fact that the samples from the third campaign were located at approximately 5 m intervals. While this omnidirectional variogram shows a fairly clear structure, there is still some erraticness in the sample variogram values. There are several small jumps from one lag to the next, creating a jagged appearance.

The number of pairs, their average separation distance, and the sample variogram value for each lag is given in Table 7.1. These results give us an indication that the jagged appearance of our variogram may be due to our choice of lag increment and tolerance. From the fourth lag onward, the number of sample pairs increases and decreases regularly, with the lags centered on multiples of 10 m containing more sample pairs than the others. The problem is that the 5 m spacing from the third sampling campaign was only in an east-west direction. This means that while the lags centered on multiples of 10 m contain samples that are representative of many directions and of the entire area, the alternate lags are restricted predominantly to east-west pairs from areas that received additional sampling.

A better choice of lag increment might be 10 m, with a 5 m tolerance on the distance. The omnidirectional variogram calculated using these parameters is shown in Figure 7.3, with the accompanying details in Table 7.2. The structure in this variogram is similar to that seen in our initial attempt, but the jagged appearance is now gone. For the

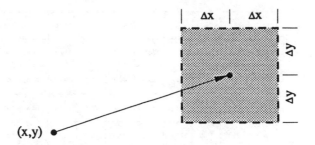

Figure 7.4 An illustration of how two dimensional data can be grouped to form a variogram surface. For any one block all samples falling within the shaded portion will be paired with the sample at (x,y). The variogram values of all such pairs are averaged and plotted as shown in Figure 7.5. A regular grid of these variogram averages forms the variogram surface.

remainder of our analysis of the spatial continuity, we will continue to use a lag increment of 10 m and a tolerance of 5 m.

Finding the Anisotropy Axes

Having found an acceptable omnidirectional variogram, we can now try to study directional anisotropies in the variogram. In many data sets the data values along certain direction are more continuous than along others. A display that quickly reveals directional anisotropies is a contour map of the sample variogram surface. Since contouring programs typically require data on a rectangular grid, a contour map of the variogram surface is easier to construct if the tolerance on $\mathbf{h}$ is defined in a rectangular coordinate system, such as the one shown in Figure 7.4, rather than the conventional polar coordinate system shown in Figure 7.1. In calculating the variogram value for pairs of points separated by the vector $\mathbf{h} = (h_x, h_y)$, we group together all pairs whose separation in the x direction is $h_x \pm \Delta x$ and whose separation in the y direction is $h_y \pm \Delta y$.

Figure 7.5 shows the variogram calculations for pairs whose separation distance in the east-west direction, h_x, is less than 50 m and whose separation in the north-south direction, h_y, is also less than 50 m. The data pairs have been grouped into 100 lags, with the a lag increment of 10 m and a lag tolerance of ± 5 m in both directions. These values

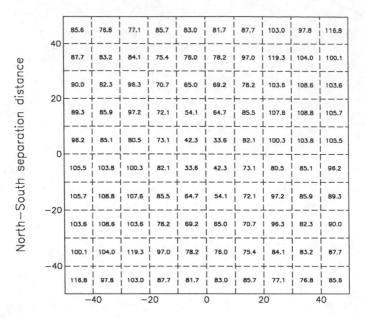

Figure 7.5 A posting of grouped sample variogram values for lag tolerances as shown in Figure 7.4. The variogram values have been grouped into 100 lags with a lag increment of 10 m and a lag tolerance of ± 5 m in both directions.

have been contoured in Figure 7.6. There is a clear anisotropy, with the variogram surface rising rapidly along the N76°E direction, and slowly along the N14°W direction [3].

Despite their effectiveness, contour maps of the variogram surface are not commonly used in practice [4]. For very erratic data sets, the variogram values within the various rectangular lags may be too erratic to produce a useful contour map. Furthermore, many practitioners have neither the software nor the hardware required for displays such as the one shown in Figure 7.5. The conventional approach to finding the directions of maximum and minimum continuity consists essentially of trying to trace only one of the contour lines.

Nine directional variograms are shown in Figure 7.7. For each of these directional variograms, an angular tolerance of ±45 degrees is large enough to give us an interpretable structure. The variogram

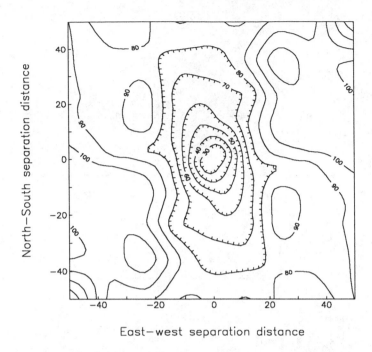

Figure 7.6 A Contour map of the variogram values grouped in Figure 7.5. The contour values are in thousands of parts per million squared.

reaches a sill above $80,000$ ppm^2 along each of these directions, so we will try to trace this particular contour line. For each of the directions, we find the distance at which the variogram reaches $80,000$ ppm^2 by linear interpolation between the two closest points; these nine distances are shown in Figure 7.7a-i. Plotting these distances on a rose diagram, as in Figure 7.8, amounts to showing the distance in several directions to the $80,000$ ppm^2 contour line. Such diagrams often show an anisotropy that appears elliptical; Figure 7.9 shows that the ranges shown in Figure 7.7 are fit quite well by an ellipse whose major axis is N10°W and whose minor axis is N80°E.

The choice of which contour line to trace is not crucial. Any value of $\gamma(h)$ for which the distance is unique and easily interpolated is adequate. The complete contour map in Figure 7.6 shows that a similar result would have been reached with any contour line from $50,000$ to $80,000$ ppm^2.

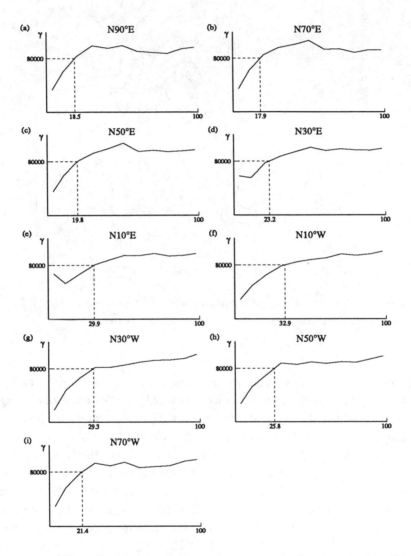

Figure 7.7 Nine directional sample variograms each showing the lag distance corresponding to a variogram value of 80,000 ppm^2.

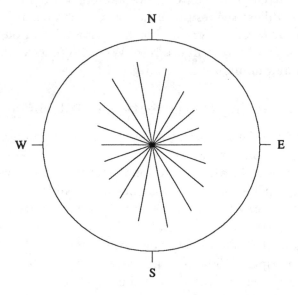

Figure 7.8 A rose diagram of the nine ranges shown in Figure 7.7.

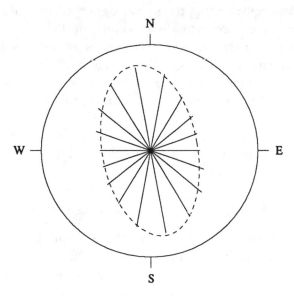

Figure 7.9 An ellipse fit to the rose diagram of the nine ranges shown in Figure 7.8. The major and minor axes of the ellipse represent the axes of a geometric anisotropy.

Since the attempt to contour the sample variogram surface was successful, we will use the results of that approach rather those from the more conventional rose diagram. For the remainder of our analysis of the spatial continuity, we will use N14°W and N76°E as the directions of maximum and minimum continuity.

Choosing the Directional Tolerance

Having discovered the directions of maximum and minimum continuity, the remaining parameter we need to choose is the angular or directional tolerance. When calculating directional variograms, we would ideally like to use as small an angular tolerance as possible to limit the blurring of the anisotropy that results from combining pairs from different directions. Unfortunately, too small a directional tolerance often gives us so few pairs that the directional variogram is too erratic to serve as a useful description. The best approach is to try several tolerances and use the smallest one that still yields good results.

Figure 7.10 shows the N14°W and N76°E variograms with four different tolerances; Table 7.4 provides the details of the calculations for an angular tolerance of ± 40 degrees. For any particular lag, the number of pairs contributing to the variogram calculation increases as the directional tolerance increase as shown in Table 7.3. In the first lag, for example, there are no pairs separated by less than 5 m in the N14°W direction when the tolerance is only ±10 degrees; as the angular tolerance is increased to ±20 degrees, one pair appears in this first lag; and as it is widened to ±40 degrees, this first lag includes six pairs.

An angular tolerance of ±40 degrees appears to be large enough to allow well-defined directional variograms while still preserving the evident anisotropy.

Sample Variograms for U

We now turn to the secondary variable, U, and try to describe its spatial continuity. Our first attempt, an omnidirectional variogram, is shown in Figure 7.11, along with the accompanying details in Table 7.5. This is not a very satisfying description of the spatial continuity of U; even though there is some visible increase of $\gamma(h)$ with distance, it is slight and somewhat erratic. Despite the contribution of 124 pairs, the value of $\gamma(h)$ for the first lag is very high. Furthermore, the dip that

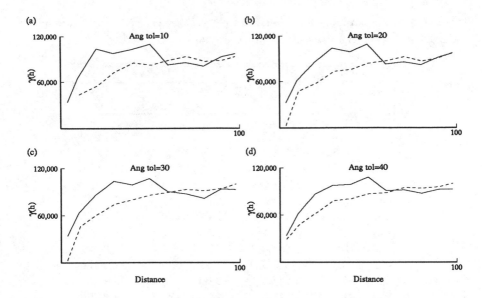

Figure 7.10 Directional sample variograms of V for various angular tolerances. In each plot, the variogram in the N76°E direction is shown by the solid line and in the N14°W direction by the dashed line. The number of pairs for each angular tolerance is given in Table 7.4. Details of the variogram shown in (d) are in Table 7.4.

Table 7.3 Table showing the number of pairs in the directional sample variograms of V for increasing angular tolerances*.

±10°		±20°		±30°		±40°	
N76°E	N14°W	N76°E	N14°W	N76°E	N14°W	N76°E	N14°W
22	0	57	1	66	1	76	6
168	176	352	370	532	543	674	714
165	281	710	681	871	911	1,058	1,152
327	363	631	682	936	1,051	1,386	1,521
251	445	806	966	1,099	1,405	1,453	1,798
338	554	692	1,071	1,103	1,605	1,521	2,071
411	709	893	1,476	1,269	2,000	1,744	2,646
388	783	809	1,505	1,259	2,178	1,706	2,703
494	854	923	1,658	1,416	2,396	1,879	2,999
457	768	915	1,542	1,380	2,193	1,828	2,804
234	314	459	634	687	890	957	1,250

*See Figure 7.10.

Table 7.4 Details of the directional sample variograms of V with an angular tolerance of $\pm$ 40 degrees as shown in Figure 7.10(d).

N76°E			N14°W		
No. of Pairs	Lag	$\gamma(h)$	No. of Pairs	Lag	$\gamma(h)$
Angular Tol = $\pm$ 40°					
76	3.6	34,154.8	6	3.7	29,249.0
674	10.5	62,228.4	714	11.3	48,228.6
1,058	19.9	86,521.1	1,152	20.6	62,705.8
1,386	29.7	97,758.3	1,521	30.5	78,342.6
1,453	40.2	98,921.3	1,798	40.8	80,920.5
1,521	50.0	108,078.7	2,071	50.4	86,908.9
1,744	60.1	90,870.2	2,646	60.6	88,282.6
1,706	70.1	91,668.4	2,703	70.4	95,068.3
1,879	80.1	87,278.7	2,999	80.5	94,065.4
1,828	90.0	92,427.5	2,804	90.2	95,929.5
957	97.7	92,742.5	1,250	97.8	100,023.3

occurs around 50–60 m presents a confusing story. This characteristic, often referred to as a *hole effect*, suggests that samples separated by 50 m are actually more similar than those separated by only 30 m.

There are certain natural phenomena for which a hole effect variogram is to be expected. If there are discrete lenses of high values, as in many ore deposits, a hole effect may be seen on the variogram at roughly the average spacing between adjacent lenses. Hole effects also occur in data sets in which there is a natural cyclicity or repetition. The vertical variogram of porosity or permeability from a well drilled through sedimentary rock often shows hole effectsthat can be attributed to cycles of facies changes.

When one sees a hole effect on a sample variogram, it is useful to check maps of the data set to see if there is an obvious cause. In the case of the U variable for Walker Lake, there is no evident explanation for the hole effect seen on Figure 7.11, and it is reasonable to conclude that this is undesirable noise in our summary of spatial continuity.

Before we proceed with our analysis of the spatial continuity of U, we need to investigate the reasons for the poor behavior of the

Table 7.5   Omnidirectional sample variogram for U^*.

No. of Pairs	Lag	$\gamma(h)$	Lag Mean	Lag Variance
124	3.6	494,439.5	802.1	725,323.6
1,986	10.8	511,155.6	678.1	652,400.0
2,852	20.4	563,576.2	656.3	631,980.0
2,890	30.0	615,467.1	655.8	689,183.6
3,126	40.2	623,923.8	602.8	694,152.3
3,352	50.2	531,279.6	537.0	565,290.7
3,850	60.3	546,761.8	545.1	570,845.2
4,026	70.2	588,416.7	577.3	597,686.1
4,008	80.2	596,601.6	587.9	601,619.8
3,728	90.2	672,284.1	647.7	663,917.6
1,730	97.7	673,677.4	666.0	659,787.1

*See Figure 7.11.

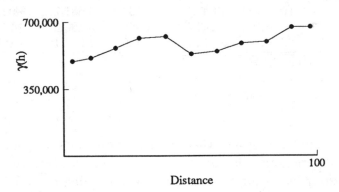

Figure 7.11   Omnidirectional sample variogram for U with a 10 m lag. The lag statistics are given in Table 7.5.

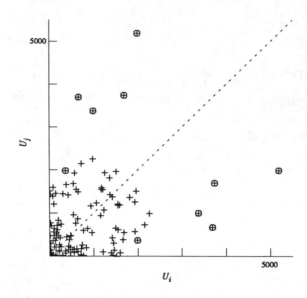

Figure 7.12 The omnidirectional lag 1 h-scatterplot of U.

Table 7.6 The 10 largest differences from the lag 1 h-scatterplot of U^*.

U_i	U_j	$\|U_i - U_j\|$	U_i	U_j	$\|U_i - U_j\|$
360.4	1,987.0	1,626.5	3,371.4	997.3	2,374.1
1,987.0	360.4	1,626.5	3,691.7	664.1	3,027.6
3,738.9	1,696.3	2,042.5	664.1	3,691.7	3,027.6
1,696.3	3,738.9	2,042.5	1,992.1	5,190.0	3,197.9
997.3	3,371.4	2,374.1	5,190.0	1,992.1	3,197.9

*See Figure 7.12.

omnidirectional variogram; there is little point in trying to calculate directional variograms that are likely to be even more erratic. If we can identify the reasons for the omnidirectional variogram's lack of clear structure, we may be able to adapt the traditional variogram calculation to reflect our new insights. Each point on the sample variogram is actually a summary of a particular h-scatterplot. A good place to begin our investigation, therefore, is the h-scatterplots for each lag.

Figure 7.12 shows the h-scatterplot for the first lag. For this partic-

Table 7.7 Omnidirectional sample variogram of U with pairs containing sample values greater than 5,000 ppm removed*.

No. of Pairs	Lag	$\gamma(h)$	No. of Pairs	Lag	$\gamma(h)$
122	3.6	418,719.1	3,822	60.3	480,825.8
1,972	10.8	453,294.6	3,986	70.2	489,074.1
2,832	20.4	500,586.1	3,974	80.2	506,808.8
2,858	30.0	510,753.0	3,688	90.2	572,382.5
3,090	40.2	524,893.4	1,718	97.7	598,922.1
3,334	50.2	479,926.6			

*See Figure 7.13.

ular lag, we are interested in understanding why its variogram value is so high. Since $\gamma(h)$ is the moment of inertia about the 45 degree line, we can focus our attention on the points that are farthest away from this line. The 10 pairs of values that plot farthest from this line, and which therefore make major contributions to the large variogram value, have been circled in Figure 7.12. These pairs have also been listed in Table 7.6. It is important to note that the pairs that contribute most to the variogram value do not always correspond to the largest data values. For example, there are samples with U values larger than the 1,987 ppm sample that appears in Table 7.6, but they do not appear in this table since they are paired with more similar values.

In an attempt to make the variogram a clearer description of the spatial continuity, we can try removing the particular samples that contribute most to the large variogram value in the first lag. The result of removing the largest of these samples, the 5,190 ppm value, is shown in Figure 7.13, with the accompanying details in Table 7.7. The removal of this particular sample reduces the number of pairs in the first lag by only two, but reduces the variogram value by nearly 20%. A similar effect can be seen in the other lags. The overall effect of the removal of this high value from the sample data set is a small but noticeable improvement in the shape of the variogram.

The removal of other erratic sample values does little to improve the appearance of the variogram. Figure 7.14 shows the omnidirectional U variogram calculated without the five samples with the largest U

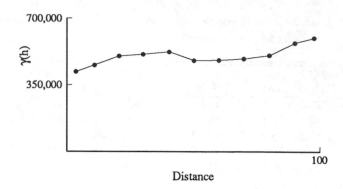

Figure 7.13 Omnidirectional sample variogram of U with pairs containing sample values greater than 5,000 ppm removed. The lag statistics are given in Table 7.7.

Table 7.8 Omnidirectional sample variogram of U with pairs containing sample values greater than 3,000 ppm removed*.

No. of Pairs	Lag	$\gamma(h)$	No. of Pairs	Lag	$\gamma(h)$
116	3.6	276,799.6	3,680	60.3	315,242.9
1,882	10.8	290,115.9	3,840	70.3	331,924.9
2,704	20.4	328,055.9	3,812	80.2	336,060.6
2,718	30.0	343,956.8	3,520	90.2	392,316.2
2,936	40.2	337,972.1	1,628	97.7	405,636.3
3,190	50.2	304,634.7			

* See Figure 7.14.

values: the 5,190 ppm sample at $(73, 199)$, the 3,739 ppm sample at $(51, 260)$, the 3,692 ppm sample at $(60, 151)$, the 3,637 ppm sample at $(39, 181)$, the 3,371 ppm sample at $(78, 159)$. While there is a small improvement in the appearance of the variogram for short distances, the overall appearance of the sample variogram is still quite close to that of a pure nugget effect.

Removing samples from the data set is rarely satisfying; there are no rules that specify which samples need to be removed, so the procedure may seem arbitrary. One also may worry that the removal of a particular sample results in the loss of all pairs which involved that

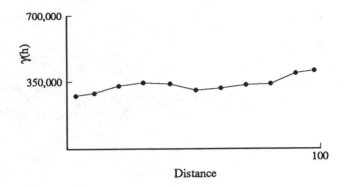

Figure 7.14 Omnidirectional sample variogram of U with pairs containing sample values greater than 3,000 ppm removed. The lag statistics are given in Table 7.8.

Table 7.9 Omnidirectional sample variogram of U with 10% of the largest pair differences removed*.

No. of Pairs	Lag	$\gamma(h)$	No. of Pairs	Lag	$\gamma(h)$
112	3.6	242,609.5	3,466	60.3	201,567.9
1,787	10.8	197,504.0	3,626	70.2	222,236.5
2,567	20.4	235,272.9	3,607	80.2	227,732.3
2,601	30.0	255,042.3	3,355	90.2	289,415.6
2,813	40.2	244,652.8	1,557	97.7	315,174.8
3,017	50.2	203,036.0			

* The plot is shown in Figure 7.15.

sample. While some of these may have been erratic, others may have been truly representative of the spatial continuity.

A less arbitrary procedure, and one that retains the useful contribution of large sample values, is to remove only the most erratic pairs from each particular lag. For example, we could decide to examine each h-scatterplot and to remove the 10% of the sample pairs that are farthest from the 45 degree line. If the variogram as defined in Equation 7.1 is seen as the mean of the distribution of squared differences for each h-scatterplot, the procedure previously described can be seen as the truncated mean of the lower 90% of this distribution. Such

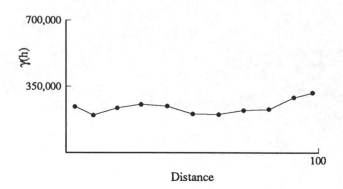

Figure 7.15 Omnidirectional sample variogram of U with 10% of the largest pair differences removed. The lag statistics are given in Table 7.9.

truncated or trimmed means are often used in statistics to reduce the adverse effects of erratically high values.

Figure 7.15 shows the effect of this procedure on the omnidirectional variogram of the U samples; the details of the calculations are given in Table 7.9. It is clear from Figure 7.15 that this procedure does little to clarify the structure of the variogram for this particular data set.

Neither the removal of particular samples nor the removal of particular pairs of samples does much to improve the appearance of our omnidirectional U variogram. This should not be taken as evidence that these procedures are not useful; there are many data sets for which these procedures can significantly improve the appearance of an erratic sample variogram. For this particular data set, the problem is not with these procedures, but with the variogram itself; there are many data sets for which the sample variogram is simply inappropriate as a descriptive tool. The variogram works well for data sets in which there is no proportional effect or in which the sample locations are not preferentially clustered. Table 7.5 provides a good example of the problems that a clustered, heteroscedastic data set presents. The final two columns in this table give the mean and variance of all the sample values that contribute to each lag. We can see that there are considerable fluctuations in these lag means and variances. The mean and variance of the sample U values are highest for the pairs that contribute to the first lag. The lags with the lowest lag mean and variance are the 50 and 60 m lags, the same two lags that produce the apparent hole effect on the sample variogram in Figure 7.11.

The reason for the fluctuating lag means becomes apparent if we consider the sampling strategy. The first lag contains sample pairs that are less than 5 m apart; until the third sampling campaign, the closest samples were approximately 10 m apart. All of the sample pairs that appear on the h-scatterplot for the first lag must therefore contain at least one sample from this final campaign, causing the samples that contribute to this first lag to be preferentially located in anomalously high areas. For the second lag, where the sample pairs are between 5 and 10 m apart, not only are there many more pairs that fall into this lag, but also these pairs are more representative of the entire area.

This discrepancy between the sample pairs that appear on different h-scatterplots is a common occurrence in practice. The most closely spaced samples are often the result of some final sampling program aimed specifically at delineating anomalously high areas. This creates problems for the sample variogram if a proportional effect exists; the average squared difference within a given lag will be influenced by the statistics of the pairs that contribute to each lag.

Relative Variograms

The dependence of the value of $\gamma(\mathbf{h})$ on the mean of the data values for each h-scatterplot leads us in practice to consider alternatives to $\gamma(\mathbf{h})$ which take account of this changing mean. These "relative variograms" scale $\gamma(\mathbf{h})$ to some local mean value [5]. There are three types of relative variograms that are often used to produce clearer descriptions of the spatial continuity; any one of these three could be used for purely descriptive purposes.

Local Relative Variograms. The most natural way to consider accounting for the effect of the local mean is to define separate regions and treat the data within each region as a separate population. One may find that the variograms for each population are similar in shape, with the local mean determining the actual magnitude of each separate variogram. In such cases, a single variogram, $\gamma_{LR}(\mathbf{h})$, can be used to describe the overall spatial continuity with the understanding that the values of $\gamma_{LR}(\mathbf{h})$ should be scaled by the local mean to obtain the actual local variogram. Once we have sample variograms from separate regions, the local relative variogram can be calculated by:

$$\gamma_{LR}(\mathbf{h}) = \frac{\sum_{i=1}^{n} N_i(\mathbf{h}) \frac{\gamma_i(\mathbf{h})}{m_i^2}}{\sum_{i=1}^{n} N_i(\mathbf{h})} \tag{7.2}$$

$\gamma_1(\mathbf{h}), \ldots, \gamma_n(\mathbf{h})$ are the local variograms from the n separate regions; $m_1, \ldots, m_n$ are the local mean values within each of the regions; and $N_1(\mathbf{h}), \ldots, N_n(\mathbf{h})$ are the numbers of sample pairs on the h-scatterplots from each region. This equation scales each local variogram by the square of the local mean then combines them by taking into account the number of sample pairs on which each local variogram is based.

The commonly observed linear relationship between the local mean and the local standard deviation leads to the common assumption that the local variogram is proportional to the square of the local mean. If the relationship between the local mean and the local standard deviation is something other than linear, one should consider scaling the local variograms by some function other than m_i^2.

General Relative Variograms. One of the drawbacks of the local relative variogram is that the local variograms from each separate region are based on smaller populations. The use of fewer data may cause the local variograms to be as erratic as the original overall sample variogram. Unlike the local relative variogram, the general relative variogram does not require the definition of smaller populations. Instead, the moment of inertia for each h-scatterplot is adjusted using the mean of the data values that appear on that particular h-scatterplot:

$$\gamma_{GR}(\mathbf{h}) = \frac{\gamma(\mathbf{h})}{m(\mathbf{h})^2} \tag{7.3}$$

The numerator, $\gamma(\mathbf{h})$, is the same value that we calculated earlier using Equation 7.1 and the denominator, $m(\mathbf{h})$, is the mean of all the data values that are used to calculate $\gamma(\mathbf{h})$:

$$m(\mathbf{h}) = \frac{1}{2N(\mathbf{h})} \sum_{(i,j)|\mathbf{h}_{ij}\approx\mathbf{h}} v_i + v_j = \frac{m_{+\mathbf{h}} + m_{-\mathbf{h}}}{2} \tag{7.4}$$

If we calculate $m(\mathbf{h})$ as we are calculating $\gamma(\mathbf{h})$, then the computation of the general relative variogram requires little additional effort should $\gamma(\mathbf{h})$ turn out to be erratic.

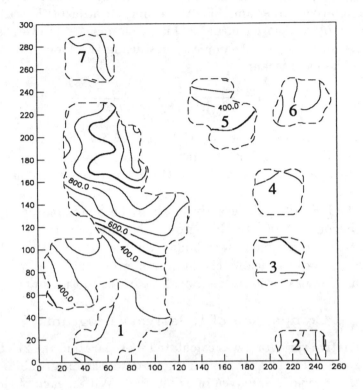

Figure 7.16 The seven sample regions of Walker lake used for calculating local relative variograms.

Table 7.10 Table giving the local means for the seven local regions of U^*.

Region	Local mean	Region	Local mean
1	579	5	428
2	704	6	868
3	540	7	1,263
4	264		

*See Figure 7.16.

Pairwise Relative Variogram. The third relative variogram that often helps to produce a clearer display of the spatial continuity is the pairwise relative variogram. Like the others, it adjusts the variogram calculation by a squared mean. This adjustment, however, is done separately for each pair of sample values, using the average of the two values as the local mean:

$$\gamma_{PR}(\mathbf{h}) = \frac{1}{2N(\mathbf{h})} \sum_{(i,j)|\mathbf{h}_{ij}\approx\mathbf{h}} \frac{(v_i - v_j)^2}{(\frac{v_i+v_j}{2})^2} \tag{7.5}$$

The difference between this and the equation for the variogram that we gave in Equation 7.1 is the denominator, which serves to reduce the influence of very large values on the calculation of the moment of inertia.

If v_i and v_j are both exactly 0, then the denominator in Equation 7.5 becomes 0 and $\gamma_{PR}(\mathbf{h})$ becomes infinite. For this reason it is advisable to choose some lower bound for the denominator. Pairs of values whose mean is below this minimum either can be dropped from the calculation or can have their local mean set to this lower bound.

Comparison of Relative Variograms

A local relative variogram was calculated by separating the 275 U samples into seven regions as shown on Figure 7.16; the mean values within each of these regions are given in Table 7.10. Within each of these regions, we have separately calculated the omnidirectional variogram; these local variograms are shown in Figures 7.17, with the accompanying details in Table 7.11. With the exception of the largest region, which contains most of the samples from the Wassuk Range anomaly, none of the local omnidirectional variograms can be calculated beyond the third lag; this is a result of the fact that the separate regions we have chosen to define are isolated clusters of samples. Although we depend entirely on the Wassuk Range samples for the definition of the variogram for large distances, there is still hope that this separation of samples into different populations will improve the definition of the variogram for short distances.

Combining the nine local variograms using the formula given in Equation 7.2 results in the local relative variogram shown in Figure 7.18a. Though this is slightly better than the original sample variogram, the improvement is marginal and the local relative variogram still appears quite erratic for short distances.

Table 7.11 Local omnidirectional sample variograms of U from the seven regions of Walker lake*.

No. of Pairs	Lag	$\gamma(h)$	No. of Pairs	Lag	$\gamma(h)$
		Region 1			Region 2
90	3.6	51,374.1	8	3.7	553,244.1
1,548	10.9	525,686.6	46	10.8	417,831.2
2,404	20.4	575,266.6	32	19.8	492,707.3
2,732	30.1	639,998.0	4	26.2	8,359.7
2,966	40.2	629,798.1			
3,072	50.2	527,483.1			
3,344	60.3	549,758.8			
3,358	70.2	573,168.6			
3,120	80.2	555,525.0			
2,662	90.1	617,799.5			
1,086	97.7	652,286.8			
		Region 3			Region 4
8	3.7	434,167.0	2	4.1	2,015.0
74	10.8	218,114.9	50	10.9	43,533.3
80	20.0	233,527.2	48	20.2	36,953.8
20	27.3	123,746.8	10	27.9	46,943.5
		Region 5			Region 6
2	3.2	107,360.0	6	3.1	281,039.4
84	10.7	181,468.2	80	10.5	623,550.6
108	19.6	245,486.3	74	19.9	559,025.8
78	29.5	204,877.7	22	28.8	303,400.8
54	40.1	178,301.7			
16	48.3	272,527.2			
		Region 7			
8	3.0	661,691.5			
104	10.5	949,303.7			
106	20.3	1134,673.2			
22	27.9	201,530.2			

* The plot is shown in Figure 7.17.

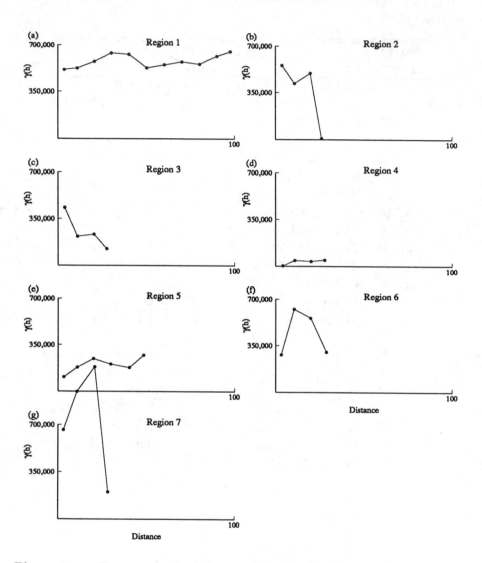

Figure 7.17 The seven local relative omnidirectional variograms calculated from each of the regions shown in Figure 7.16. The lag statistics are given in Table 7.11.

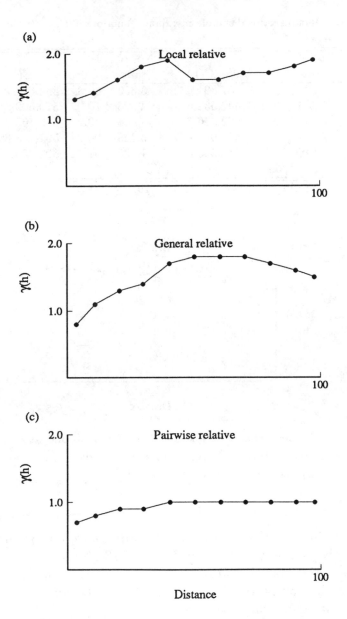

Figure 7.18 Three omnidirectional relative variograms of the U samples. The composite local relative variogram is shown in (a), the general relative variogram is shown in (b), and the pairwise relative variogram in (c).

Table 7.12 Omnidirectional sample covariance functions of U. Note the covariance is given as $\sigma^2 - C(h)^*$.

No. of Pairs	Lag	$\sigma^2 - C(h)$	No. of Pairs	Lag	$\sigma^2 - C(h)$
124	3.6	357,891.4	3,850	60.3	564,696.2
1,986	10.8	447,532.6	4,026	70.2	579,509.6
2,852	20.4	520,376.6	4,008	80.2	583,763.3
2,890	30.0	515,060.4	3,728	90.2	597,143.3
3,126	40.2	518,553.6	1,730	97.7	602,666.8
3,352	50.2	554,765.2			

*The plot is shown in Figure 7.19.

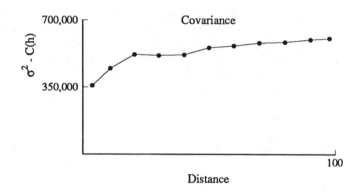

Figure 7.19 Omnidirectional sample covariance functions of U at a 10 m lag and an angular tolerance of $\pm$ 10 degrees. Note the covariance has been plotted as $\sigma^2 - C(h)$. The lag statistics are given in Table 7.12.

The general relative variogram shown in Figure 7.18b and the pairwise relative variogram shown in Figure 7.18c both provide adequate displays of the spatial continuity. There are still some small fluctuations in the variogram, but we can clearly see how the continuity changes with increasing distance.

The Covariance Function and the Correlogram

The other two tools we have not yet tried are the covariance function and the correlogram that we introduced in Chapter 4. Like the relative

Table 7.13 Omnidirectional sample correlogram of U^*.

No. of Pairs	Lag	$1 - \rho(h)$	No. of Pairs	Lag	$1 - \rho(h)$
124	3.6	0.68	3,850	60.3	0.95
1,986	10.8	0.78	4,026	70.2	0.98
2,852	20.4	0.89	4,008	80.2	0.99
2,890	30.0	0.89	3,728	90.2	1.01
3,126	40.2	0.89	1,730	97.7	1.02
3,352	50.2	0.93			

*The plot is shown in Figure 7.20.

variograms described in the previous section, these two are likely to be more resistant to erratic values since they account for the lag means [in the case of $C(\mathbf{h})$] and the lag variances [in the case of $\rho(\mathbf{h})$] Neither of these is commonly used in geostatistics and our use of them here raises the question of how spatial continuity should conventionally be displayed. Time series analysts and classical statisticians would likely prefer to see covariance functions, which decrease with distance, rather than variograms, which typically increase with distance. We will defer to the geostatistical convention, however, and continue to show variogramlike displays which start at the origin and increase with increasing distance. In this section, and throughout the remainder of the book, we will plot the covariance function in the form of the variogram as $C(0) - C(\mathbf{h})$ and the correlogram as $\rho(0) - \rho(\mathbf{h})$. For covariance functions, the value at $|\mathbf{h}| = 0$ is simply the sample variance σ^2; for correlation functions, the value at $|\mathbf{h}| = 0$ is 1.

The omnidirectional sample covariance function is shown in Figure 7.19, with the details of the calculations given in Table 7.12. Figure 7.20 shows the omnidirectional sample correlogram, with the details of the calculations given in Table 7.13. Both of these produce a clearer picture of the spatial continuity than does the sample variogram.

We now have four tools that can adequately describe the spatial continuity of the sample U values: the general relative variogram, the pairwise relative variogram, the covariance function, and the correlation function. As we proceed to describe the spatial continuity in

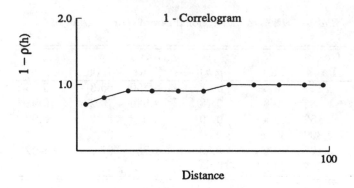

Figure 7.20 Omnidirectional sample correlogram of U. Note the correlation coefficient has been plotted as $1-\rho(h)$. The lag statistics are given in Table 7.13.

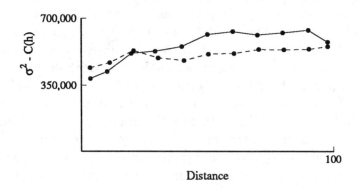

Figure 7.21 Directional sample covariance functions of U. Note the covariance function has been plotted as $\sigma^2-C(h)$. The dashed line represents N14°W and the solid N76°E. The lag statistics are given in Table 7.14.

different directions, we could continue with any one of these. We have chosen here to use the covariance function, largely because it is the particular function we will require later when we tackle the problem of local estimation. Had we wanted only to provide a description of the spatial continuity, there would have been no reason to choose $C(\mathbf{h})$ over any of the other three tools.

Table 7.14 Directional sample covariance functions for U^*.

N76°E			N14°W		
No. of Pairs	Lag	$\sigma^2 - C(h)$	No. of Pairs	Lag	$\sigma^2 - C(h)$
Angular Tol = ± 40°					
54	3.6	385,364.2	5	3.5	441,331.0
427	10.2	421,267.6	475	11.2	467,790.3
578	19.9	517,143.2	635	20.7	532,065.4
594	29.3	528,872.5	742	30.5	493,747.6
537	39.8	553,211.2	835	40.7	479,879.6
501	50.0	616,159.8	1,020	50.4	513,523.1
514	60.1	630,748.3	1,242	60.6	517,122.5
493	69.9	613,809.9	1,329	70.4	537,606.5
486	79.8	624,710.9	1,371	80.4	535,925.6
471	90.0	638,308.3	1,234	90.3	539,035.9
271	97.7	576,078.8	521	97.7	553,724.9

*The plot is shown in Figure 7.21.

Directional Covariance Functions for U

Having decided to use the covariance function to describe the spatial continuity of U, we should now try to find the directions of maximum and minimum continuity. As we did for V earlier in this chapter, we could find these directions either through a contour map of the covariance surface or through a rose diagram of the range of the covariance function in different directions. For this particular data set, the directions of maximum and minimum continuity are the same for both variables. This is not always the case, however, and when dealing with several variables, one should repeat the analysis of the axes of anisotropy for each variable.

Figure 7.21 shows the covariance function along the N14°W direction and the N76°E direction, using an angular tolerance of ±40 degrees for each direction; the details of the calculations are provided in Table 7.14. Along the N76°E direction, the structure is clear; in the perpendicular direction, however, there is considerable noise. Increasing the angular tolerance beyond ±45 degrees is not desirable since there will then be some overlap between the two directional calcula-

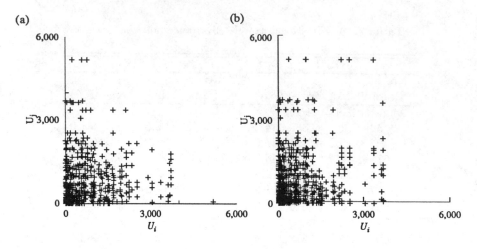

Figure 7.22 h-scatterplots from the N14°W directional variogram of U shown in Figure 7.21. (a) is at lag 3 and (b) at lag 5.

tions, with some of the pairs that contribute to the calculation of the covariance function in the direction of maximum continuity also contributing to the calculation in the direction of minimum continuity. In some situations this may be necessary. Before we take this step, however, let us see if we can improve the appearance of the directional covariance functions by removing particular samples or particular pairs of samples.

In Figure 7.22 we show the h-scatterplots for the third lag and the fifth lag of the N14°W directional covariance function of Figure 7.21 since these are the two lags that appear most aberrant. As we noted when we were trying to improve the appearance of the omnidirectional variogram, the 5,190 ppm sample probably contributes significantly to the covariance estimate. The directional covariance functions calculated without this sample are shown in Figure 7.23, with the details of the calculations shown in Table 7.15. The removal of this single sample does not completely sort out the problem we have with the N14°W covariance function since the third lag appears too high.

The other possibility we considered earlier was the removal of the 10% of the pairs that contribute most to the calculation within each lag. The effect of this procedure is shown in Figure 7.24, with the

Table 7.15 N14°W sample covariance function of U with pairs differing more than 5,000 ppm removed*.

No. of Pairs	Lag	σ^2 - C(h)	No. of Pairs	Lag	σ^2 - C(h)
Angular Tol = ± 40°					
5	3.5	366,751.1	1,231	60.6	469,437.0
472	11.2	404,769.8	1,312	70.4	473,866.1
631	20.7	449,793.3	1,357	80.4	463,043.1
735	30.5	422,736.8	1,221	90.3	482,475.1
829	40.7	442,807.3	518	97.7	477,219.5
1,015	50.4	453,468.3			

*The plot is shown in Figure 7.23.

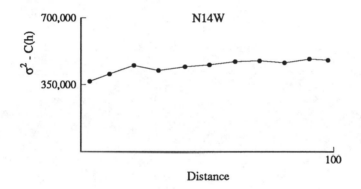

Figure 7.23 N14°W directional sample covariance function of U with pairs differing more than 5,000 ppm removed. The lag statistics are given in Table 7.15.

accompanying details in Table 7.16. This finally produces an interpretable structure in both the principle directions.

Cross-Variograms

Like the variogram for spatial continuity of a single variable, the cross-variogram is what geostatisticians traditionally use for describing the cross-continuity between two variables. The omnidirectional cross-variogram for U and V is shown in Figure 7.25. The structure is

Table 7.16 Directional sample covariance functions of U with 10% of the largest differences between pairs removed*.

N76°E			N14°W		
No. of Pairs	Lag	$\sigma^2 - C(h)$	No. of Pairs	Lag	$\sigma^2 - C(h)$
Angular Tol = ± 40°					
54	3.6	468,158.0	5	3.5	441,330.0
427	10.2	473,501.0	475	11.2	513,677.0
578	19.9	559,600.0	635	20.7	535,898.0
594	29.3	575,238.0	742	30.5	523,894.0
537	39.8	606,368.0	835	40.7	546,787.0
501	50.0	609,570.0	1,020	50.4	551,377.0
514	60.1	620,964.0	1,242	60.6	556,343.0
493	69.9	613,306.0	1,329	70.4	562,268.0
486	79.8	604,503.0	1,371	80.4	568,298.0
471	90.0	627,283.0	1,234	90.3	576,646.0
271	97.7	585,535.0	521	97.7	572,683.0

*The plot is shown in Figure 7.24.

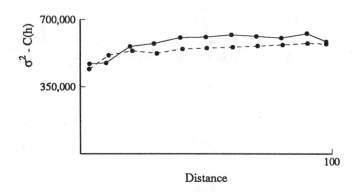

Figure 7.24 Directional sample covariance functions of U with 10% of the largest differences between pairs removed. The direction of the dashed line is N14°W, the solid N76°E. The lag statistics are given in Table 7.16.

clear enough on this cross-variogram to warrant trying to construct directional cross-variograms.

Figure 7.26 shows the cross-variograms calculated along the directions of maximum continuity (N14°W) and minimum continuity (N76°E) with an angular tolerance of ±40 degrees. Although the N14°W cross-variogram is reasonably well behaved, the perpendicular direction presents a somewhat unsatisfactory structure. The first lag contains only five points, so we should not pay too much attention to that value. Even after ignoring the first lag, the remaining points are still a bit disappointing; without the point at 10 m, the cross-variogram would appear to be a nugget effect. It is worth investigating the possibility of improving the clarity of these cross-variograms through the same procedures that we used earlier to improve the U variogram and covariance functions.

In our earlier investigations into the effect of the 5,091 ppm U sample we noticed that the U variogram was slightly improved if this sample was removed; the cross-variogram also benefits slightly from the the removal of this sample. Further analysis of the h-scatterplots reveals that the comparison of type 1 samples with type 2 samples produces many erratic pairs. It seems reasonable, therefore, to include only those pairs for which both samples are of the same type in our calculations. Figure 7.27 shows the directional cross-variograms calculated without the 5,091 ppm sample and without pairs that involve different types. This summary of the spatial cross-continuity between the U values and the V values is now fairly clear. With the exception of the very first lag, which still contains only five points, the cross-variogram along the N14°W direction in Figure 7.27 is clearly defined. It now rises gradually rather than reaching the sill quickly and flattening out as it did in our first attempt.

Summary of Spatial Continuity

With all of the different avenues we have explored in this chapter, using and adapting tools for each new problem, it is possible that the actual descriptions of the spatial continuity in the sample data set have become lost along the way. To conclude the chapter, let us briefly review the best descriptions we found.

For the spatial continuity of V, we were able to use the variogram. Figure 7.28a shows the sample V variograms calculated along the two

Table 7.17   Omnidirectional sample cross-variogram*.

No. of Pairs	Lag	$\gamma_{UV}(h)$	No. of Pairs	Lag	$\gamma_{UV}(h)$
124	3.6	80,057.5	3,850	60.3	124,551.9
1,986	10.8	88,042.5	4,026	70.2	123,121.5
2,852	20.4	111,262.7	4,008	80.2	120,822.8
2,890	30.0	118,806.8	3,728	90.2	129,019.6
3,126	40.2	124,865.1	1,730	97.7	134,559.0
3,352	50.2	114,747.0			

*The plot is shown in Figure 7.25.

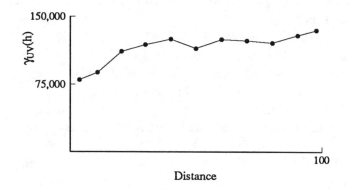

Figure 7.25   Omnidirectional sample cross-variogram at a 10 m lag. The lag statistics are given in Table 7.17.

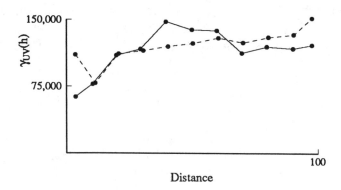

Figure 7.26   Directional sample cross variograms of U and V. The direction of the dashed line is N14°W, the solid N76°E.

Table 7.18 Directional sample cross-variograms with pairs differing more than 5,000 ppm and different types removed*.

N14°W			N76°E		
No. of Pairs	Lag	$\gamma_{UV}(h)$	No. of Pairs	Lag	$\gamma_{UV}(h)$
Angular Tol $= \pm 40°$					
5	3.5	110,186.8	63	3.6	68,274.6
583	11.4	78,175.6	527	10.2	80,763.9
776	20.7	104,249.0	708	19.9	113,423.8
1,002	30.7	113,271.7	860	29.6	124,999.5
1,109	40.7	112,725.0	781	39.9	145,441.8
1,391	50.6	118,516.5	854	50.2	131,384.5
1,573	60.5	124,032.6	770	60.0	134,935.7
1,795	70.4	122,094.0	869	69.9	110,911.0
1,847	80.3	125,040.5	782	79.6	117,344.8
1,774	90.3	131,875.6	790	90.1	117,189.9
751	97.6	149,300.6	410	97.6	117,553.4

*The plot is shown in Figure 7.27.

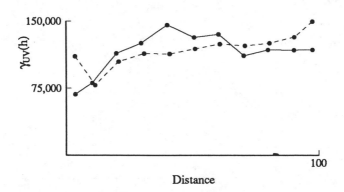

Figure 7.27 Directional sample cross variograms of U and V with pairs differing more than 5,000 ppm removed. Pairs of different types were also removed. N14°W is represented by the dashed line and N76°E by the solid. The lag statistics are given in Table 7.18.

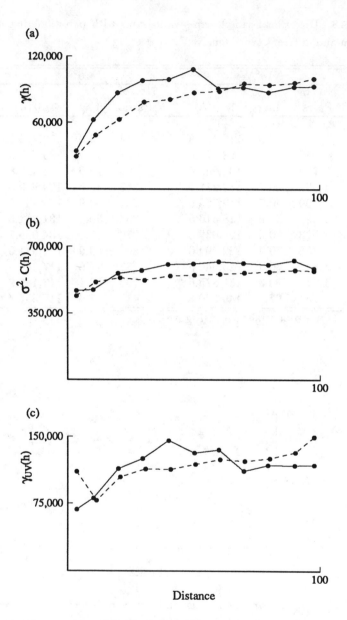

Figure 7.28 A summary of the final directional sample variograms of V is shown in (a), of U in (b) and of the cross-variogram between V and U in (c).

principle directions, N14°W and N76°E, with a ±40 degree angular tolerance on each direction. For the spatial continuity of U and for the spatial cross-continuity of U and V, we used the same two principle directions and the same directional tolerance.

For the spatial continuity of U, the covariance produced a better description than did the variogram. The directional covariances were still somewhat erratic, so we removed the 10% of the pairs that contributed most heavily to the calculations within each lag. The resulting directional covariances are shown in Figure 7.28b.

For the spatial cross-continuity between U and V, we found the cross-variogram to be adequate if we removed the largest U value and also removed all pairs which involved different sample types. These directional cross-variograms are shown in Figure 7.28c.

Notes

[1] For a three-dimensional data set, the direction of variogram computation is traditionally specified using two parameters. The first parameter defines the horizontal direction and is usually referred to as the horizontal angle, the second parameter is the vertical angle. Typically, tolerances are given for both angles: an angular tolerance on the horizontal angle as well as an angular tolerance on the vertical angle. In three dimensions the combination of these tolerances describes an elliptical or flattened cone. A datum at the apex of the cone is paired at the lag distance (plus or minus the lag tolerance) with all data found within the cone. In cases where a three-dimensional anisotropy is suspected to be unusually oriented with reference to the coordinate system of the data, one can apply a coordinate transformation to the data before computing the sample variogram. The axes of the new or transformed coordinated system are made to align with the suspected directions of the anisotropy. This enables a straightforward computation of the directional sample variograms along the axes of the anisotropy.

[2] The variogram cloud can also be useful for gaining some insight into the reasons for a messy sample variogram. The cloud is made by plotting each squared difference $(v_i - v_j)^2$ versus the separation distance h_{ij}. The cloud formed by these points may reveal extreme outlier points that dominate the estimation of the sample variogram. It may also reveal that the distribution of $(v_i - v_j)^2$ for

any one lag is severely skewed, in which case the arithmetic average of $(v_i - v_j)^2$ may provide a poor estimate of the sample variogram for that lag. Sometimes the relationship between $\gamma(h)$ and h can be made clearer by running a resistant smoother, of the type discussed in the Notes to Chapter 3, over the variogram cloud.

[3] The apparent rotation of the axes of anisotropy near the origin is an artifact of the contouring program caused by the relatively coarse grid of values being contoured. The $50,000$ ppm^2 contour line, which is the first one that crosses more than the central four lags, is the first reliable indication of anisotropy.

[4] Several practitioners used contour maps of the covariance function or the correlogram in the early 1960s. The same approach to presenting the variogram surface was used in the following article:

Rendu, J. , "Kriging for ore valuation and mine planning," *Engineering and Mining Journal*, vol. 181, no. 1, pp. 114–120, 1980.

[5] Though much has been written on *robust* or *resistant* variography, little attention has been given to the various types of relative variograms that are often used in practice. Journel and Huijbregts (1978) and David (1977) discuss the local relative variogram in some detail. No mention is made, however, of the general and pairwise relative variograms. Despite the fact that the general and pairwise relative variograms are not well understood theoretically, they are well known in practice, due largely to David, who has pioneered the use of these variograms and, through the success of his applications, has succeeded in popularizing variograms of this form.

Further Reading

Cressie, N. and Hawkins, D. M. , "Robust estimation of the variogram: I," *Mathematical Geology*, vol. 12, no. 2, pp. 115–125, 1980.

David, M. , *Geostatistical Ore Reserve Estimation*. Amsterdam: Elsevier, 1977.

Isaaks, E. H. and Srivastava, R. M. , "Spatial continuity measures for probabilistic and deterministic geostatistics," *Mathematical Geology*, vol. 20, no. 4, pp. 313–341, 1988.

Journel, A. G. and Huijbregts, C. J. , *Mining Geostatistics*. London: Academic Press, 1978.

Omre, H. , *Alternative Variogram Estimators in Geostatistics*. PhD thesis, Stanford University, 1985.

Srivastava, R. and Parker, H. , "Robust measures of spatial continuity," in *Third International Geostatistics Congress*, (Armstrong, M. et al, ed.), D. Reidel, Dordrecht, Holland, 1988.

Verly, G. , David, M. , Journel, A. G. , and Marechal, A. , eds., *Geostatistics for Natural Resources Characterization*, Proceedings of the NATO Advanced Study Institute, South Lake Tahoe, California, September 6-17, D. Reidel, Dordrecht, Holland, pp. 1-140, 1983.

8

ESTIMATION

The focus of the last six chapters has been purely descriptive as we dealt with the first of the goals we posed in Chapter 1. This second part of the book looks at the remaining goals, all of which involve estimation. We are no longer interested merely in describing the sample data set; we now want to use the sample information to predict values in areas we have not sampled.

As the list of goals in Chapter 1 shows, there are many different types of estimation problems. In the remaining chapters we will look at several different estimation methods. Each of these is useful for a particular type of problem, so it is important to understand which methods are applicable to which types of problems. In the approach taken here, we consider the following three features of an estimation problem:

- Do we want a global or local estimate?

- Do we want to estimate only the mean or the complete distribution of data values?

- Do we want estimates of point values or of larger block values?

The answers to these questions will dictate what methods are appropriate. In this chapter we will elaborate on these features, giving practical examples of the different types of problems. We will also introduce the basic tools that will be needed for the various estimation methods.

Chapter 9 is a rather key chapter. In this chapter we elaborate on the use of models in estimation. Particular attention is given to the introduction of probability models and their applications. Chapter 10 looks at global estimation, which requires only some appropriate declustering method. In Chapter 11 we will turn to local point estimation, which requires us to account not only for the clustering but also for the distance to our sample locations. In Chapters 12 and 13 we will look at some of the practical aspects of the geostatistical estimation method known as *kriging*. Chapter 14 deals with various strategies for selecting data to be used for local estimation. Chapter 15 deals with cross validation and Chapter 16 describes the fitting of variogram models to the sample variogram. In Chapter 17 we will see how the kriging approach can easily be adapted to include information from other variables. In Chapter 18 we will use several methods to estimate local distributions. In Chapter 19 we will look at ways of adjusting our estimates to account for the fact that average values over large areas tend to have less variability than average values over small areas. In Chapter 20 we will look at ways of assessing the accuracy of our various estimates.

Weighted Linear Combinations

The methods we discuss all involve weighted linear combinations:

$$estimate = \hat{v} = \sum_{i=1}^{n} w_i \cdot v_i \qquad (8.1)$$

$v_1, \ldots, v_n$ are the n available data values and w_i is a weight assigned to the value v_i. These weights are usually standardized so that they sum to one, though this is not a requirement of all estimation methods. Throughout the remainder of the book, we will use the small hat ˆ to denote an estimate.

Different approaches to assigning the weights to the data values give rise to many different methodologies. Some of these methods are based on common sense notions of which data values are most important; others are based on statistical theory. As we will see, these two are not incompatible; much of what makes good common sense also happens to make good statistical sense.

Weighted linear combinations are not the only way of combining data values to obtain estimates. There are estimation methods that use more complicated functions, but many of these involve more complicated mathematics than we intend to get into in this book. We will, however, look at some more advanced estimation methods that involve weighted linear combinations of transformed data values. Later we will discuss the practical and theoretical reasons for working with transformed data values. For the moment, we will simply outline the general form of such methods.

If we transform our data values and combine them with some linear combination, then we get some transformed estimate:

$$transformed\ estimate = \hat{t} = \sum_{i=1}^{n} w_i \cdot T(v_i) \qquad (8.2)$$

The only difference between this and Equation 8.1 is that we have transformed the individual data values using the function $T(v)$. For example, this function could be a polynomial, such as $T(v) = v^2$. It could also be a much more complicated expression; the Fourier analysis approach to estimation, for example, transforms the data values using a series of sines and cosines. $T(v)$ need not have an analytical expression; it could be any consistent and repeatable procedure for mapping the original data values into a different set of transformed values. It may be quicker to transform the data values using some graphical method rather than to calculate some high-order polynomial that accomplishes the same thing.

We rarely want transformed estimates, so we have to use a back transform to restore our estimate to some meaningful value. If we had transformed all of our data values by taking their logarithms, then their weighted linear combination would give us an estimate of a logarithm; we would rather have an estimate of the original value, so we use some back transform to turn our transformed estimate into something more useful:

$$estimate = \hat{v} = B(\hat{t}) \qquad (8.3)$$

$B(t)$ is the back transform we use and is not necessarily the inverse of $T(v)$. In our example with the logarithms, we would not use $B(t) = e^t$ since this produces a biased result [1]. Later, when we look at some of the estimation methods that require transformations, we will look at this issue of the back transform in more detail. Like the original transformation, our back transform need not have an actual analytical expression; it could be a graphical procedure or any other consistent and repeatable method for restoring the transformed estimate to an estimate of the original value.

Global and Local Estimation

Estimation over a large area within which we have many samples is *global* estimation. In *local* estimation, we are considering a smaller area, one in which there are few samples; in such situations we often use nearby samples located outside the area being estimated. In our Walker Lake study, an estimate of the mean value of U over the entire map area would be considered a global estimate since we have a lot of samples within the area being estimated. On the other hand, an estimate of the U mean within a particular 10 x 10 m^2 area would be considered a local estimate because there would be very few (perhaps even zero) samples within the 10 x 10 m^2 area and we would likely have to use nearby samples from outside the area to get a reasonable estimate.

In practical situations there is usually some target area over which samples are collected. Whether we are studying ore grades in a mineral deposit, pollutant concentrations in a toxic waste site, or soil strengths over the area of a proposed building, we have a large area that serves as the area of interest. Global estimation is commonly used at a very early stage in most studies to obtain some characteristics of the distribution of data values over the whole area of interest.

A single global estimate rarely satisfies the goals of a study and we usually also require a complete set of local estimates. For example, in planning a mine it is not sufficient to know the overall average grade, one also needs detailed local information on the ore grades. In a pollution study, an estimate of the overall concentration does not give us the information we need to decide which specific localities have unacceptably high concentrations. In many fluid flow studies, a global estimate of the permeability is rather meaningless since flow is controlled by the extremely high and the extremely low permeability zones; local estimates are required to reveal the presence of such zones.

Global estimation is fairly straightforward if the samples are located on a regular grid or are located randomly. Unfortunately, this is not typical in practice and, as our description in the previous chapter showed, the statistics from a clustered sample data set may not be representative of the exhaustive data set. Global estimation methods should therefore account for the possibility of clustering. Groups of clustered samples should have their weights reduced to account for the fact that they are not representative of as large an area as are unclustered samples.

Local estimates need to account for the distance from the point(s) we are estimating to the individual sample locations. Some samples will be much closer than others and the values at these closer locations should be given more weight than the values that are further away. Local estimates, like global ones, are also affected by clustered sampling. A group of closely spaced samples with similar values contains redundant information; our local estimate would likely improve if we could locate samples uniformly over the small area we are estimating. The weights assigned by local estimation methods should account for both the distance to the samples and also for the possible redundancy between samples.

Means and Complete Distributions

Though there are many ways of summarizing a distribution, the mean is the statistic most commonly used. As a measure of the location of the center of the distribution it is particularly interesting. As we have seen, however, the "center" of the distribution is a slippery concept and measures other than the mean might be more relevant in certain situations. For strongly skewed distributions, such as precious metal grades or pollutant concentrations, an estimate of the median is a good supplement to an estimate of the mean. A low median coupled with a high mean warns us that the distribution is quite positively skewed, and that the overall average is due largely to a small proportion of extremely high values. In such situations, estimates of the lower and upper quartiles will further enhance our understanding of how the values are distributed.

There are some applications that call for an estimation of the variability. Many metallurgical processes are adversely affected by large fluctuations in the ore grade; estimates of the variability will help us decide if special measures are needed to reduce ore grade fluctuations. In fluid flow studies, variability is also an important factor; large fluctuations in permeability in a petroleum reservoir can make secondary recovery schemes very inefficient.

In most mining applications, a single mean value is not sufficient since there is typically some cutoff grade above which the rock is processed as ore and below which it is discarded as waste. There may even be several categories of ore and waste corresponding to different stockpiles. In such situations, we require estimates of the proportion of the distribution above and below certain cutoffs along with the mean value above and below these cutoffs.

Despite its shortcomings in certain applications, the mean does have special significance for a wide variety of practical estimation problems. If we had exhaustive sampling, our sample mean would be identical to the spatial arithmetic average. In most studies we are interested in some kind of spatial average and quite often the averaging process is indeed arithmetic. For example, grades and concentrations average arithmetically; the combined average grade of two truckloads of ore with equal weights is the arithmetic average of the two individual truckload grades. To many practitioners, this will seem like a rather obvious remark. There are a growing number of geostatistical applications, however, for which the averaging process is not arithmetic. For example, in the study of fluid flow through porous media, the permeability does not average arithmetically. The effective permeability of two blocks in series is not the arithmetic average of the two individual block permeabilities. In civil engineering applications, the soil strength of a large area is not the average of the soil strengths calculated over its subareas. In ore deposits where the density varies with grade, the average grade of several samples is not the arithmetic average of the individual sample grades.

Despite the growing number of applications for which arithmetic averaging is not relevant, the majority of current practical studies involve quantities that do average arithmetically and for these the mean remains a useful parameter to estimate. The equation we gave for the sample mean is a weighted linear combination in which the available samples are all given an equal weight of $\frac{1}{n}$. Most methods for estimating an exhaustive mean, whether local or global, use weighted linear combinations in which some of the available values receive more weight than others. As the weight assigned to a particular data value increases, the influence of that particular value on the estimate of the mean also increases. Conversely, data values with small weights have little influence on the estimate of the mean.

If we require other statistics of the whole distribution, such as the median or variance, or statistics for truncated distributions, the obvious solution is to try to estimate the complete distribution rather than just its mean. If we can successfully estimate the complete distribution then any summary, even of a truncated portion, is easy to calculate.

There are two different philosophies for estimating complete distributions; the parametric approach and the nonparametric approach. Both of these approaches typically involve weighted linear combinations of transformed data values. In the parametric methods, we make some assumptions about the type of distribution we have, then we estimate some parameters of the model we have assumed. In the nonparametric approach, we estimate some points on the cumulative histogram then interpolate between these estimated points to get a complete distribution. The disadvantage of this nonparametric approach is that it requires us to extrapolate beyond the last point we have estimated on the cumulative histogram; experience has shown that our final results depend heavily on the way we choose to do this extrapolation. The main drawback of parametric approach is that it causes us to depend very heavily on our assumptions about the model, and these assumptions are typically very difficult to verify. The problem of how we choose to extrapolate beyond the last point on the cumulative histogram is replaced by what model we choose to use for the entire distribution.

Point and Block Estimates

In any earth science study, the size of each sample is an important consideration. There is a relationship between the size or "support" of our data and the distribution of their values. We can imagine that if we used very small samples, such as rock chips, to sample a gold deposit, there could be a lot of variability between sample values. One rock chip might contain almost pure gold, another might contain nothing. If we sampled the same deposit with large truckloads, there would likely be much less variability. The mixing of high and low values that is to be expected with large samples would give us less erratic values.

We have already seen this effect when we were looking at the moving window statistics of the U values in our exhaustive data set. In Figure 5.14 we drew a contour map of the moving window means for 10 x 10 m^2 windows; in Figure 5.15 we drew a similar map for 20 x 20 m^2 windows. The larger window size produced a smoother

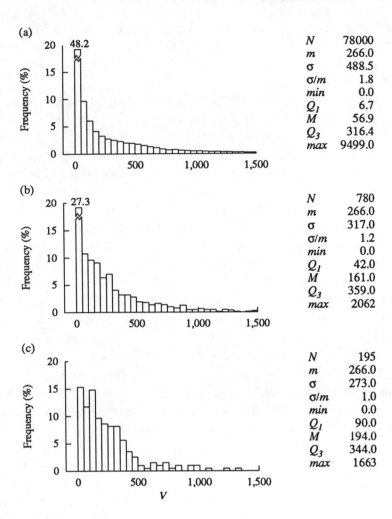

Figure 8.1 Exhaustive histograms of U for three different support volumes. The histogram of the 78,000 point U values is shown in (a); the histogram of the 780 average U values within 10 x 10 m^2 blocks is shown in (b); the histogram of the 195 average U values within 20 x 20 m^2 blocks is shown in (c).

map. If we look at the histograms of the individual moving window means for these two maps we can better understand this effect.

In Figure 8.1b we show the histogram of the 780 moving window means from our 10 x 10 m^2 windows; the histogram of the 195 individual means from our 20 x 20 m² windows are shown in Figure 8.1c. In Figure 8.1a we have included the histogram of the 78,000 original point values as a reminder of what the distribution looked like for the smallest possible support.

As the support of the data increases we notice that the maximum value decreases. The 9,500 ppm value from our exhaustive data set gets diluted by lesser values as we average the U values over larger areas. The same is true of the lowest values. In Figure 8.1a, 48% of the values fall within the lowest class on the histogram; in Figure 8.1c, where we have averaged all of the values within 20 x 20 m² squares, only 10% of the values fall within this lowest class.

Averaging values together over larger areas generally has the effect of reducing the variance of the data and of making their distribution more symmetric. From Figure 8.1 we can see that the standard deviation, the coefficient of variation and the difference between the mean and the median all decrease as the support of the data increases. Though the support of the data has an effect on the spread and the symmetry of the distribution, it does not affect the mean. For all three histograms shown in Figure 8.1 the mean value is 266 ppm.

This relationship between the support of the data and the distribution of their values has serious implications in practice. When we estimate a complete distribution from some sample data set, what we get is an estimate of the exhaustive distribution for the same support as our samples. In the example we used earlier, if we used all of our rock chip samples to estimate a complete distribution, we would be estimating the exhaustive distribution of gold grades for rock chip sized volumes in our deposit. On the other hand, if we used our truckload samples then we would be estimating the exhaustive distribution of gold grades for truckload sized volumes.

In most practical applications, the support of the samples is not the same as the support of the estimates we are trying to calculate. In feasibility studies of ore deposits, we want to know the global and local distributions of truckload sized volumes. In the actual operation of the mine, entire truckloads will be treated as either ore or waste; we will not be able to discriminate between ore and waste within a particular truckload. Our sample data set, however, is typically based on measurements from drill hole cores. If we were to plan the entire mining operation based on these core sized samples, we would be vastly overestimating the spread of the distribution of ore grades. This problem is particularly severe in deposits with strongly skewed sample grades. Purely on the basis of the sample values, we might conclude that a large enough proportion of the deposit is above our economic cutoff grade to warrant developing the prospect. When we actually mine the deposit using trucks that are much larger than our original core samples, the inevitable averaging of high grades and low grades may result in a far lower proportion of economically viable material than we originally thought.

As an example, let us treat the Walker Lake data set as a prospective ore deposit. At the feasibility study stage we may decide that 1,000 ppm is an economically viable cutoff grade. Material with a U value greater than 1,000 ppm will be processed as ore, the remainder will be discarded as waste. Let us pretend that we have been able to estimate perfectly the exhaustive distribution of the $78,000$ U values. Using this estimate of the distribution, we would conclude that about 7% of the area will eventually be processed as ore. If in the actual mining operation we are forced to classify entire 20 x 20 m^2 units as either or ore waste then we are in for a surprise: only 2% of the area will actually be rich enough to be processed as ore.

Similar problems exist in many other applications. In forecasting the performance of a petroleum reservoir, estimates are needed of the porosity and permeability of very large blocks. The only available measurements, however, may be for much smaller volumes such as core plugs. Similarly, the standard penetration tests for soil strength measurements provide information for a very small support; from these sample strength measurements one needs estimates of the soil strength over much larger areas.

It is important to evaluate the support of the sample data set and the support we intend for the final estimates. If the two are different, some correction will have to be applied to our sample support estimates. There are a variety of mathematical methods for adjusting a distribution so that its variance will be reduced while its mean remains unchanged. Unfortunately, all of these depend on unverifiable assumptions about how the distribution changes as the support increases; they also require knowledge of certain parameters that are very difficult to estimate precisely.

This problem of the discrepancy between the support of our samples and the intended support of our estimates is one of the most difficult we face in estimation. Despite the shortcomings of all currently available methods, it is better to attempt a rough correction than to ignore the problem.

Notes

[1] Perhaps this is best explained using a small example. Consider the following table of three numbers and their logarithms:

	Value	Logarithm
	3	1.0986
	6	1.7918
	12	2.4849
Averages	7	1.7918
Antilog	–	6

The average of the three numbers is 7 while the average of their logarithms is 1.7918 and the antilog of 1.7918 is 6. This clearly demonstrates that one cannot take the logarithms of a set of sample values; compute the average logarithm, (weighted or not), and then compute the antilog of the logarithmic average. The result is biased! In fact this sequence of steps will result in a bias with not only the log transformation, but with all nonlinear transformations.

Further Reading

Buxton, B. , *Coal Reserve Assessment: A Geostatistical Case Study.* Master's thesis, Stanford University, 1982.

David, M. , *Geostatistical Ore Reserve Estimation.* Amsterdam: Elsevier, 1977.

Journel, A. G. and Huijbregts, C. J. , *Mining Geostatistics.* London: Academic Press, 1978.

Parker, H. , "The volume-variance relationship: a useful tool for mine planning," in *Geostatistics*, (Mousset-Jones, P. , ed.), pp. 61–91, McGraw Hill, New York, 1980.

Parker, H. , "Trends in geostatistics in the mining industry," in *Geostatistics for Natural Resources Characterization*, (Verly, G. , David, M. , Journel, A. G. , and Marechal, A. , eds.), pp. 915–934,

NATO Advanced Study Institute, South Lake Tahoe, California, September 6-17, D. Reidel, Dordrecht, Holland, 1983.

Verly, G. and Sullivan, J. , "Multigaussian and probability krigings-application to the Jerritt Canyon deposit," *Mining Engineering*, pp. 568–574, 1985.

9

RANDOM FUNCTION
MODELS

Estimation requires a model of how the phenomenon behaves at locations where it has not been sampled; without a model, one has only the sample data and no inferences can be made about the unknown values at locations that were not sampled. One of the important contributions of the geostatistical framework is the emphasis it places on a statement of the underlying model. Unlike many earlier methods that do not state the nature of their model, geostatistical estimation methods clearly state the model on which they are based. In this chapter we address the issue of modeling. After a brief discussion of deterministic models, we will discuss probabilistic models. We will give a brief review of the essential concepts of basic probability and show an example of their application.

The Necessity of Modeling

Throughout this chapter we will be using the hypothetical example shown in Figure 9.1. In this example, we have measurements of some variable, v, at seven regularly spaced locations and are interested in estimating the unknown values of v at all of the locations we have not sampled. Though this example is one dimensional for graphical convenience, the remarks made in this chapter are not limited to one-dimensional estimation problems.

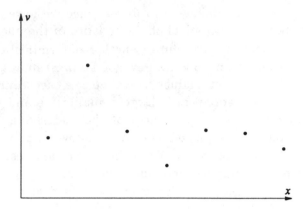

Figure 9.1 An example of an estimation problem. The dots represent seven sample points on a profile to be estimated.

The sample data set in Figure 9.1 consists of seven locations and seven v values. By itself, this sample data set tells us virtually nothing about the entire profile of v. All we know from our samples is the value of v at seven particular locations. Estimation of the values at unknown locations demands that we bring in additional information or make some assumptions.

Perhaps the most desirable information that can be brought to bear on the problem of estimation is a description of how the phenomenon was generated. In certain situations, the physical or chemical processes that generated the data set might be known in sufficient detail so that an accurate description of the entire profile can be made from only a few sample values. In such situations a deterministic model is appropriate.

Unfortunately, very few earth science processes are understood well enough to permit the application of deterministic models. Though we do know the physics or chemistry of many fundamental processes, the variables of interest in earth science data sets are typically the end result of a vast number of processes whose complex interactions we are not yet able to describe quantitatively. For the vast majority of earth science data sets, we are forced to admit that there is some uncertainty about how the phenomenon behaves between the sample locations. The random function models that we will introduce later in this chapter recognize this fundamental uncertainty and give us tools for estimating values at unknown locations once we have made some assumptions about the statistical characteristics of the phenomenon.

With any estimation procedure, whether deterministic or probabilistic, we inevitably want to know how good our estimates are. Without an exhaustive data set against which we can check our estimates, the judgment of their goodness is largely qualitative and depends to a large extent on the appropriateness of the underlying model. As conceptualizations of the phenomenon that allow us to predict what is happening at locations where we do not have samples, models are neither right nor wrong; without additional data, no proof of their validity is possible. They can, however, be judged as appropriate or inappropriate. Such a judgment, which must take into account the goals of the study and whatever qualitative information is available, will benefit considerably from a clear statement of the model.

In addition to making the nature of our assumptions clear, a clearly stated model also provides us with a constant reminder of what is real and what is modeled. With the sample data set providing a very limited view of the complete profile, there is a strong temptation to replace the frustrating reality of the estimation problem with the mathematical convenience of a model, and in so doing, to lose sight of the assumptions on which our estimation procedure is based. A typical symptom of this is the reliance on statistical hypothesis tests to test model parameters. While such tests may demonstrate that the model is self-consistent, they do not prove that the model is appropriate.

Deterministic Models

As discussed earlier, the most desirable type of estimation problem is one in which there is sufficient knowledge about the phenomenon to allow a deterministic description of it. For example, imagine that the seven sample data were measurements of the height of a bouncing ball. Knowledge of the physics of the problem and the horizontal velocity of the ball would allow us to calculate the trajectory shown in Figure 9.2. While this trajectory depends on certain simplifying assumptions, and is therefore somewhat idealized, it still captures the overall characteristics of a bouncing ball and serves as a very good estimate of the height at unsampled locations. In this particular example, we rely very heavily on our deterministic model; in fact, we could have calculated the same estimated profile with a smaller sample data set. Our determin-

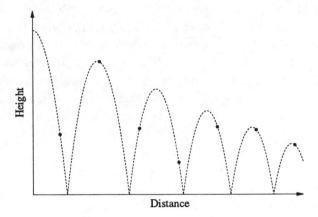

Figure 9.2 With the seven sample points shown in Figure 9.1 viewed as heights of a bouncing ball, the dashed curve shows a deterministic model of the heights at unsampled locations.

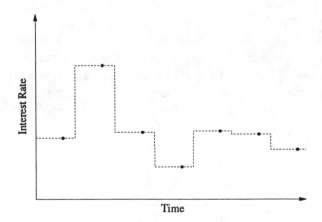

Figure 9.3 With the seven sample points shown in Figure 9.1 viewed as interest rates which change only on specific days, the dashed line shows a deterministic model of the interest rates at unsampled locations.

istic model also allows reasonable extrapolation beyond the available sampling.

Using the same seven sample data, we can imagine a scenario that would produce a very different estimated profile. We can imagine that these seven samples are interest rates at a bank, measured on the Tuesday of seven consecutive weeks. Combining this with the knowledge that the bank adjusts the interest rate only once a week, on Thursdays, we can produce the estimated profile shown in Figure 9.3. Like our previous example with the bouncing ball, accurate estimation is made possible by our knowledge of the context of the data set. Unlike the previous example, we depend on all of our sample data and our knowledge of the phenomenon is not good enough to allow us to extrapolate very far beyond the available samples.

From these two examples, it is clear that deterministic modeling is possible only if the context of the data values is well understood. The data values, by themselves, do not reveal what the appropriate model should be.

Probabilistic Models

For the seven sample data shown in Figure 9.1 we can imagine many other contexts for which a deterministic description of the complete profile would be possible. Unfortunately, few earth science applications are understood in sufficient detail to permit a deterministic approach to estimation. There is a lot of uncertainty about what happens at unsampled locations. For this reason, the geostatistical approach to estimation is based on a probabilistic model that recognizes these inevitable uncertainties.

In a probabilistic model, the available sample data are viewed as the result of some random process. From the outset, it should be clear that this model conflicts with reality. The processes that actually do create an ore deposit, a petroleum reservoir or a hazardous waste site are certainly extremely complicated, and our understanding of them may be so poor that their complexity appears as random behavior to us, but this does not mean that they are random; it simply means that we are ignorant.

Unfortunately, our ignorance does not excuse us from the difficult task of making predictions about how apparently random phenomena behave where we have not sampled them. Though earth science data are not, in fact, the result of random processes, this conceptualization does turn out to be a useful one for the problem of estimation. Though the word *random* often connotes "unpredictable," it turns out that viewing our data as the outcome of some random process does help us with the problem of predicting unknown values. Not only does it give us estimation procedures that, in practice, have sometimes proved to be very good, it also gives us some ability to gauge the accuracy of our estimates and to assign confidence intervals to them.

To take a simple but familiar example, consider the problem of estimating the sum of two dice. With a single die able to show only the numbers from 1 to 6, the sum of two dice must be in the range from 2 to 12. While an estimate of "somewhere from 2 to 12" is not very satisfying it is at least a safe start and is certainly better than avoiding the problem by claiming total ignorance. We can go beyond this safe statement, however, since some outcomes are more likely than others. A probability model in which the numbers from 1 to 6 all have an equal probability of appearing on a single die allows us to predict that 7 is the most likely outcome for the sum of two dice. Were we to use 7 as an estimate, the probability model could tell us that we would be exactly right about 17% of the time, and that we would be off by more than two only 33% of the time. If, for some reason, we preferred to use 10 as our estimate, the probability model could tell us that we would be exactly correct less than 9% of the time and would be off by more than two nearly 60% of the time.

In this example with the dice, we benefit considerably from our knowledge of the details of the random process that is generating each outcome; namely that we are dealing with dice and that a single die shows numbers only in the range from 1 to 6, each with equal probability. In the actual practice of estimation we are handicapped by not knowing the details of the random process. In fact, as we have already noted, there is no random process that is generating our sample data: no dice are being thrown behind the scenes, no coins are being tossed, and no cards are being shuffled and dealt. Having chosen to view our data as the outcome of some random process, we are responsible for defining the hypothetical random process that might have conceivably generated our data.

It is possible in practice to define a random process that might have conceivably generated any sample data set. The application of the most commonly used geostatistical estimation procedures, however, does not require a complete definition of the random process; as we will see shortly, it is sufficient to specify only certain parameters of the random process.

In the next few sections we will make the notion of a random process more precise and discuss the probabilistic tools most frequently used in geostatistics: the mean and variance of a linear combination of random variables. Though the theory of geostatistics is traditionally presented in terms of continuous random variables, the following presentation avoids this approach for the following reasons. First, for readers who are already familiar with these concepts, a review of the basic theory of continuous random variables is gratuitous and boring. Second, for readers who are unfamiliar with these concepts, a rapid overview of the theory of continuous random variables constitutes cruel and unusual punishment; as several of the references at the end of the chapter will testify, the topic is a book in itself. For the purposes of understanding random processes and how the mean and variance of a linear combination of random variables can help us in the problem of estimation, it is sufficient to reach these concepts through the more easily understood discrete approach. Those readers who are approaching these concepts for the first time, and who understand the following discrete presentation, are encouraged to read Appendix B, which provides the continuous analogy. The material in this appendix is not necessary for understanding how geostatistical estimation works, but it may provide a background for understanding the traditional geostatistical jargon.

Random Variables

A *random variable* is a variable whose values are randomly generated according to some probabilistic mechanism. The throwing of a die, for example, produces values randomly from the set $\{1, 2, 3, 4, 5, 6\}$. Similarly, if we designate the "heads" side of a coin as 0 and the "tails" side as 1, then tossing a coin produces numbers randomly from the set $\{0, 1\}$.

Throughout the remaining chapters, it will be important to keep clear the distinction between a random variable and the actual outcomes of a random variable. To maintain this distinction, we will use the following notation: upper case letters, such as V, will denote random variables while lower case letters, such as v, will denote outcomes. A further distinction is needed between the set of possible outcomes that a random variable might have, denoted $\{v_{(1)}, \ldots, v_{(n)}\}$, and the outcomes that are actually observed, denoted $v_1, v_2, v_3, \ldots$ With a single die, for example, we can view the resulting throws as a random variable called D. The six possible outcomes of D, each having equal probability, are:

$$d_{(1)} = 1, \quad d_{(2)} = 2, \quad d_{(3)} = 3, \quad d_{(4)} = 4, \quad d_{(5)} = 5, \quad d_{(6)} = 6$$

The parentheses around the subscript will serve as a reminder that these symbols refer to the set of n possible values that the random variable can take. As we throw the die and record each result, we will get a sequence of observed outcomes. Suppose that we observe a sequence beginning with the following outcomes:

The first observed outcome in this sequence is $d_1 = 4$, the second is $d_2 = 5$, the tenth is $d_{10} = 6$, and the sixteenth is $d_{16} = 4$. Without the parentheses around the subscript, the lower case letters will refer to the outcomes actually observed through several repetitions of the probabilistic mechanism.

Each possible outcome has a probability of occurrence and any random variable, V, can be completely defined by stating its set of possible outcomes, $\{v_{(1)}, \ldots, v_{(n)}\}$ and the set of corresponding probabilities, $\{p_1, \ldots, p_n\}$. The probabilities $p_1, \ldots, p_n$ must sum to one.

For the random variable we have defined as D, the result of throwing a single die, each of the outcomes in the set $\{1, 2, 3, 4, 5, 6\}$ has a probability of $\frac{1}{6}$.

The possible outcomes of a random variable need not all have equal probabilities. For example, we can define a random variable that we will call L whose outcomes are produced by throwing two dice and taking the larger of the two values. Adding a second die to our previous sequence of throws:

and choosing the larger value in each pair gives us the following outcomes of L:

$$4, 5, 4, 3, 2, 4, 5, 5, 5, 6, 6, 3, 5, 6, 5, 4, \ldots$$

L has the same set of possible outcomes as D, the result of throwing a single die; the possible values of L are $\{1, 2, 3, 4, 5, 6\}$. While the outcomes of D are all equally probable, common sense tells us that some outcomes of L are more likely than other. For example, 1 is not

Table 9.1 The probability distribution of the random variable L, whose outcomes are generated by throwing two dice and taking the larger of the two values.

$l_{(i)}$	1	2	3	4	5	6
p_i	$\frac{1}{36}$	$\frac{3}{36}$	$\frac{5}{36}$	$\frac{7}{36}$	$\frac{9}{36}$	$\frac{11}{36}$

a very likely outcome since it will occur only if both dice show 1s; the chances of this happening are only $\frac{1}{36}$. The complete probability distribution for the random variable L is given in Table 9.1; as we noted earlier, the sum of the probabilities is 1.

Functions of Random Variables

Since the outcomes of a random variable are numerical values, it is possible to define other random variables by performing mathematical operations on the outcomes of a random variable. For example, we earlier defined the random variable D, which was the result of throwing a single die. Beginning with D, we can define another random variable called $2D$, whose outcomes are generated by throwing a single die and doubling the results. Using the same sequence of throws of a die we gave earlier, the outcomes of $2D$ would be

$$8, 10, 6, 6, 4, 8, 6, 10, 10, 12, 12, 4, 10, 4, 2, 8, \ldots$$

The set of possible outcomes of $2D$ is $\{2, 4, 6, 8, 10, 12\}$, each of which has a probability of $\frac{1}{6}$ of occurring.

As another example, consider the random variable $L^2 + L$. The outcomes of this random variable would be generated by first generating an outcome of the random variable L that we defined earlier and then adding it to its own square. Beginning with the sequence of pairs of throws that we used earlier:

we would generate outcomes of $L^2 + L$ by first generating the outcomes of L:

$$4, 5, 4, 3, 2, 4, 5, 5, 5, 6, 6, 3, 5, 6, 5, 4, \ldots$$

then performing the prescribed mathematical operations, to produce the following sequence of outcomes:

$$20, 30, 20, 12, 6, 20, 30, 30, 30, 42, 42, 12, 30, 42, 30, 20, \ldots$$

The six possible outcomes of $L^2 + L$ are $\{2, 6, 12, 20, 30, 42\}$ and the corresponding set of probabilities is $\{\frac{1}{36}, \frac{3}{36}, \frac{5}{36}, \frac{7}{36}, \frac{9}{36}, \frac{11}{36}\}$. The random variable $L^2 + L$ has the same set of probabilities as the random variables from which it is defined since its definition involves only monotonic functions.

With random variables that are a defined as monotonic functions of another random variable, it is fairly straightforward to define the set of possible outcomes and the corresponding probabilities. If the n possible outcomes of the random variable V are $\{v_{(1)}, \ldots, v_{(n)}\}$ and the corresponding probabilities are $\{p_1, \ldots, p_n\}$, then the random variable $f(V)$ also has n possible outcomes, $\{f(v_{(1)}), \ldots, f(v_{(n)})\}$, with the same set of corresponding probabilities, $\{p_1, \ldots, p_n\}$.

In addition to creating new random variables by performing mathematical operations on the outcomes of a single random variable, we can also create new random variables by performing mathematical operations on the outcomes of several random variables. A simple example that we have already discussed briefly is the summing of the results of two dice. With the sequence of throws of two dice we used earlier:

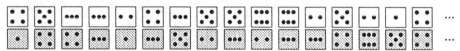

the outcomes of the random variable T could be determined by summing the values on the two dice:

$$5, 9, 7, 6, 3, 7, 8, 7, 8, 8, 9, 5, 9, 8, 6, 8, \ldots$$

The possible outcomes of T are $\{2, 3, 4, 5, 6, 7, 8, 9, 10, 11, 12\}$. As with the variable L that we discussed earlier, some of the outcomes of T are more likely than others. The only way of getting the dice to sum to 2, for example, is to have 1s on both of them; the probability of this unlikely event is $\frac{1}{36}$. There are a total of six possible combinations which sum to 7, and the probability of this event is $\frac{6}{36}$. The complete probability distribution for the possible outcomes of T is given in Table 9.2.

We should note here that the random variable T is not the same as the random variable $2D$ we discussed earlier. The outcomes of $2D$ are

Table 9.2 The probability distribution of the random variable T, whose outcomes are generated by throwing two dice and adding the two values.

$t_{(i)}$	2	3	4	5	6	7	8	9	10	11	12
p_i	$\frac{1}{36}$	$\frac{2}{36}$	$\frac{3}{36}$	$\frac{4}{36}$	$\frac{5}{36}$	$\frac{6}{36}$	$\frac{5}{36}$	$\frac{4}{36}$	$\frac{3}{36}$	$\frac{2}{36}$	$\frac{1}{36}$

generated by taking a *single* random variable, D, and doubling it. The outcomes of T are generated by taking *different* random variables, D_1 and D_2, and summing them. Though D_1 and D_2 are generated by the same random mechanism, their outcomes are not necessarily the same. As a function of a single random variable, $2D$ has the same number of possible outcomes as the random variable from which it is defined. T, on the other hand, has several more outcomes than either of the random variables from which it is defined.

In combining different random variable to create new ones, we are not limited to addition. For example, we could define a random variable called DH, whose outcomes would be generated by throwing a die, tossing a coin (heads = 0, tails = 1), and multiplying the two results. For this particular random variable, it is not too difficult to specify the complete set of possible outcomes and their probabilities; there are seven possible outcomes, $\{0, 1, 2, 3, 4, 5, 6\}$ with the following set of probabilities: $\{\frac{1}{2}, \frac{1}{12}, \frac{1}{12}, \frac{1}{12}, \frac{1}{12}, \frac{1}{12}, \frac{1}{12}\}$.

For random variables defined as functions of other random variables, it becomes difficult to define the complete set of possible outcomes and their corresponding probabilities, particularly if the function is complicated. Fortunately, for the problems we will be tackling, we will never have to deal with anything more complicated than a sum of several random variables.

Parameters of a Random Variable

The set of outcomes and their corresponding probabilities is sometimes referred to as the *probability law* or *probability distribution* of a random variable. If this probability distribution is known, one can calculate many parameters that describe interesting features of the random variable. Each of the descriptive statistics we discussed in the first section of this book has a corresponding parameter. For example, there is a maximum value for any random variable, as well as a minimum; it will also have a mean and a standard deviation. If the set of possible outcomes is large enough, it will have a median and two quartiles.

There are two important remarks to be made concerning the parameters of a random variable. First, the complete distribution cannot, in general, be determined from the knowledge of only a few parameters. Second, the parameters cannot be calculated by observing the outcomes of a random variable.

The distribution cannot generally be deduced from a few parameters. In the second chapter, we discussed the use of statistics for summarizing a distribution. While statistics such as the mean and the standard deviation do describe important features of the distribution, they are only summaries, and do not tell us everything about the distribution. Two distributions may have the same mean and standard deviation and still be quite different in other respects. The same is true of random variables and their parameters. While two random variables may have the same mean and variance, their distributions need not be the same. In later chapters, where we discuss the calculation of confidence intervals, we will make use of random variables whose complete distributions can be determined from the knowledge of only a few parameters. For example, a Gaussian random variable can be completely defined by its mean and variance; a uniform random variable can be completely defined by its minimum and maximum. In general, however, a random variable is not fully described by a few parameters.

The parameters of a random variable cannot be calculated exactly by observing a few outcomes of the random variable; rather, they are parameters of a conceptual model. From a sequence of observed outcomes, all we can do is calculate sample statistics based on that particular set of data; a different set of outcomes would produce a different set of sample statistics. It is true that as the number of observed outcomes increases, the sample statistics calculated from the set of available observations tend to become more similar to their corresponding model parameters. This leads us in practice to assume that the parameters of our random variable are the same as the sample statistics we can calculate. For example, having calculated the mean of our available samples we could choose to conceptualize these samples as outcomes of a random variable whose mean is the same as our sample mean.

The step of assigning the same mean to our random variable as that of our observed data involves an assumption. There are several other random variables with different means that could also have generated the same data. To take a specific example, the 16 throws of a die that we gave earlier have a sample mean of 3.75. Were these sample data from a real data set, we would not know the details of the generating mechanism, and would typically assume that the random variable that generated these 16 sample values also has a mean of 3.75. From our knowledge of the underlying random mechanism, however, we know that in this case the mean of the random variable is, in fact, 3.5.

The important and fundamental difference between the parameters of a conceptual model and the statistics which can be calculated from a finite number of observed outcomes calls for a certain care in our notation. We will be using the same symbols for the model parameters that we used earlier for the descriptive statistics. To emphasize the difference between the two, we will use a tilde (˜) above a symbol if it refers to a model parameter. The mean of a set of observed values will still be denoted by m, for example, and if we choose to conceptualize these values as outcomes of some random variable, then the mean of the corresponding random variable will be denoted by $\tilde{m}$.

The two model parameters most commonly used in probabilistic approaches to estimation are the mean or "expected value" of the random variable and its variance. For certain random variables, the mean and the variance provide a complete description of the distribution. Even for those random variables for which the mean and variance do not provide a complete description, these two parameters often provide useful information on how the random variable behaves.

Expected Value. The expected value of a random variable is its mean or average outcome. It is the weighted average of the n possible outcomes, with each outcome being weighted by its probability of occurrence:

$$E\{V\} = \tilde{m} = \sum_{i=1}^{n} p_i v_{(i)} \tag{9.1}$$

As discussed earlier, the ˜ above the m reminds us that m is a model parameter, and not a sample statistic.

Using the information in Table 9.1, which gave the probability distribution of the random variable L that we defined earlier, the expected value of the larger of two dice throws is

$$E\{L\} = \frac{1}{36}(1) + \frac{3}{36}(2) + \frac{5}{36}(3) + \frac{7}{36}(4) + \frac{9}{36}(5) + \frac{11}{36}(6)$$
$$= 4.47$$

As we can see from this particular example, the expected value need not be one of the possible outcomes. Also, it need not be close to the most likely outcome, the mode, or to the median of our random variable.

The expected value of the sum of two random variables is the sum of their expected values:

$$E\{U + V\} = E\{U\} + E\{V\} \qquad (9.2)$$

The random variable that we called T, for example, was defined as the sum of two dice: $T = D_1 + D_2$. Using Equation 9.1 and the probabilities of the various outcomes that we gave in Table 9.2, the expected value of T is 7.0. Even if we had not known the complete probability distribution of T, we could have reached the same conclusion by using Equation 9.2. The expected value of a single throw of a die is 3.5, so the expected value of the sum of two dice throws is 7.0.

Variance. The variance of a random variable is the expected squared difference from the mean of the random variable:

$$Var\{V\} = \tilde{\sigma}^2 = E\{[V - E\{V\}]^2\} \qquad (9.3)$$

An alternate expression for the variance can be derived by expanding the square inside the curly brackets:

$$Var\{V\} = E\{V^2 - 2V E\{V\} + E\{V\}^2\}$$

and using the fact that the expected value of a sum is equal to the sum of the expected values:

$$\begin{aligned} Var\{V\} &= E\{V^2\} - E\{2V E\{V\}\} + E\{E\{V\}^2\} \\ &= E\{V^2\} - 2E\{V\}E\{V\} + E\{V\}^2 \\ &= E\{V^2\} - E\{V\}^2 \qquad (9.4) \end{aligned}$$

If the probability law is known, the variance of a random variable can be expressed as

$$Var\{V\} = \sum_{i=1}^{n} p_i v_{(i)}^2 - \left(\sum_{i=1}^{n} p_i v_{(i)}\right)^2 \qquad (9.5)$$

Using this equation and the information given in Table 9.1, the variance of the random variable L is 1.97.

As with other model parameters, the variance of a random variable need not be the same as the sample variance calculated from a sequence of actual observations. For example, the sequence of 16 outcomes we gave earlier for the random variable T, the sum of two dice, has a variance of 2.68, while the corresponding model parameter is 5.83. The particular outcomes we saw in our sequence were much less variable than the probability model would have predicted.

Joint Random Variables

Random variables may also be generated in pairs according to some probabilistic mechanism; the outcome of one of the variables may influence the outcome of the other. For example, we could create a pair of random variables, L and S, by throwing two dice, letting the outcome of L be the larger of the two values shown on the dice and the outcome of S be the smaller. The sequence of throws of two dice given earlier:

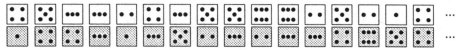

results in the following pairs of (L, S) values:

$$(4,1), (5,4), (4,3), (3,3), (2,1), (4,3), (5,3), (5,2),$$
$$(5,3), (6,2), (6,3), (3,2), (5,4), (6,2), (5,1), (4,4), \ldots$$

There is some relationship between the value of L and the value of S. As the value of L gets smaller, the range of possible values for S also gets smaller. A complete definition of these two random variables should include a description of how they *jointly* vary.

Pairs of joint discrete random variables, (U, V), may be completely defined by stating all the possible outcomes,

$$\{(u_{(1)}, v_{(1)}), \ldots, (u_{(1)}, v_{(m)}), \ldots, (u_{(n)}, v_{(1)}), \ldots, (u_{(n)}, v_{(m)})\}$$

along with the corresponding probabilities, which must sum to 1:

$$\{p_{11}, \ldots, p_{1m}, \ldots, p_{n1}, \ldots, p_{nm}\}$$

where there are n possible outcomes for U and m possible outcomes for V.

Table **9.3** The joint probability distribution for the pair of random variables (L, S), whose outcomes are determined by throwing two dice.

		Possible Outcomes of S					
		1	2	3	4	5	6
	1	$\frac{1}{36}$	0	0	0	0	0
Possible	2	$\frac{2}{36}$	$\frac{1}{36}$	0	0	0	0
Outcomes	3	$\frac{2}{36}$	$\frac{2}{36}$	$\frac{1}{36}$	0	0	0
of L	4	$\frac{2}{36}$	$\frac{2}{36}$	$\frac{2}{36}$	$\frac{1}{36}$	0	0
	5	$\frac{2}{36}$	$\frac{2}{36}$	$\frac{2}{36}$	$\frac{2}{36}$	$\frac{1}{36}$	0
	6	$\frac{2}{36}$	$\frac{2}{36}$	$\frac{2}{36}$	$\frac{2}{36}$	$\frac{2}{36}$	$\frac{1}{36}$

Table 9.3 completely defines the joint variables (L, S) that we discussed earlier. For any particular pair of outcomes, $(l_{(i)}, s_{(j)})$, this table gives its probability of occurrence.

The notion of a pair of joint random variables can be extended to any number of random variables. For example, we can imagine defining a triple of random variables (L, M, S), whose outcomes were determined by throwing three dice, letting the outcomes of L and S be the largest and smallest values and letting the outcome of M be the remaining middle value. The probability law of these joint random variables would consist of a set of all possible triples, $\{\ldots, (l_{(i)}, m_{(j)}, s_{(k)}), \ldots\}$, and a corresponding set of probabilities, $\{\ldots, p_{ijk}, \ldots\}$. As with the univariate and bivariate probabilities we discussed earlier, the set of trivariate probabilities would sum to 1. The multivariate probability distribution of n joint random variables consists of a set of all possible n-tuples and a corresponding set of probabilities of occurrence that sum to one.

Marginal Distributions

The knowledge of the joint distribution of two random variables allows the calculation of the univariate or marginal distribution of each random variable. The distribution of one variable ignoring the effect of the second is usually referred to as the *marginal distribution*. For discrete random variables (U, V), the probability law of U by itself can be calculated from the joint probability law in the following manner:

$$P\{U = u_{(i)}\} = p_i = \sum_{j=1}^{m} p_{ij} \tag{9.6}$$

If the joint probability distribution is tabulated in a form similar to that shown in Table 9.3, Equation 9.6 corresponds to summing across the rows or down the columns. Using Table 9.3, we can calculate the marginal probability that L is 5 by summing all of the joint probabilities in the fifth row:

$$
\begin{aligned}
P\{L = 5\} = p_5 &= \sum_{j=1}^{6} p_{5j} \\
&= \frac{2}{36} + \frac{2}{36} + \frac{2}{36} + \frac{2}{36} + \frac{1}{36} + 0 \\
&= \frac{9}{36}
\end{aligned}
$$

This procedure gives us a marginal probability of $p_5 = \frac{9}{36}$, which agrees with the information given earlier in Table 9.1.

Conditional Distributions

The knowledge of the joint distribution of two random variables also allows us to calculate the univariate distribution of one variable given a particular outcome of the other random variable. For example, with our discrete random variables L and S, we might be interested in knowing the probability that the outcome of L is 3 if we already have the information that the outcome of S is also 3.

The conditional probabilities of one random variable given a particular value of another can be calculated from the following axiom of conditional probability:

$$P\{U = u | V = v\} = \frac{P\{U = u \text{ and } V = v\}}{P\{V = v\}} \tag{9.7}$$

The numerator in this equation is given by the joint probability law; the denominator can be calculated using the equations given in the previous section on marginal distributions. The idea behind Equation 9.7 is that the joint probability $P\{U = u \text{ and } V = v\}$ needs to be rescaled since our probability law no longer consists of all possible pairs of (U, V) outcomes, but only those pairs that have v as the outcome of the second variable. The division by $P\{V = v\}$ guarantees that the sum of the probabilities is still 1.

For discrete random variables, conditional probabilities can be calculated in the following manner:

$$P\{U = u_{(i)}|V = v_{(j)}\} = \frac{p_{ij}}{\sum_{k=1}^{n} p_{kj}} \qquad (9.8)$$

With the joint probabilities tabulated in the form given in Table 9.3, Equation 9.8 corresponds to dividing a particular entry by the sum of all other entries in the same row or column. For example, returning to the problem we posed earlier the conditional probability that L is 3 given that S is 3 can be calculated by dividing p_{33} by the sum of the entries in the third column:

$$
\begin{aligned}
P\{L = 3|S = 3\} &= \frac{p_{33}}{\sum_{k=1}^{6} p_{k3}} \\
&= \frac{\frac{1}{36}}{\frac{2}{36} + \frac{2}{36} + \frac{2}{36} + \frac{1}{36} + 0 + 0} \\
&= \frac{1}{7}
\end{aligned}
$$

The conclusion is that if two dice are rolled and we somehow discover that the smaller of the two values is 3, then the probability of the larger one also being 3 is $\frac{1}{7}$.

Parameters of Joint Random Variables

Joint random variables are completely defined by their joint probability law, which includes the set of possible pairs and the set of corresponding probabilities of occurrence. The knowledge of the probability law of joint random variables allows the calculation of several parameters that are useful in describing how the two variables are related. As with the univariate parameters we discussed earlier, the knowledge of bivariate parameters does not generally allow the complete probability law to be deduced. Also, these bivariate parameters are not the same as the corresponding statistics that can be calculated from a finite set of observations.

Covariance. The covariance between two random variables is defined as follows:

$$Cov\{UV\} = \tilde{C}_{UV} = E\{(U - E\{U\})(V - E\{V\})\} \quad (9.9)$$
$$= E\{UV\} - E\{U\}E\{V\} \quad (9.10)$$

If the joint probability law is known, then the covariance can be calculated by the following equation:

$$Cov\{UV\} = \sum_{i=1}^{n}\sum_{j=1}^{m} p_{ij}u_{(i)}v_{(j)} - \sum_{i=1}^{n} p_i u_{(i)} \sum_{j=1}^{m} p_j v_{(j)} \quad (9.11)$$

The marginal probabilities of the n possible outcomes of U and the m possible outcomes of V can be calculated using Equation 9.6. Using the joint probability distribution for the random variables L and S given in Table 9.3, the first term in Equation 9.11 is 12.25, the expected value of L is 4.47, and the expected value of S is 2.53. The covariance between the random variables L and S is 0.941. As we noted earlier, this parameter need not be identical to the sample covariance calculated from an observed series of outcomes. For example, the sample covariance of the 16 (L, S) pairs that we gave earlier is 0.219. If we continued throwing the pair of dice and produced many more outcomes of (L, S) pairs, the sample covariance calculated on the observed outcomes would tend to approach the corresponding parameter.

The covariance between a random variable and itself, $\tilde{C}_{UU}$, is identical to its variance, $\tilde{\sigma}_U^2$.

The covariance of a random variable is influenced by the magnitude of the possible outcomes. If we defined a pair of random variables $(10L, 10S)$, whose outcomes are determined simply by multiplying the outcomes of (L, S) by 10, the covariance of $(10L, 10S)$ is 941, which is 100 times greater than the covariance of (L, S).

Correlation Coefficient. It is often desirable to have a parameter that describes how two random variables jointly vary yet is not affected by a simple rescaling of the values. The *correlation coefficient* provides such a parameter, and is defined as follows:

$$\tilde{\rho}_{UV} = \frac{\tilde{C}_{UV}}{\sqrt{\tilde{\sigma}_U^2 \tilde{\sigma}_V^2}} \quad (9.12)$$

The variances $\tilde{\sigma}_U^2$ and $\tilde{\sigma}_V^2$ can be calculated from their corresponding marginal distributions using Equation 9.5.

Continuing with the covariance calculation for L and S given earlier, the variance of L is 1.97, and the variance of S is also 1.97, giving us a correlation coefficient of 0.478.

If the possible pairs of outcomes of two random variables all fall on a straight line when plotted on a scatterplot, then the two random variables have a correlation coefficient of $+1$ or -1 depending on the slope of the line. For two random variables whose outcomes are completely independent of one another, the correlation coefficient is 0. Though the correlation coefficient is a parameter that summarizes the strength of the *linear* relationship, it is often used as a parameter of the bivariate distribution of joint random variables that are not linearly related. The correlation coefficient of 0.478 between L and S agrees with our earlier intuition that these two random variables are somehow related. If the outcome of one of them is large, the outcome of the other one also tends to be large.

Weighted Linear Combinations of Random Variables

Weighted linear combinations of random variables are particularly important for estimation since, as we discussed in the previous chapter, all of the estimators we will be discussing are weighted linear combinations of the available sample data. In probabilistic approaches to estimation, where the available sample data are viewed as outcomes of random variables, the estimate is therefore an outcome of a random variable that is created by a weighted linear combination of other random variables.

To completely describe the distribution of a random variable that is created by combining other random variables, we would need to know the multivariate distribution of all the random variables involved in the linear combination. Even without the knowledge of this multivariate distribution, we are able to describe certain parameters of the linear combination by knowing certain parameters of the random variables involved in the combination.

Expected Value of a Linear Combination. Equation 9.2 can be generalized to include not only sums of two random variables but also weighted sums of any number of random variables. The expected value of a weighted linear combination of random variables is equal to the weighted linear combination of the individual expected values:

$$E\{\sum_{i=1}^{n} w_i V_i\} = \sum_{i=1}^{n} w_i E\{V_i\} \tag{9.13}$$

Equation 9.2 is a particular case of this in which $n = 2$ and $w_1 = w_2 = 1$.

Earlier, we gave a simple example of this equation using the random variable T defined as the result of throwing two dice and summing the results. In this particular example, the random variables involved in the linear combination, D_1 and D_2, were independent; the outcome of one die did not influence the outcome of the other. Even if there is some dependence between the variables involved in the linear combination, Equation 9.13 is still valid. As an example of this, consider the two random variables L and S. We have already seen that there is some correlation between these two; larger values of L tend to be associated with larger values of S. Before we try to apply Equation 9.13, we can use our common sense to figure out what the expected value of the sum of L and S should be. To generate outcomes of L and S we threw two dice, designating the larger one as the outcome of L and the smaller one as the outcome of S. When we sum the two outcomes, it really does not matter which one is the larger and which one is the smaller; the values on both dice get added together regardless of which one is larger. Following this argument, the random variable we create by summing the outcomes of L and S is the same as the random variable T that we defined simply as the result of throwing two dice and summing the two values. The expected value of $L + S$ should therefore be the same as the expected value of T.

Earlier, when we calculated the covariance between L and S, we had to calculate their expected values: $E\{L\} = 4.47$ and $E\{S\} = 2.53$. Using these two values in Equation 9.13 brings us to the result that the expected value of $L+S$ is 7.0. This result agrees with our common sense conclusion, and provides one example of the fact that Equation 9.13 is valid even if the variables being summed are correlated.

Variance of a Weighted Linear Combination. The variance of a random variable that is created by a weighted linear combination of other random variables is given by the following equation:

$$Var\{\sum_{i=1}^{n} w_i \cdot V_i\} = \sum_{i=1}^{n} \sum_{j=1}^{n} w_i \cdot w_j \cdot Cov\{V_i V_j\} \tag{9.14}$$

As an example of the application of this equation, let us try to calculate the variance of T and also of $L + S$. While T is created by summing two independent random variables, D_1 and D_2, $L + S$ is created by summing two dependent random variables. In the previous section we showed that these two random variables are, in fact, the same. The variance we calculate for T using Equation 9.14 should be the same as the variance we calculate for $L + S$.

For the simple addition of two random variables, Equation 9.14 becomes

$$Var\{U + V\} = Cov\{UU\} + Cov\{UV\} + Cov\{VU\} + Cov\{VV\}$$

For the random variable $T = D_1 + D_2$, the two random variables involved in the sum are independent:

$$Cov\{D_1 D_2\} = Cov\{D_2 D_1\} = 0$$

As we noted earlier, the covariance between a random variable and itself is identical to the variance.

$$Cov\{D_1 D_1\} = Var\{D_1\}$$
$$Cov\{D_2 D_2\} = Var\{D_2\}$$

Despite their outcomes being different, D_1 and D_2 are both produced by the same random mechanism, the throwing of a single die, and their variances are equal. Their variance, calculated using Equation 9.5 with the probabilities all equal to $\frac{1}{6}$, is 2.917. Substituting these results into Equation 9.14, the variance of T is 5.83. This can be checked by calculating the variance of T directly using Equation 9.5 and the probability distribution given in Table 9.2.

For the random variables L and S, we have already calculated the following results:

$$Cov\{LS\} = Cov\{SL\} = 0.941$$
$$Cov\{LL\} = Var\{L\} = 1.971$$
$$Cov\{SS\} = Var\{S\} = 1.971$$

Substituting these results into Equation 9.14, the variance of $L + S$ is 5.83, which agrees with our previous calculations.

Random Functions

A random function is a set of random variables that have some spatial locations and whose dependence on each other is specified by some probabilistic mechanism. For example, we can take a series of points regularly spaced in one dimension and define a random function $V(x)$ that is generated by the following probabilistic mechanism:

$$V(0) \;=\; \begin{cases} 0 & \text{with probability } \tfrac{1}{2} \\ 1 & \text{with probability } \tfrac{1}{2} \end{cases} \qquad (9.15)$$

$$V(x) \;=\; \begin{cases} V(x-1) & \text{with probability } \tfrac{3}{4} \\ 1 - V(x-1) & \text{with probability } \tfrac{1}{4} \end{cases} \qquad (9.16)$$

All values of $V(x)$ are either 0 or 1. The random function begins at either 0 or 1 with equal probability. At each subsequent location, the value of the function has a 75% chance of staying the same and a 25% chance of switching to the other value. The decision to stay the same or switch could be decided by flipping two coins and switching only if both of them showed heads.

As with random variables, which have several possible outcomes, random functions also have several possible outcomes or "realizations." Figure 9.4 shows three of the possible realizations of the random function $V(x)$ that we defined earlier. Though each of these realizations are different in their details, they have similar characteristics due to the fact that they were all generated by the procedure described by Equations 9.15 and 9.16.

The random variables $V(0), V(1), V(2), \ldots, V(x_i), \ldots$ all have the same univariate probability law: the set of possible outcomes of $V(x_i)$ is $\{0, 1\}$, with both outcomes being equally probable. This is due to the fact that Equation 9.15 gives us an equal chance of getting a 0 or a 1 at the first location and that Equation 9.16 does not discriminate between 0s and 1s—it gives us an equal chance of staying at 0 as it does of staying at 1.

The pairs of random variables at consecutive locations:

$$(V(0), V(1)), (V(1), V(2)), \ldots, (V(x), V(x+1)), \ldots$$

all have the same joint probability law; the set of possible outcomes of $(V(x), V(x+1))$ is $\{(0,0), (0,1), (1,0), (1,1)\}$, with the corresponding

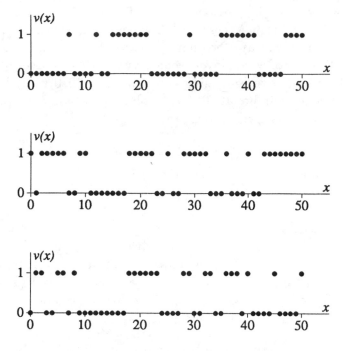

Figure 9.4 Three realizations of the random function $V(x)$.

set of probabilities being $\{\frac{3}{8}, \frac{1}{8}, \frac{1}{8}, \frac{3}{8}\}$. This common joint distribution between consecutive pairs is a consequence of Equation 9.16, which makes the outcomes $(0,0)$ and $(1,1)$ three times as likely as the outcomes $(0,1)$ and $(1,0)$.

Equation 9.16 actually does a lot more than impose a common joint distribution between consecutive pairs; it also imposes a common joint distribution between pairs of values separated by any specified distance. For example, the pairs separated by a distance of two units, $(V(x), V(x+2))$, have a particular joint distribution, as can be seen by considering the possible combinations of three consecutive values, $(V(x), V(x+1), V(x+2))$:

$$(0,0,0), (0,0,1), (0,1,0), (0,1,1)$$
$$(1,0,0), (1,0,1), (1,1,0), (1,1,1)$$

Since staying at the same value is three times as likely as switching to the other value, some of these possible combinations of three consecutive values are more likely than others. The two least likely ones are $(0,1,0)$ and $(1,0,1)$ since these both involve two switches. Three times as likely as these two are the combinations $(0,0,1)$, $(0,1,1)$, $(1,0,0)$, and $(1,1,0)$, each of which involves only one switch. Three times more likely still are the two remaining combinations, $(0,0,0)$ and $(1,1,1)$, which involve no switches at all. The probabilities associated with each of the combinations given above are:

$$p = \tfrac{9}{32} \quad : \quad (0,0,0),(1,1,1)$$
$$p = \tfrac{3}{32} \quad : \quad (0,0,1),(0,1,1),(1,0,0),(1,1,0)$$
$$p = \tfrac{1}{32} \quad : \quad (0,1,0),(1,0,1)$$

Looking at these eight possible outcomes of three consecutive values, we find that two of them start with 0 and end with 0, $(0,0,0)$ and $(0,1,0)$. Summing the probabilities of these two events, we reach the conclusion that the pair of random variables $(V(x), V(x+2))$ has a probability of $\tfrac{10}{32}$ of being $(0,0)$. The other possible outcomes of $(V(x), V(x+2))$ have the following probabilities:

$$P\{(V(x), V(x+2)) = (0,1)\} \quad = \quad \frac{6}{32}$$
$$P\{(V(x), V(x+2)) = (1,0)\} \quad = \quad \frac{6}{32}$$
$$P\{(V(x), V(x+2)) = (1,1)\} \quad = \quad \frac{10}{32}$$

These results completely define the probability law of the joint random variables $(V(x), V(x+2))$. Even at a separation of two units, there is a greater chance that two values will be the same than be different.

If we repeat this type of analysis for other pairs of random variables $(V(x), V(x+h))$, we find that Equation 9.16 implicitly states the joint probability distribution for pairs of random variables separated by any distance. For any value of h, the outcomes $(0,0)$ and $(1,1)$ are more likely to occur than the outcomes $(0,1)$ and $(1,0)$. For $h = 1$, we saw that the probability of staying at the same value was $\tfrac{3}{4}$; at $h = 2$, it decreased slightly to $\tfrac{20}{32}$. As h increases, the probability of staying at the same value decreases, asymptotically approaching $\tfrac{1}{2}$.

For this random function $V(x)$ that we have been using as an example, the univariate probability law does not depend on the location x; at all locations, 0 and 1 have an equal probability of occurring.

Similarly, the bivariate probability law of $V(x)$ and $V(x + h)$ does not depend on the location x, but only on the separation h; regardless of their locations, all pairs of random variables separated by a particular distance h have the same joint probability distribution. This independence of the univariate and bivariate probability laws from the location x is referred to as *stationarity*. There are random functions that are not stationary and there are geostatistical techniques that make use of such random function models. The most widely used geostatistical estimation procedures, however, including the ones we will be discussing in later chapters, use stationary random function models.

Parameters of a Random Function

From a complete description of its probabilistic generating mechanism, we can calculate several parameters that describe interesting features of a random function. If the random function is stationary, then the univariate parameters we discussed earlier, the expected value and the variance, can be used to summarize the univariate behavior of the set of random variables. For the random function $V(x)$ that we used as an example in the previous section, 0 and 1 both had an equal probability of occurring at all locations. Using Equation 9.1, the expected value of $V(x)$ is 0.5; using Equation 9.5, its variance is 0.25.

The other parameters that are commonly used to summarize the bivariate behavior of a stationary random function are its covariance function, $\tilde{C}(h)$, its correlogram, $\tilde{\rho}(h)$, and its variogram, $\tilde{\gamma}(h)$. For stationary random functions, these three parameters are related by a few simple expressions. As with the univariate and bivariate parameters we discussed earlier, it is important to keep in mind that the descriptive statistics that can be calculated from a sample data set are not the same as the conceptual model parameters. In the case of the covariance function, the correlogram and the variogram, the simple expressions that describe the relationships between the parameters of a random function are not valid for the corresponding descriptive statistics.

The covariance function is the covariance between random variables separated by a distance h:

$$\begin{aligned} \tilde{C}_V(h) &= Cov\{V(x) \cdot V(x + h)\} \\ &= E\{V(x) \cdot V(x + h)\} - E\{V(x)\}E\{V(x + h)\} \end{aligned}$$

For stationary random functions, this can be expressed as

$$\tilde{C}_V(h) = E\{V(x) \cdot V(x+h)\} - E\{V(x)\}^2 \tag{9.17}$$

The covariance between random variables at identical locations is the variance of the random function:

$$\tilde{C}_V(0) = Cov\{V(x) \cdot V(x)\} = Var\{V(x)\} \tag{9.18}$$

The correlogram is the correlation coefficient between random variables separated by a specified distance:

$$\tilde{\rho}_V(h) = \frac{Cov\{V(x) \cdot V(x+h)\}}{\sqrt{Var\{V(x)\} \cdot Var\{V(x+h)\}}} = \frac{\tilde{C}_V(h)}{\tilde{C}_V(0)} \tag{9.19}$$

The correlation coefficient between random variables at identical locations is 1:

$$\tilde{\rho}_V(0) = \frac{Cov\{V(x) \cdot V(x)\}}{Var\{V(x)\}} = 1 \tag{9.20}$$

The variogram is half the expected squared difference between random variables separated by a specified distance:

$$\tilde{\gamma}_V(h) = \frac{1}{2}E\{[V(x) - V(x+h)]^2\} \tag{9.21}$$

Using the properties of the expected value that we discussed earlier, this can also be expressed as

$$\tilde{\gamma}_V(h) = \frac{1}{2}E\{V(x)^2\} + \frac{1}{2}E\{V(x+h)^2\} - E\{V(x) \cdot V(x+h)\}$$

For stationary random functions, $E\{V(x)^2\} = E\{V(x+h)^2\}$, which allows us to rewrite the previous equation as

$$\tilde{\gamma}_V(h) = E\{V(x)^2\} - E\{V(x) \cdot V(x+h)\}$$

The equation is unchanged if we subtract $E\{V(x)\}^2$ from the first term and add it to the second:

$$\tilde{\gamma}_V(h) = E\{V(x)^2\} - E\{V(x)\}^2 - E\{V(x) \cdot V(x+h)\} + E\{V(x)\}^2$$

Using Equation 9.4, the variogram can be written as

$$\begin{aligned}\tilde{\gamma}_V(h) &= Var\{V(x)\} - [E\{V(x) \cdot V(x+h)\} - E\{V(x)\}^2] \\ &= \tilde{C}_V(0) - \tilde{C}_V(h) \end{aligned} \tag{9.22}$$

Since the expected squared difference between samples at identical locations is 0, the value of $\tilde{\gamma}(0)$ is always 0.

For the majority of random functions used in practical geostatistics, the pairs of widely separated random variables are independent of one another. The covariance function and the correlogram, therefore, eventually reach 0 while the variogram eventually reaches a maximum value, usually referred to as the *sill*. This sill value of the variogram is also the variance of the random function, which allows Equation 9.22 to be expressed as

$$\tilde{C}_V(h) = \tilde{\gamma}_V(\infty) - \tilde{\gamma}_V(h) \qquad (9.23)$$

For the stationary random function models most frequently used in the practice of geostatistics, the covariance function, the correlogram and the variogram provide exactly the same information in a slightly different form. The correlogram and the covariance have the same shape, with the correlogram being scaled so that its maximum value is 1. The variogram also has the same shape as the covariance function, except that it is upside-down; while the covariance starts from a maximum of $\tilde{\sigma}^2$ at $h = 0$ and decreases to 0, the variogram starts at 0 and increase to a maximum of $\tilde{\sigma}^2$. Figure 9.5 shows these three summaries for the random function $V(x)$ that we used earlier as an example.

By summarizing the joint distribution of pairs as a function of distance, the variogram (or the covariance or the correlogram) provide a measurement of the spatial continuity of the random function. As we noted when we first discussed the random function $V(x)$, it has a tendency to stay at the same value rather than to switch values. This tendency is controlled by the probability values given in Equation 9.16. We can define a random function called $U(x)$ that is similar to $V(x)$ but has a greater tendency to stay at the same value:

$$U(0) = \begin{cases} 0 & \text{with probability } \frac{1}{2} \\ 1 & \text{with probability } \frac{1}{2} \end{cases}$$

$$U(x) = \begin{cases} U(x-1) & \text{with probability } \frac{7}{8} \\ 1 - U(x-1) & \text{with probability } \frac{1}{8} \end{cases}$$

Figure 9.6 shows three realizations of this random function. Comparing this to the realizations of $V(x)$ shown earlier in Figure 9.4, we see that

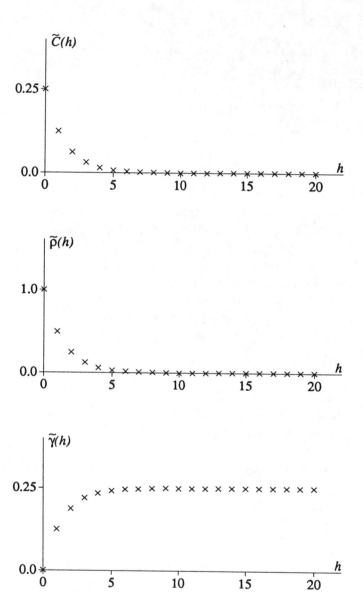

Figure 9.5 Three measures of the bivariate behavior of the random function $V(x)$.

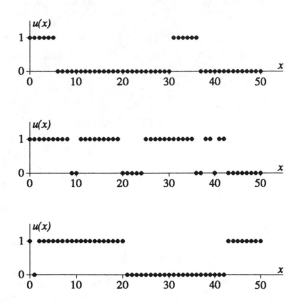

Figure 9.6 Three realizations of the random function $U(x)$.

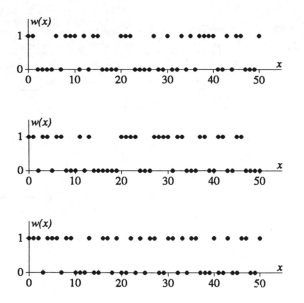

Figure 9.7 Three realizations of the random function $W(x)$.

by increasing the probability that $U(x) = U(x - 1)$, we have created a random function that is more continuous.

We can define another function called $W(x)$ that is also similar to $V(x)$ but has less of a tendency to stay at the same value:

$$W(0) = \begin{cases} 0 & \text{with probability } \frac{1}{2} \\ 1 & \text{with probability } \frac{1}{2} \end{cases}$$

$$W(x) = \begin{cases} W(x - 1) & \text{with probability } \frac{1}{2} \\ 1 - W(x - 1) & \text{with probability } \frac{1}{2} \end{cases}$$

Figure 9.7 shows three realizations of $W(x)$. Comparing this to the realizations of $V(x)$ and $U(x)$, we see that by decreasing the probability that $W(x) = W(x - 1)$, we have created a random function that is less continuous.

Figure 9.8 shows the variograms of the three random functions $V(x)$, $U(x)$, and $W(x)$. We have already observed that these random functions have different spatial continuities. Figure 9.8 shows that these differences in the spatial continuity are captured by the variogram. The variogram for $W(x)$, the least continuous of the three, rises immediately, reaching the maximum value of 0.25 at $h = 1$; the variogram for $U(x)$, the most continuous of the three, rises much more slowly. Random functions for which closely spaced values may be quite different will have variograms that rise quickly from the origin; random functions for which closely spaced values are very similar will have variograms that rise much more slowly.

The Use of Random Function Models in Practice

We are now in a position to discuss how random function models are typically used in practice. As we noted earlier, a random function is purely a conceptual model that we choose to use because we do not yet have accurate deterministic models. Having decided to use this type of conceptualization, we are responsible for defining the random process that might have created our observed sample values.

We could try to define the probabilistic generating mechanism such as the one we gave in Equation 9.15 and 9.16 for the random function $V(x)$. From the probabilistic generating mechanism we could generate many realizations, each one of which would be a possible reality

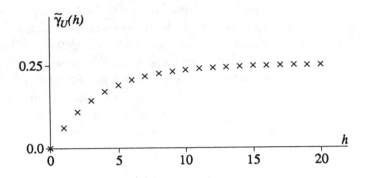

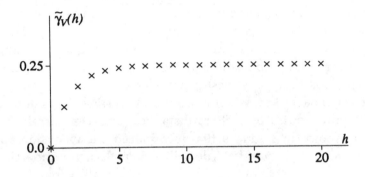

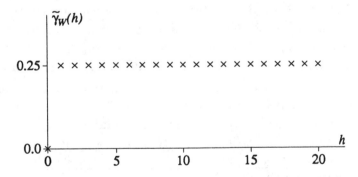

Figure 9.8 Variograms for the three random functions $U(x)$, $V(x)$, and $W(x)$.

that is consistent with our available data and with the random function model we have chosen. In practice, however, we usually do not bother to completely describe the probabilistic generating mechanism for several reasons. First, the complete definition of the probabilistic generating mechanism is very difficult even in one dimension. The examples of random processes that we used earlier were sufficient to convey the important concepts, but are not able to adequately model all of the possible data sets we might encounter. Second, for many of the problems we typically encounter, we do not need to know the probabilistic generating mechanism. As we will see in the following section, it is possible to tackle many estimation problems with a knowledge of only a few parameters.

In the practice of geostatistics we usually adopt a stationary random function as our model and specify only its covariance or variogram. Returning to the sample data set we showed at the beginning of this chapter in Figure 9.1, we could choose to view this as the outcome of any one of a number of random processes. Figure 9.9 shows three possible variogram models and Figure 9.10 shows realizations of random functions that have these particular variogram parameters. The variogram model shown in Figure 9.9a describes a random function that is quite erratic over short distances, and the corresponding realization shown in Figure 9.10a does indeed show considerable short scale variability. Figure 9.9b describes a random function that is less erratic than the first, and the corresponding realization shown in Figure 9.10b is less erratic than the first one. Figure 9.9c, on the other hand, describes a random function that is extremely continuous; the variogram is tangential to the x-axis at the origin, and rises very slowly. The corresponding realization, shown in Figure 9.10c, is very smooth, and gently undulates through the available sample data points.

If we restrict ourselves to the seven available sample values, then any one of the realizations shown in Figure 9.10 is a plausible version of the true profile; all pass through the available sample data points. The key to successful estimation is to choose a variogram or a covariance that captures the pattern of spatial continuity that we believe the exhaustive profile should have. If the seven available samples were measurements of some quantity that is known to fluctuate wildly, such as the concentration of some pollutant or the grade of a precious metal, then we would prefer the random function that generated the outcome shown in Figure 9.10a. On the other hand, if we are dealing with a

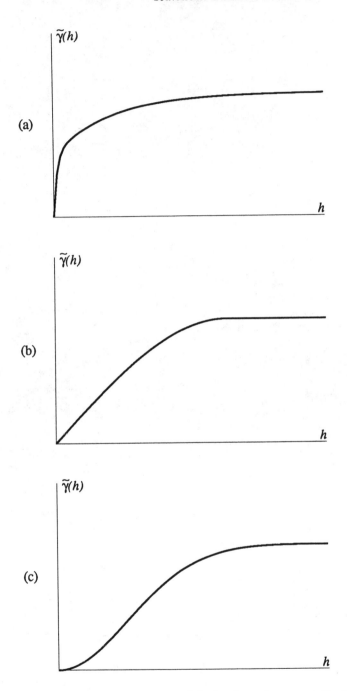

Figure 9.9 Three possible variogram models for the unknown profile from which the samples in Figure 9.1 were drawn.

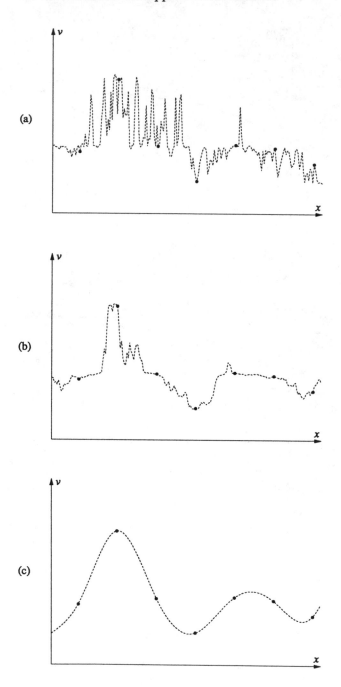

Figure 9.10 Realizations corresponding to each of the three variogram models shown in Figure 9.9.

quantity that is known to be very continuous, such as the elevation of an oil-water contact or the thickness of a coal seam, then we would prefer the random function that generated the outcome shown in Figure 9.10c.

The choice of a variogram or a covariance model is an important step in geostatistical estimation procedures. When we present the method of ordinary kriging in Chapter 12, we will discuss how the variogram model is typically chosen in practice and show how this choice affects the resulting estimates. For the moment, we will simply note that the choice of a particular variogram model directly implies a belief in a certain type of spatial continuity. Of the many characteristics of an earth science data set, the pattern of spatial continuity is among the most important for the problem of estimation. If the phenomenon is very continuous, then estimates based on only the closest samples may be quite reliable. On the other hand, if the phenomenon is erratic, then estimates based on only the closest available samples may be very unreliable; for such an erratic phenomenon, good estimates may require the use of many more sample data beyond the closest ones.

An Example of the Use of a Probabilistic Model

Since the use of probabilistic models is likely to be unfamiliar to many readers, we will conclude this chapter with an example of how they can be used to answer a specific estimation problem.

For example, we may wish to estimate the complete profile of the phenomenon whose samples are shown in Figure 9.1. At every point where we do not have a sample, we could estimate the unknown true value using a weighted linear combination of the seven available sample:

$$\hat{v} = \sum_{j=1}^{7} w_j \cdot v_j$$

Furthermore, suppose the set of weights is allowed to change as we estimate unknown values at different locations. The question we will address in this example is how to choose the weights so that the average estimation error is 0.

If we define the error, r, of any particular estimated value to be the difference between the estimated value and the true value at that same location:

$$\text{Error of the } i\text{th estimate} = r_i = \hat{v}_i - v_i \qquad (9.24)$$

then the average error of a set of n estimates is

$$Average\ error = m_r = \frac{1}{n}\sum_{i=1}^{n} r_i = \frac{1}{n}\sum_{i=1}^{n} \hat{v}_i - v_i \qquad (9.25)$$

Even though we have been able to define the average error with Equation 9.25, this does not help us very much in addressing the problem of how to weight the samples so that the average error is 0. When we set Equation 9.25 to 0, we cannot make any progress since it it involves quantities that we do not know, namely the true values $v_1, \ldots, v_n$.

The probabilistic solution to this problem consists of conceptualizing the unknown exhaustive profile as the outcome of a random process and solving the problem for our conceptual model. For any point at which we attempt to estimate the unknown value, our model is a stationary random function that consists of eight random variables, one for the value at each of the seven sample locations, $V(x_1), \ldots, V(x_7)$, and one for the unknown value at the point we are trying to estimate, $V(x_0)$. Each of these random variables has the same probability law; at all eight locations, the expected value of the random variable is $E\{V\}$. Any pair of random variables has a joint distribution that depends only on the separation between the two points and not on their locations. The covariance between pairs of random variables separated by a particular distance, h, is $\tilde{C}_V(h)$.

Every value in this model is seen as the outcome of random variables; the seven samples are outcomes of random variables, as is the unknown true value. Our estimate is also a random variable since it is a weighted linear combination of the random variables at the seven available sample locations:

$$\hat{V}(x_0) = \sum_{i=1}^{7} w_i \cdot V(x_i)$$

Similarly, the *estimation error*, defined as the difference between the estimate and the random variable modeling the true value, is also a random variable:

$$R(x_0) = \hat{V}(x_0) - V(x_0)$$

By substituting the previous equation which expressed $\hat{V}(x_0)$ in terms of other random variables, we can express $R(x_0)$ solely in terms of the eight random variables in our random function model:

$$R(x_0) = \sum_{i=1}^{7} w_i \cdot V(x_i) \quad - \quad V(x_0) \qquad (9.26)$$

The error that we make when we estimate the unknown value at x_0 is an outcome of the random variable $R(x_0)$.

If we perform estimation at n locations, then the average of the n errors is itself a random variable:

$$Average \; error = A = \frac{1}{n} \sum_{i=1}^{n} R_i \qquad (9.27)$$

In this equation, the random variables $R_1, \ldots, R_n$ are the errors at the n locations at which we are calculating estimates. We can set the expected value of A to 0 by using Equation (9.13) which gave us a formula for calculating the expected value of a linear combination of random variables. Applying this equation to the random variable A as defined in Equation 9.27 leads to the following result:

$$E\{A\} = E\{\frac{1}{n} \sum_{i=1}^{n} R_i\} = \frac{1}{n} \sum_{i=1}^{n} E\{R_i\}$$

One way to guarantee that the expected value of A is zero is to insist that each of the random variables $R_1, \ldots, R_n$ have an expected value of 0. We can ensure that each of these errors has an expected value of 0 by applying the formula for the expected value of a linear combination to Equation 9.26:

$$E\{R(x_0)\} \quad = \quad E\{\sum_{i=1}^{7} w_i \cdot V(x_i) \quad - \quad V(x_0)\}$$

$$= \quad \sum_{i=1}^{7} w_i E\{V(x_i)\} \quad - \quad E\{V(x_0)\}$$

We have already assumed that the random function is stationary, which allows us to express all of the expected values on the right hand side as $E\{V\}$:

$$E\{R(x_0)\} = \sum_{i=1}^{7} w_i E\{V\} \quad - \quad E\{V\}$$

The expected value of the error at any particular location, $E\{R(x_0)\}$ is often referred to as the *bias*. Setting this expected value to 0 to ensure unbiasedness results in the following conclusion:

$$E\{R(x_0)\} = 0 \ = \ E\{V\}\sum_{i=1}^{7} w_i \ - \ E\{V\}$$

$$E\{V\}\sum_{i=1}^{7} w_i \ = \ E\{V\}$$

$$\sum_{i=1}^{7} w_i \ = \ 1 \tag{9.28}$$

The conclusion, then, is that if we want unbiased estimates, we should restrict ourselves to weighted linear combinations whose weights sum to one.

This result, often referred to as the *unbiasedness condition*, makes such obvious common sense that it is often taken for granted. For example, all of the estimation methods presented in Chapter 11 use weighted linear combinations in which the weights sum to 1. Though none of these methods makes clear exactly why this is being done, the preceding demonstration makes it clear that the use of this condition can be justified by a conceptual model in which the values are viewed as outcomes of a stationary random process.

Figure 9.11 reviews the steps in the derivation of the unbiasedness condition. There are two remarks that were made earlier that should be emphasized again here.

First, though the problem arose from the practical intention of having the actual average error of a particular set of estimates be 0, it has been solved in the realm of a random function model. Whether or not it works in reality depends on the appropriateness of the model.

Second, even if a stationary random function is an appropriate model, the actual average error may not be 0. Our unbiasedness condition has been based on setting an expected value to 0. A random variable with an expected value of 0 can still produce a set of actual outcomes whose sample mean is not 0. The parameters of a random variable need not be identical to the corresponding statistics calculated from a set of observed outcomes.

In the case studies we present in the following chapters, we will see that even though an estimation method makes use of the unbiasedness

Reality	Conceptual Model
• a sample data set consisting of seven data:	• a stationary random function consisting of eight random variables:

$$v_1, \ldots, v_7$$

$$V(x_0), V(x_1), \ldots, V(x_7)$$

| • estimate is a specific value calculated by a weighted linear combination of the seven data: | • estimate is a random variable; some of its parameters are known since it is a linear combination of known random variables: |

$$\hat{v}_0 = \sum w_i \cdot v_i$$

$$\hat{V}(x_0) = \sum w_i \cdot V(x_i)$$

| • estimation error is a specific but unknown quantity, | • estimation error is a random variable; some of its parameters are known since it is a linear combination of known random variables: |

$$r_0 = \hat{v}_0 - v_0$$

$$R(x_0) = \hat{V}(x_0) - V(x_0)$$

Problem: How to weight samples so that average error is 0? $\Rightarrow$ Problem: How to weight random variables $V(x_1) \ldots V(x_7)$ so that expected value of the average estimation error is 0?

$$\Downarrow$$

Solution: $\sum w_i = 1$ $\Leftarrow$ Solution: $\sum w_i = 1$

Figure 9.11   An outline of the use of a probabilistic model. The problem of producing estimates whose average error is 0 is translated into a similar problem in the probabilistic model. The solution derived from the probabilistic model is then assumed to apply to reality.

condition given in Equation 9.28, it can still produce estimates that are actually quite biased.

Further Reading

Box, G. and Jenkins, G. , *Time Series Analysis forecasting and control.* Oakland, California: Holden-Day, 1976.

Gardner, W. A. , *Introduction to Random Processes.* New York: Macmillan, 1986.

Matheron, G. F. , "Estimer et choisir," *Les Cahiers du Center de Morphologie Mathématique de Fontainebleau,* Fascicule 7, École Supérieure des Mines de Paris, 1978.

Olkin, I. , Gleser, L. , and Derman, C. , *Probability Models and Applications.* New York: Macmillan, 1980.

10

GLOBAL ESTIMATION

When we calculated the sample statistics in Chapter 6 we noticed that
the naive sample mean was a very poor estimate of the exhaustive
mean. Further analysis revealed that the sampling strategy caused our
samples to be preferentially located in areas with high V values. For
example, more than 125 of the 470 samples, for example, fall within the
Wassuk Range anomaly. While it is likely that these samples give good
information on the exhaustive mean within this anomaly, they are not
representative of the remaining area. Unfortunately, the remaining
area is not as densely sampled as the Wassuk Range anomaly. To
obtain a good estimate of the exhaustive mean, we will need to find
some way of weighting individual samples so that the clustered ones
do not have an undue influence on our estimate.

We have already seen in Table 6.2 that if we use only the 195
samples from the initial campaign, our sample mean is much closer
to the actual exhaustive value than if we include the samples from
the second and third campaigns. A weighted linear combination that
gives equal weight to the first 195 sample values and no weight to the
last 275 produces a better estimate of the global mean than one that
weights all 470 sample equally.

This method of giving no weight to the clustered samples has two
drawbacks. First, it completely ignores useful information. We should
not discard the measurements from our 275 clustered samples; instead,
we should try to find some way to moderate their influence. Second,
it may not always be possible to identify those samples that should

be kept and those that should be discarded. We are fortunate with
the Walker Lake sample data set in being able to identify the three
sampling campaigns. In practice, we may not be able to find a natural
subset of samples that completely covers the area on a pseudo regular
grid.

In this chapter we look at two declustering methods that are generally applicable to any sample data set. In both methods we use a
weighted linear combination of all available sample values to estimate
the exhaustive mean. By assigning different weights to the available
samples, we can effectively decluster the data set. The first method,
called the *polygonal method,* assigns a polygon of influence to each
sample. The areas of these polygons are then used as the declustering
weights. The second method, called the *cell declustering method,* uses
the moving window concept to calculate how many samples fall within
particular regions or cells. The declustering weight assigned to a sample is inversely proportional to the number of other samples that fall
within the same cell. Following a detailed description of how these two
methods are implemented, we will see how both methods perform on
the Walker Lake data set.

Polygonal Declustering

Each sample in our data set has a polygon of influence within which
it is closer than any other sample. Figure 10.1 shows the locations
of some arbitrary samples. The shaded area shows the polygon of
influence for the 328 ppm sample located near the center of this area.
Any point within the shaded region is closer to the 328 ppm sample
than to any other.

Figure 10.2 shows how the boundaries of the polygon of influence
are uniquely defined. The perpendicular bisector of a line segment is a
line on which points are equidistant from either end of the line segment;
points on either side of the perpendicular bisector have to be closer to
one end or the other. The perpendicular bisectors between a sample
and its neighbors form the boundaries of the polygon of influence.

The edges of the global area require special treatment. A sample
located near the edge of the area of interest may not be completely
surrounded by other samples and the perpendicular bisectors with its
neighbors may not form a closed polygon. Figure 10.3a shows an example where the perpendicular bisectors between the 85 ppm sample and

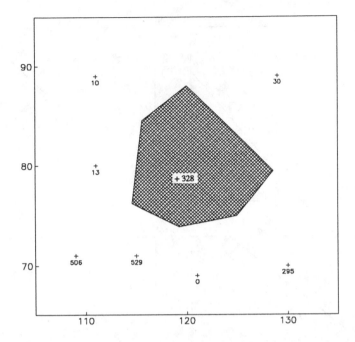

Figure 10.1 An example showing the polygon of influence of a sample

its three neighboring samples do not form a closed region. One solution
is to choose a natural limit, such as a lease boundary or a geologic contact, to serve as a boundary for the entire area; this can then be used
to close the border polygons. In Figure 10.3b we use the rectangular
boundaries of the map area as the natural limit of our area of interest.
An alternative in situations where a natural boundary is not easy to
define is to limit the distance from a sample to any edge of its polygon
of influence. This has the effect of closing the polygon with the arc of
a circle. In Figure 10.3c we see how the polygon of influence is closed
if it is not allowed to extend more than 10 m from the 85 ppm sample.

By using the areas of these polygons of influence as weights in our
weighted linear combination, we accomplish the declustering we require. Clustered samples will tend to get small weights corresponding
to their small polygons of influence. On the other hand, samples with
large polygons of influence can be thought of as being representative
of a larger area and are therefore entitled to a larger weight.

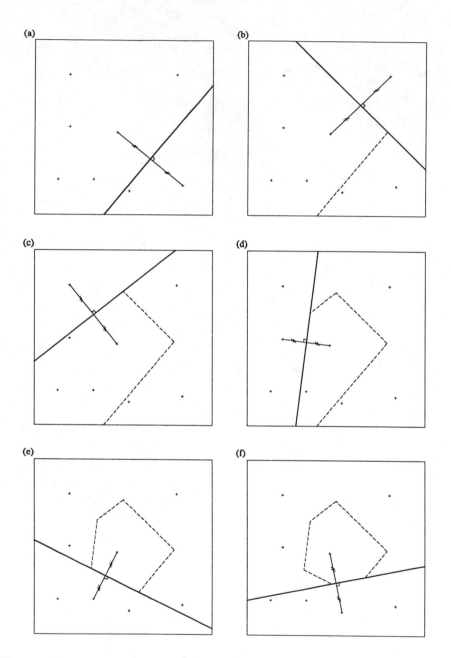

Figure 10.2 Construction of a polygon of influence using the method of perpendicular bisectors. Figures (a) to (f) show the steps in constructing a region within which the central sample is closer than any other sample.

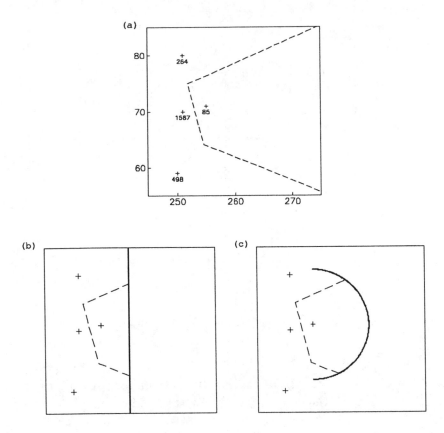

Figure 10.3 Defining the polygon on the border of the global area. (a) shows a polygon that cannot be closed by the method of perpendicular bisectors between data points. Alternatively the polygon can be closed by a natural limit such as the lease boundary in (b) or by limiting the distance from a sample to the edge of a polygon as shown in (c).

Cell Declustering

In the cell declustering approach, the entire area is divided into rectangular regions called *cells*. Each sample receives a weight inversely proportional to the number of samples that fall within the same cell. Clustered samples will generally receive lower weights with this method because the cells in which they are located will also contain several other samples.

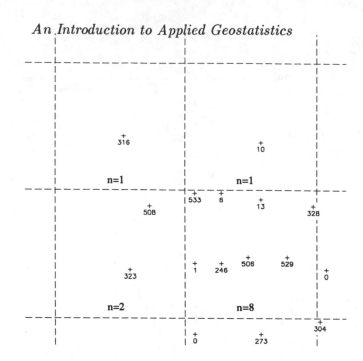

Figure 10.4 An example of cell declustering.

Figure 10.4 shows a grid of such cells superimposed on a number of clustered samples; the dashed lines show the boundaries of 20 x 20 m^2 cells. Each of the two northernmost cells contains only one sample, so both of these samples receive a weight of 1. The southwestern cell contains two samples, both of which receive a weight of $\frac{1}{2}$. The southeastern cell contains eight samples, each of which receives a weight of $\frac{1}{8}$.

Since all samples within a particular cell receive equal weights and all cells receive a total weight of 1, the cell declustering method can be viewed as a two step procedure. First, we use our samples to calculate the mean value within moving windows, then we take these moving window means and use them to calculate the mean of the global area.

The estimate we get from this cell declustering method will depend on the size of the cells we choose to use. If the cells are very small, then each sample will fall into a cell of its own and all samples will therefore receive equal weights of 1. If the cells are as large as the entire global area, all samples will fall into the same cell and will again receive equal weights. Somewhere between these two extremes we must find an appropriate medium.

If there is an underlying pseudo regular grid, then the spacing of this grid usually provides a good cell size. In our Walker Lake example, the sampling grid from the first program suggests that 20 x 20 m^2 cells would adequately decluster our data. If the sampling pattern does not suggest a natural cell size, a common practice is to try several cell sizes and to pick the one that gives the lowest estimate of the global mean. This is appropriate if the clustered sampling is exclusively in areas with high values. In such cases, which are common in practice, we expect the clustering of the samples to increase our estimate of the mean, so we are justified in choosing the cell size that produces the lowest estimate.

Comparison of Declustering Methods

Having discussed how the two declustering methods are implemented, we can now try them both on our Walker Lake sample data set. We will estimate the global V mean using the 470 V samples. From our description of the exhaustive data set in Chapter 5 we know that the true value is 278 ppm. Though the case study here aims only at the estimation of the global mean, we will see later that the declustering weights we calculate here can also be used for other purposes. In Chapter 18, we will take a look at estimating an entire distribution and its various declustered statistics.

The polygons of influence for the Walker Lake sample data set are shown in Figure 10.5. In this figure we have chosen to use the rectangular boundaries of the map area to close the border polygons.

For the cell declustering method we must choose an appropriate cell size. Since we are using rectangular cells we can vary both the east-west width and the north-south height of our cells. In Figure 10.6 we have contoured the estimated global means we obtain using cells of different sizes. The minimum on this map occurs for a cell whose east-west dimension is 20 m and north-south dimension is 23.08 m. We are justified in choosing this minimum over all other possibilities since our clustered samples are all located in areas with high V values. This 20 x 23 m^2 cell size also nearly coincides with the spacing of the pseudo regular grid from the first sampling campaign.

For the first 20 samples, Table 10.1 gives details of the calculation

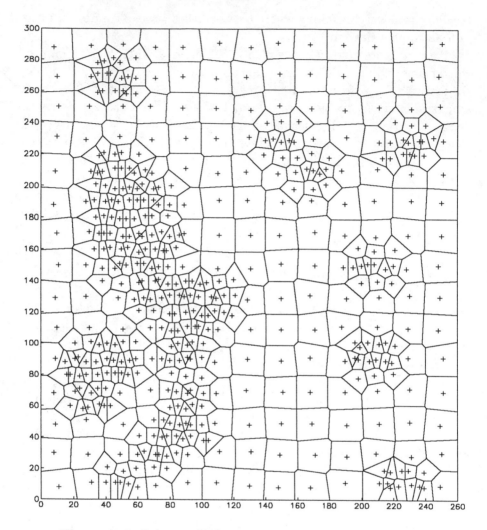

Figure 10.5 Polygon of influence map for the 470 V sample data.

of the weights for the polygonal method and for the cell declustering method using a cell size of 20 x 23 m². Our estimated global mean will be a weighted linear combination of the 470 sample values:

$$Estimated\ Global\ Mean = \frac{\sum_{i=1}^{470} w_i \cdot v_i}{\sum_{i=1}^{470} w_i} \qquad (10.1)$$

This is the same as the general equation we presented in the previous

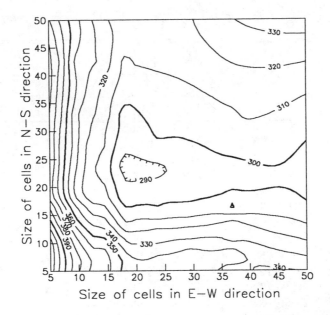

Figure 10.6 Contour map showing the relation between the global declustered mean and the declustering cell size. From this map it can be seen that a cell size of 20 m east-west and approximately 23 m north-south yields the lowest global mean.

chapter, with the denominator acting as a factor that standardizes the weights so that they sum to 1. For the polygonal approach, $\Sigma w_i = 78,000$ since the total map area is $78,000$ m^2; for the cell declustering approach, $\Sigma w_i = 169$ since the global area is covered by 169 20 x 23 m^2 cells.

Using the areas of the polygons of influence as our declustering weights we get an estimate of 276.8 ppm. Using the weights from the cell declustering method with a cell size of 20 x 23 m^2 we get an estimate of 288 pm.

For sample data sets that have an underlying pseudo regular grid and in which clustered sampling occurs only in areas with high or low values, the cell declustering method usually performs well. The estimate of 288 ppm obtained by this method is quite close to the actual value of 276.9 ppm.

In this particular study, the polygonal method performs extremely well. We should be somewhat humble, however, about our 276.8 ppm

Table 10.1 Declustering weights for the first 20 samples from the sample data set

No.	X	Y	V	POLYGONS		CELLS	
				Area of Polygon	$\frac{w_i}{\Sigma w_i}$	Number of Samples in Same Cell	$\frac{w_i}{\Sigma w_i}$
1	11	8	0	399	0.0051	1	0.0059
2	8	30	0	370	0.0047	1	0.0059
3	9	48	224	319	0.0041	3	0.0020
4	8	68	434	249	0.0032	3	0.0020
5	9	90	412	236	0.0030	4	0.0013
6	10	110	587	343	0.0044	1	0.0015
7	9	129	192	372	0.0048	1	0.0059
8	11	150	31	398	0.0051	1	0.0059
9	10	170	389	382	0.0049	1	0.0059
10	8	188	175	350	0.0045	1	0.0059
11	9	209	188	424	0.0054	1	0.0059
12	10	231	82	390	0.0050	2	0.0030
13	11	250	81	389	0.0050	2	0.0030
14	10	269	124	392	0.0050	1	0.0059
15	8	288	188	417	0.0053	1	0.0059
16	31	11	29	279	0.0036	1	0.0059
17	29	29	78	375	0.0048	1	0.0059
18	28	51	292	264	0.0034	6	0.0010
19	31	68	895	58	0.0007	6	0.0010
20	28	88	703	57	0.0007	10	0.0006
⋮	⋮	⋮	⋮	⋮	⋮	⋮	⋮

estimate. This remarkable accuracy is a peculiarity of the Walker Lake data set and we should not expect similar luck in all situations.

The polygonal method has the advantage over the cell declustering method of producing a unique estimate. In situations where the sampling does not justify our choosing the minimum of our various cell declustered estimates, the choice of an appropriate cell size becomes awkward.

An interesting case study that sheds further light on these two methods is the estimation of the U global mean. Using the 275 U

samples we can repeat the calculation of the declustering weights for both methods. In this case, the true value is 266 ppm. The polygonal estimate is 338 ppm while the minimum cell declustering estimate is 473 ppm. In this example both methods fare poorly because there are large portions of the map area with preferentially low values that have no U samples. Neither of these methods can hope to replace actual samples; all they do is make intelligent use of the available samples. It is worth noting that in this case, where there is no underlying pseudo regular grid that covers the area, the cell declustering approach produces a considerably poorer estimate than the polygonal approach.

Declustering Three Dimensional Data

The methods we have presented here work well with two-dimensional data sets. For the declustering of three-dimensional data sets, there are several possible adaptations of these tools.

If the data are layered, then one may be able to separate the data into individual layers and then use a two-dimensional declustering methods on each layer. If the data set cannot easily be reduced to several two-dimensional data sets, it is possible to use the three-dimensional version of either of the methods discussed here.

For the cell declustering approach, the cells become rectangular blocks whose width, height and depth we must choose. If the appropriate dimensions of such blocks are not obvious from the available sampling, one can still experiment with several block dimensions in an attempt to find the one that minimizes (or maximizes) the estimate of the global mean. In three dimensions, however, this procedure is more tedious and less difficult to visualize than in two.

The three-dimensional analog of the polygonal approach consists of dividing the space into polyhedra within which the central sample is closer than any other sample. The volume of each polyhedron can then be used as a declustering weight for the central sample. An alternative approach, which is easier to implement though usually more computationally expensive, is to discretize the volume into many points, and to assign to each sample a declustering weight that is proportional to the number of points which are closer to that sample than to any other.

A final alternative, one whose two-dimensional version we have not yet discussed, is to use the global kriging weights as declustering weights. In Chapter 20 we will show how these weights can be obtained by accumulating local kriging weights. If a good variogram model can be chosen, this final alternative has the advantage of accounting for the pattern of spatial continuity of the phenomenon.

Further Reading

Hayes, W. and Koch, G. , "Constructing and analyzing area-of-influence polygons by computer," *Computers and Geosciences*, vol. 10, pp. 411–431, 1984.

Journel, A. , "Non-parametric estimation of spatial distributions," *Mathematical Geology*, vol. 15, no. 3, pp. 445–468, 1983.

11

POINT ESTIMATION

In the previous chapter we looked at some ways of estimating a mean value over a large area within which there are many samples. Though this is necessary at an early stage in most studies it is rarely a final goal. We often also need estimates for much smaller areas; we may even need to estimate unknown values at specific locations. For such local estimation problems we still use weighted linear combinations, but our weights now need to account not only for possible clustering but also for the distance to the nearby samples. In this chapter we look at some methods for point estimation and check their results with the true values from the exhaustive data set, using some of our descriptive statistical tools to help us judge the performance of the different methods. Summarizing a complete set of point estimates with a single statistic provides an index that helps us decide which one is best. We will see that while some methods perform very well according to some criteria, they may not do as well according to other criteria. In the next chapter we will look at ordinary kriging, an estimation method that is designed to give the best estimates for one of the statistical criteria we look at in this chapter.

For each of the point estimation methods we describe in the following sections, we will show the details of the estimation of the V value at 65E,137N. No sweeping conclusions should be drawn from this single example; it is presented only to provide a familiar common thread through our presentation of various methods. Once we have looked at

Table 11.1   Distances to sample values in the vicinity of 65E,137N

Sample No.	X	Y	V	Distance from 65E,137N
1	225	61	139	477	4.5
2	437	63	140	696	3.6
3	367	64	129	227	8.1
4	52	68	128	646	9.5
5	259	71	140	606	6.7
6	436	73	141	791	8.9
7	366	75	128	783	13.5

the details of how each method is implemented we will compare their performances at several locations.

The V values at sample locations near 65E,137N are shown in Figure 11.1 and listed in Table 11.1. The variability of these nearby sample values presents a challenge for estimation. We have values ranging from 227 to 783 ppm; our estimated value therefore, can cover quite a broad range depending on how we choose to weight the individual values.

In the following sections we will look at four quite different point estimation methods. Two of them, the polygonal method and the local sample mean method, are adaptations of the global declustering techniques we discussed in the last chapter. As we will see, these are both extreme versions of the family of inverse distance methods. The fourth method we will examine is a geometric technique known as *triangulation*. Through a comparison of these four simple methods we can begin to understand the important factors in point estimation.

Polygons

The polygonal method of declustering that we looked at in the last chapter can easily be applied to point estimation. We simply choose as an estimate the sample value that is closest to the point we are trying to estimate. Table 11.1 shows the distances from the 65E,137N

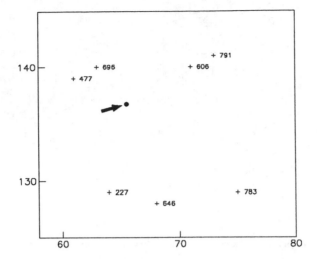

Figure 11.1 The data configuration shown in this figure is used to illustrate several point estimation methods in the following sections. The goal is to estimate the value of V at the point 65E,137N, located by the arrow, from the surrounding seven V data values.

to the each of the sample locations shown in Figure 11.1. The sample at 63E,140N is closest, so our polygonal estimate of the V value at 65E,137N is 696 ppm. This polygonal estimator can be viewed as a weighted linear combination that gives all of the weight to the closest sample value.

Polygonal estimates of the V value at other points near 65E,137N will also be 696 ppm. As long as the points we are estimating fall within the same polygon of influence, the polygonal estimate remains unchanged. As soon as we encounter a point in a different polygon of influence, the estimate jumps to a different value. Figure 11.2 shows how the polygonal estimates near 65E,137N form a discontinuous surface of plateaus.

Triangulation

Discontinuities in the estimated values are usually not desirable. This is not to say that real values are never discontinuous; indeed, as our indicator maps (Figure 5.10) of the exhaustive data set showed, the true

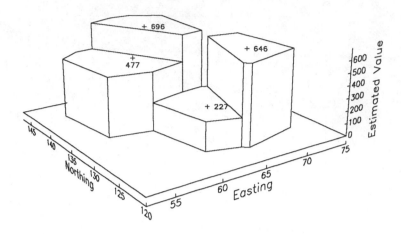

Figure 11.2 A perspective view showing the discontinuities inherent in polygonal estimates.

values can change considerably over short distances. The discontinuities that some estimation methods produce are undesirable because they are artifacts of the estimation procedure and have little, if anything, to do with reality. The method of triangulation overcomes this problem of the polygonal method, removing possible discontinuities between adjacent points by fitting a plane through three samples that surround the point being estimated. The equation of a plane can be expressed generally as

$$z = ax + by + c \tag{11.1}$$

In our example, where we are trying to estimate V values using coordinate information, z is the V value, x is the easting, and y is the northing. Given the coordinates and the V value of three nearby samples, we can calculate the coefficients a, b and c by solving the following system of equations:

$$
\begin{aligned}
ax_1 + by_1 + c &= z_1 \\
ax_2 + by_2 + c &= z_2 \\
ax_3 + by_3 + c &= z_3
\end{aligned}
\tag{11.2}
$$

From Figure 11.1 we can find three samples that nicely surround the point being estimated: the 696 ppm, the 227 ppm, and the 606 ppm

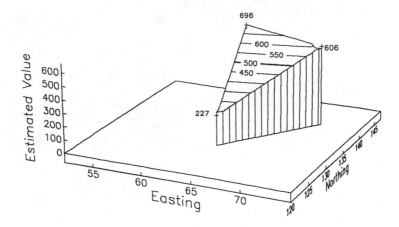

Figure 11.3 An illustration of the estimation plane obtained by triangulation. This method overcomes the problem of discontinuities by fitting a plane through three samples that surround the point being estimated.

samples. Using the data for these three samples, the set of equations we need to solve is

$$63a + 140b + c = 696$$
$$64a + 129b + c = 227 \qquad (11.3)$$
$$71a + 140b + c = 606$$

The solution to these three simultaneous equations is

$$a = -11.250 \qquad b = 41.614 \qquad c = -4421.159 \qquad (11.4)$$

which gives us the following equation as our triangulation estimator:

$$\hat{v} = -11.250x + 41.614y - 4421.159 \qquad (11.5)$$

This is the equation of the plane that passes through the three nearby samples we have chosen; in Figure 11.3 we show the contours of the estimated V values that this equation produces. Using this equation we can now estimate the value at any location simply by substituting the appropriate easting and northing. Substituting the coordinates x = 65 and y = 137 into our equation gives us an estimate of 548.7 ppm at the location 65E,137N.

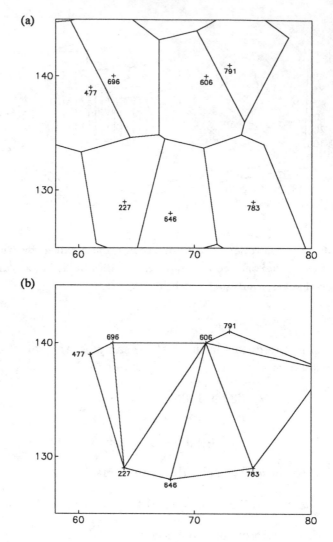

Figure 11.4 (a) The polygons of influence and (b), Delaunay triangles for the data configuration given in Figure 10.1.

This method of estimation clearly depends on which three nearby samples we use to define our plane. There are several ways we could choose to triangulate our sample data set. One particular triangulation, called the *Delaunay triangulation*, is fairly easy to calculate and has the nice property that it produces triangles that are as close to equilateral as possible. This Delaunay triangulation and our earlier polygons of influence are geometrically related. Figure 11.4a shows the polygons of influence in the vicinity of 65E,137N; we show the Delaunay triangulation of the same samples in Figure 11.4b. Three sample locations form a Delaunay triangle if their polygons of influence share a common vertex [1]. For example, the polygons of influence for the 696 ppm, 227 ppm and 606 ppm samples share a common vertex near the center of Figure 11.4a and therefore also form one of the Delaunay triangles shown in Figure 11.4b.

Though one can use an equation like Equation 11.5 to produce estimates for any location, it is unwise to do so beyond the boundaries of the triangle that connects the three samples that were originally used to calculate the coefficients of the plane. This entails that triangulation is typically not used for extrapolation purposes; we estimate values only at those locations that fall within one of our Delaunay triangles. Around the edges of the area of interest we are typically unable to use triangulation.

Though we solved a system of linear equations to derive our triangulation estimate of 548.7 ppm, we could also express it as a weighted linear combination of the three sample values. For the triangulation procedure, the weights assigned to each value can be directly calculated from the geometry of the three samples and the point being estimated.

Figure 11.5 shows the location of three samples, designated I, J, and K; the sample values at these locations are, respectively, v_I, v_J, and v_K. The point O at which we require an estimate is contained within the triangle IJK. Instead of solving three simultaneous equations and substituting the coordinates of O into our solution, we can directly calculate the triangulation estimate at O with the following equation:

$$\hat{v}_O = \frac{A_{OJK} \cdot v_I + A_{OIK} \cdot v_J + A_{OIJ} \cdot v_K}{A_{IJK}} \tag{11.6}$$

The As represent the areas of the triangles given in their subscripts.

Our triangulation estimate, therefore, is a weighted linear combination in which each value is weighted according to the area of the opposite triangle. This weighting agrees with the common intuition that closer points should receive greater weights. In Figure 11.5 we can see that as the point O gets closer to I, the area of triangle OJK will increase, giving the value at I more influence. Dividing the individual weights by A_{IJK} has the effect of normalizing the weights so

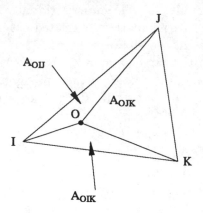

Figure 11.5 An example showing how three nearest data can be weighted by triangular areas to form a point estimate. The data are located at the corners of the triangle. The data value at I is weighted by the triangle area A_{OJK}, at J by area A_{OIK}, and at K by A_{OIJ}.

that they sum to one. Using Equation 11.6, the estimate of the V value at 65E,137N would be written as

$$\hat{v} = \frac{(22.5)(696) + (12.0)(227) + (9.5)(606)}{44} = 548.7 \text{ ppm} \qquad (11.7)$$

Local Sample Mean

In the polygonal method, where we use only the nearest sample, and in the triangulation method, where we use only three of the nearby samples, we ignore much of the nearby information. In Figure 11.1 we saw several other samples in the neighborhood of 65E,137N. There are a variety of estimation methods that give nonzero weights to all of the nearby samples, thus incorporating more of the available information in the estimate. A simplistic approach to incorporating the information from nearby samples is to weight them all equally, using the sample mean as the estimate. This equal weighting of nearby samples was the first step in the cell declustering method we looked at in the previous chapter. In practice, such simple averaging is rarely used as a point estimation method. We include it here because it will later serve as a useful reference when we explore other statistical approaches to point estimation.

Table 11.2 Inverse distance weighting calculations for sample values in the vicinity of 65E,137N

Sample No.	X	Y	V	Distance from 65E,137N	$1/d_i$	$\frac{1/d_i}{\Sigma 1/d_i}$
1	225	61	139	477	4.5	0.2222	0.2088
2	437	63	140	696	3.6	0.2778	0.2610
3	367	64	129	227	8.1	0.1235	0.1160
4	52	68	128	646	9.5	0.1053	0.0989
5	259	71	140	606	6.7	0.1493	0.1402
6	436	73	141	791	8.9	0.1124	0.1056
7	366	75	128	783	13.5	0.0741	0.0696
					$\Sigma 1/d_i =$	1.0644	

The mean of the seven nearby samples shown in Figure 11.1 is 603.7 ppm. This estimate is much higher than either our polygonal estimate or our triangulation estimate. As we noted when we first discussed the sample mean in Chapter 2, it is heavily influenced by extreme values. The two samples with V values greater than 750 ppm in the eastern half of Figure 11.1 receive more than 25% of the total weight and therefore have a considerable influence on our estimated value. None of these high sample values was given any weight in our polygonal and triangulation estimates.

Inverse Distance Methods

An improvement on naively giving equal weight to all samples is to give more weight to the closest samples and less to those that are farthest away. One obvious way to do this is to make the weight for each sample inversely proportional to its distance from the point being estimated:

$$\hat{v} = \frac{\sum_{i=1}^{n} \frac{1}{d_i} v_i}{\sum_{i=1}^{n} \frac{1}{d_i}} \qquad (11.8)$$

$d_1, \ldots, d_n$ are the distances from each of the n sample locations to the point being estimated and $v_1, \ldots, v_n$ are the sample values.

Table 11.3 The effect of the inverse distance exponent on the sample weights and on the V estimate.

	V	$\dfrac{1/d_i^p}{\Sigma 1/d_i^p}$					
		$p = 0.2$	$p = 0.5$	$p = 1.0$	$p = 2.0$	$p = 5.0$	$p = 10.0$
1	477	0.1564	0.1700	0.2088	0.2555	0.2324	0.0106
2	696	0.1635	0.1858	0.2610	0.3993	0.7093	0.9874
3	227	0.1390	0.1343	0.1160	0.0789	0.0123	<.0001
4	646	0.1347	0.1260	0.0989	0.0573	0.0055	<.0001
5	606	0.1444	0.1449	0.1402	0.1153	0.0318	0.0019
6	791	0.1364	0.1294	0.1056	0.0653	0.0077	<.0001
7	783	0.1255	0.1095	0.0696	0.0284	0.0010	<.0001
$\hat{v}$(in ppm)		601	598	594	598	637	693

In Table 11.2 we show the weight that each of our seven samples near 65E,137N gets in an inverse distance method. For each sample we show $1/d_i$, the reciprocal of its distance from 65E,137N; by dividing each weight by $\sum_{i=1}^{n}(1/d_i)$ we standardize the weights so that they sum to one.

The nearest sample, the 696 ppm sample at 63E,140N, receives about 26% of the total weight, while the farthest sample, the 783 ppm sample at 75E,128N, receives less than 7%. A good example of the effect of the inverse distance weighting can be found in a comparison of the weights given to the 477 ppm sample and the 791 ppm sample. The 791 ppm sample at 73E,141N is about twice as far away from the point we are trying to estimate as the 477 ppm sample at 61E,139N; the 791 ppm sample therefore receives about half the weight of the 477 ppm sample. Using the weights given in Table 11.2 our inverse distance estimate of the V value at 65E,137N is 594 ppm.

The inverse distance estimator we gave in Equation (11.8) can easily be adapted to include a broad range of estimates. Rather than using weights that are inversely proportional to the distance, we can make the weights inversely proportional to any power of the distance:

$$\hat{v}_1 = \frac{\sum_{i=1}^{n} \frac{1}{d_i^p} v_i}{\sum_{i=1}^{n} \frac{1}{d_i^p}} \qquad (11.9)$$

Different choices of the exponent p will result in different estimates.

Table 11.3 shows the effect of p on the weights and on the resulting estimates. As we decrease p, the weights given to the samples become more similar. for example, with $p = 0.2$ the closest sample receives nearly 16% of the total weight and the farthest sample receives about 13%. As we increase p, the individual weights become more dissimilar. The farthest samples suffer most, receiving a smaller proportion of the total weight, while the nearest samples become more influential. For $p = 5.0$, the closest sample gets 71% of the total weight while the farthest samples get virtually no weight. For progressively larger values of p the closest sample would receive a progressively larger percentage of the total weight.

The general inverse distance estimator given in Equation 11.9 offers considerable flexibility. As p approaches 0 and the weights become more similar, our inverse distance estimate approaches the simple average of the nearby sample values. As p approaches ∞ , the inverse distance estimate approaches the polygonal estimate, giving all of the weight to the closest sample. In Table 11.3 we can see that our estimate approaches the local sample mean of 604 ppm as p decreases; as p increases, our inverse distance estimate approaches our polygonal estimate of 696 ppm. Traditionally, the most common choice for the inverse distance exponent is 2. Inverse distance squared estimates are not necessarily better than estimates from inverse distance methods that use some exponent other than 2. The choice of p is arbitrary, and the traditional popularity of 2 is due, in part, to the fact that this choice involves fewer calculations and can therefore be computed very efficiently.

Search Neighborhoods

We have not yet confronted the issue of what counts as a "nearby" sample. In our example of the estimation of the V value at 65E,137N we simply accepted the samples that appeared on Figure 11.1. The choice of a search neighborhood that controls which samples are included in the estimation procedure is an important consideration in statistical approaches to point estimation and is discussed in detail in Chapter 14.

For the case studies we perform in this chapter, we use a circular search neighborhood with a radius of 25 m. All samples that fall within

25 m of the point we are estimating will be included in the estimation procedure.

Estimation Criteria

Before we compare the results from our different point estimation methods we should consider how we are going to decide which method is best overall. For estimates of a single value, such as our estimates of the global mean given in the previous chapter or the estimates at the single location 65E,137N that we calculated earlier in this chapter, it is fairly easy to decide which method worked best since one method will produce an estimate that is closest to the true value [2]. When we are comparing sets of estimates from several locations, however, it is very unlikely that one method will produce the best estimate at all locations. It is necessary, therefore, to have some criteria for comparing sets of point estimates from several locations.

In the case studies we present here, our comparisons will be based on some of the univariate and bivariate descriptive tools we discussed in Chapters 2 and 3. The univariate distribution of our estimates will be important in evaluating a set of estimates since it seems reasonable to expect that a good estimation method will produce estimated values whose distribution is similar to the distribution of true values. We will also look at the univariate distribution of our errors; as we will see shortly, there are several properties that we hope our error distribution will have. Put simply, we would like its center, its spread, and its skewness all to be as close to 0 as possible. The bivariate distribution of our estimated values and the true values will also help us to judge how well our estimation method has performed. A perfect estimation method would give us estimates that always match the true value; for such ideal estimates, the scatterplot of true and estimated values would plot as a straight line. We can judge various estimation methods by how close their scatterplots come to this ideal.

Univariate Distribution of Estimates. If we have a set of estimates at several locations it is natural to compare their distribution to the distribution of the true values at the same locations. We would like the distribution of our estimated values to be similar to the distribution of true values. Typically, we compare their means; for some applications it may make more sense to compare other summary statistics. For example, in the estimation of permeabilities it may make more sense

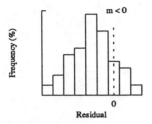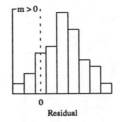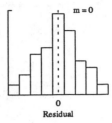

Figure 11.6 Biased hypothetical distributions of error. In (a) the distribution of estimation errors indicates a negative bias, in (b), a positive bias, and in (c), no bias.

to compare the medians of the true and estimated distributions. In some studies it is important that the variability of the estimates match that of the true values; in such cases we should compare the standard deviations or the variances. We can also construct q-q plots of the two distributions; these often reveal subtle differences that are hard to detect with only a few summary statistics.

Univariate Distribution of Errors. At every location where we perform point estimation we will have a true value, v, and an estimated value, $\hat{v}$. We will define the error at each location to be the difference between the estimate and the true value:

$$error = r = \hat{v} - v \qquad (11.10)$$

If r is positive, then we have overestimated the true value; if r is negative then we have underestimated the true value. We will often refer to these errors as *residuals*.

In Figure 11.6 we show histograms for three hypothetical error distributions. The distribution shown in Figure 11.6a has a negative mean, reflecting a general tendency toward underestimation. The opposite case, shown in Figure 11.6b, is a positive mean resulting from an abundance of overestimates. In Figure 11.6c the overestimates and underestimates are balanced and the mean is 0. The mean of the error distribution is often referred to as the *bias* and a reasonable goal for any estimation method is to produce unbiased estimates. As we have noted before, the mean is not the only measure of center. Ideally, one would like the median and the mode of the error distribution also to be 0.

Figure 11.7 Skewed distribution of estimation error. The error distribution in (a) is positively skewed in contrast to the more or less symmetric distribution shown in (b).

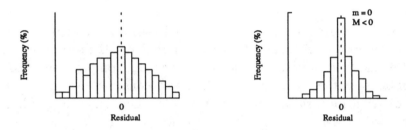

Figure 11.8 The error distribution in (a) shows a greater spread or variance about the mean error, than (b).

A mean error of 0 could be the result of many small underestimates combined with a few very large overestimates; Figure 11.7a shows such a distribution. We typically prefer to have a more symmetric distribution and an estimation method that produced the error distribution in Figure 11.7b would be preferable to the one that produced the error distribution in Figure 11.7a. The median error serves as a good check on the symmetry. If both the mean and the median are close to 0, then not only do our overestimates and underestimates balance, but they are also fairly symmetric in their magnitudes. An appreciable difference between the mean and the median warns us that the magnitude of our overestimates is likely not the same as our underestimates.

Another feature we hope to see in our error distribution is a small spread. In Figure 11.8 both error distributions are centered on 0 and are symmetric. The distribution shown in Figure 11.8a, however, has errors that span a greater range than those in Figure 11.8b. The vari-

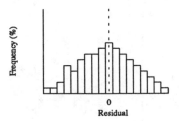
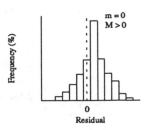

Figure 11.9 The unbiased error distribution in (a) shows a relatively large spread or variance about the mean error. The distribution in (b) shows much less variance, however it is slightly biased. Often an estimator can be improved by reducing the spread in the distribution of errors at the cost of introducing a slight bias.

ance or the standard deviation are both good yardsticks for assessing the spread of our error distribution.

The goals of minimum spread and a center close to 0 are not independent and there will be times in practice when we have to trade one off against the other. Figure 11.9 shows two possible error distributions; one has a mean of 0 but a large variance, the other has a low variance but a slightly positive mean. In such situations we might be willing to accept a small bias in return for less variable estimates. Two summary statistics that incorporate both the bias and the spread of the error distribution are the mean absolute error and the mean squared error:

$$Mean\ Absolute\ Error = MAE = \frac{1}{n}\sum_{i=1}^{n}|r| \qquad (11.11)$$

$$Mean\ Squared\ Error = MSE = \frac{1}{n}\sum_{i=1}^{n}r^2 \qquad (11.12)$$

The *MSE* can be related to other statistics of the distribution of our errors:

$$MSE = variance + bias^2 \qquad (11.13)$$

So far, we have discussed desirable properties of the entire error distribution. We would also like to see these properties hold for any range of estimated values. For example, if we separated our estimates into two groups, high values and low values, we would hope that the

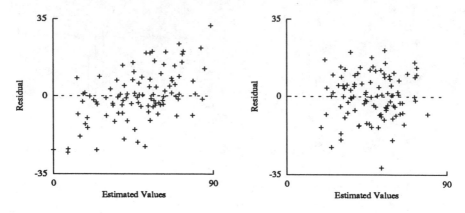

Figure 11.10 In (a) the estimation error or residuals are globally unbiased, however they are conditionally biased. For some ranges of estimates the average of the residuals will not be equal to 0. In (b) the estimates are globally unbiased as well as conditionally unbiased; for any range of estimates, the positive residuals balance the negative residuals.

bias would still be close to 0 within each group. Ideally, we would like to be able to subdivide our estimates into many different groups and have an unbiased error distribution within each group. This property is called *conditional unbiasedness* and is quite difficult to achieve in practice. A set of estimates that is conditionally unbiased is also globally unbiased. As we will see in our case studies, the reverse is not true; most estimation methods, even those that are globally unbiased, are guilty of overestimation or underestimation for some range of values.

One way of checking for conditional bias is to plot the errors as a function of the estimated values. Figure 11.10 shows such plots for two sets of estimates; both are globally unbiased since the mean of all the errors is 0. In Figure 11.10a the estimates are conditionally biased; the lowest estimated values tend to be too low while the highest ones tend to be too high. In Figure 11.10b the estimates are conditionally unbiased; for any range of estimates the overestimates balance the underestimates and the mean error within the range is 0.

Bivariate Distribution of Estimated and True Values. A scatterplot of true versus predicted values provides additional evidence on how well an estimation method has performed. The best possible esti-

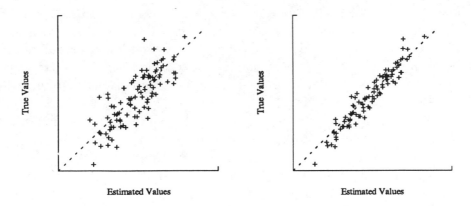

Figure 11.11 Two sets of estimates with differing spreads about the main diagonal. The estimates shown on the scatterplot in (b) are preferable to those shown in (a) since they generally fall closer to the diagonal on which perfect estimates would plot.

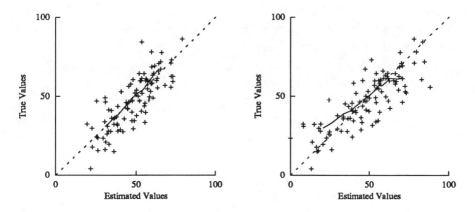

Figure 11.12 Conditional bias on a scatterplot of true versus estimated values. The estimates in (a) are preferable to those in (b) since their conditional expectation curve falls closer to the diagonal line.

mates would always match the true values and would therefore plot on the 45-degree line on a scatterplot. In actual practice we will always have to live with some error in our estimates, and our scatterplots of true versus estimated values will always appear as a cloud of points. Figure 11.11 shows two hypothetical scatterplots of true versus estimated values. We typically want a set of estimates that comes as close as possible to the line $v = \hat{v}$ so we would prefer the results shown in Figure 11.11b. The correlation coefficient is a good index for summarizing how close the points on a scatterplot come to falling on a straight line, and we will often make use of it in our comparisons of different point estimation methods [3].

The scatterplot also offers us another way of checking conditional bias. If the mean error is 0 for any range of estimated values, then the conditional expectation curve of true values given estimated ones will plot on the 45-degree line. In Figure 11.12 we show scatterplots of true versus estimated values for two sets of estimates; both sets are globally unbiased and both clouds have the same correlation coefficient. The conditional expectation of true values given estimated ones is shown as the solid line on each scatterplot. For the estimates shown in Figure 11.12b there is definitely some conditional bias; the true values for the very low estimates tend to be higher than their estimates, and the true values for the very highest estimates tend to be lower than their estimates. In Figure 11.12a, the conditional expectation curve plots very close to the 45-degree diagonal line for the complete range of estimated values.

Even though we rarely expect to get completely conditionally unbiased estimates, comparing the conditional expectation curve to the 45-degree line helps us understand how our estimation method performs. The causes of global bias, for example, are often made apparent by such plots.

Case Studies

The four point estimation methods we have discussed so far have been used to estimate the V value at points located on a regular square grid. The origin of this grid is 5E,5N, and the spacing between points is 10 m in both the north-south and the east-west directions. None of the 780 points on this grid coincides with a sample from our sample data set. At each point we have calculated the polygonal estimate, the triangulation estimate using the Delaunay triangulation, the mean of all samples within 25 m, and the inverse distance squared estimate using all samples within 25 m. Table 11.4 shows the resulting estimates at 20 of these points near the center of the map area.

The univariate distributions of the estimates and of the true values

Table 11.4 Twenty point estimates of V from near the center of the Walker Lake map area are tabulated for the polygonal, Delaunay triangulation, inverse distance squared, and moving average methods.

(E)	(N)	TRUE	POLY	TRIANG	LOCAL MEAN	INVERSE DIST. SQ.
⋮	⋮	⋮	⋮	⋮	⋮	⋮
135	25	186.2	29.1	235.8	239.9	154.8
135	35	349.4	29.1	93.8	187.9	84.7
135	45	89.8	75.2	128.3	190.8	109.9
135	55	122.6	75.2	214.5	273.6	199.1
135	65	465.5	399.9	367.8	334.8	369.1
135	75	461.9	399.9	445.6	430.4	408.5
135	85	762.2	243.1	370.3	344.3	314.5
135	95	491.4	243.1	254.7	318.2	264.6
135	105	356.3	0.0	114.5	278.0	106.9
135	115	214.8	0.0	39.2	184.9	117.7
135	125	327.1	244.7	159.3	185.2	191.0
135	135	195.6	244.7	183.9	138.4	156.9
135	145	368.7	185.2	160.3	106.6	155.5
135	155	15.3	185.2	98.9	100.2	126.8
135	165	89.4	26.0	23.1	83.7	53.6
135	175	146.7	26.0	18.7	45.8	42.9
135	185	0.0	0.0	6.8	41.9	20.0
135	195	31.3	0.0	81.9	102.5	52.2
135	205	63.07	63.0	162.2	290.9	223.6
135	215	383.8	383.8	351.3	390.9	408.0
⋮	⋮	⋮	⋮	⋮	⋮	⋮

are summarized in Table 11.5. For 108 of the points along the edges of the map area it was not possible to calculate the triangulation estimate since the point being estimated did not fall within any of the triangles defined by the Delaunay triangulation. The triangulation estimates should not be compared to the true values from all 780 points but rather to the true values at the 672 points where a triangulation estimate could be calculated.

The triangulation method produce estimates whose mean is lower than the mean of the true values; the local sample mean and the inverse distance squared estimates, on the other hand, have a higher

Table 11.5 Comparison of the distributions of true and estimated values for the four point estimation methods.

	True	Polygonal	Local Sample Mean	Inverse Distance Squared	True	Triangulation
n	780	780	780	780	672	672
m	283	284	334	310	294	277
σ	251	246	184	199	256	211
CV	0.89	0.87	0.55	0.64	0.87	0.76
min	0	0	0	0	0	0
Q_1	70	72	186	141	75	109
M	219	235	338	297	237	237
Q_3	446	443	468	439	462	401
max	1,323	1,393	890	1,065	1,323	1,231
$\rho_{\hat{z}z}$		0.69	0.67	0.78		0.80

mean than the true values. These differences reflect a global bias in all these methods. The particular biases we observe here are peculiarities of the Walker Lake data sets; in other situations we should not expect similar results. For example, the triangulation method does not consistently underestimate nor does the inverse distance squared method consistently overestimate. The difference between the mean estimate and the true mean will depend more on the character of the data set under study and on the available sample values than on the estimation method. The excellent polygonal estimate is rather fortunate and again we should not expect such accuracy in other situations.

Comparing the standard deviations of the true and estimated distributions, we notice that the only estimates whose variability closely matches that of the true values are the polygonal estimates; the other methods, including triangulation, produce estimates less variable than reality. Unlike the differences we noticed in the means of the estimates distributions, these differences in the standard deviation will be observed in most practical studies. This reduced variability of estimated values is often referred to as *smoothing* and is a consequence of combining several sample values to form an estimate. The polygonal estimates, which use only one sample value, are unsmoothed. The tri-

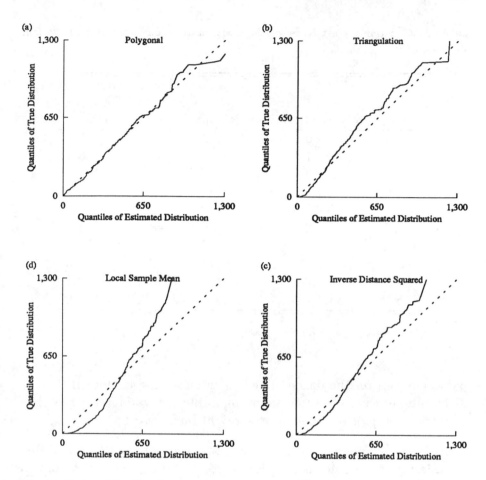

Figure 11.13 The distribution of 780 true point V values compared to the distribution of point estimates provided by the four estimation methods.

angulation estimates, which use three sample values, are less smoothed than the local sample mean or the inverse distance squared estimates, both of which give some weight to all of the nearby samples. As more sample values are incorporated in a weighted linear combination, the resulting estimates generally become less variable.

The q-q plots in Figure 11.13 help us see where the estimated distributions differ from the true one. If the true and estimated distributions are identical, then all of their quantiles will be the same and the q-q

Table 11.6 Summary statistics for the error distributions from each of the four point estimation methods.

	Polygonal	Triangulation	Local Sample Mean	Inverse Distance Squared
n	780	672	780	780
m	1.1	−16.8	51.1	26.8
σ	175.2	153.7	187.5	156.0
min	−651	−546	−618	−630
Q_1	−97	−58	−66	−65
M	0	3.9	48.9	24.4
Q_3	95	105	175	121
max	595	524	537	473
MAE	128	111	154	121
MSE	30,642	23,885	37,741	25,000

pairs will plot on the dashed line shown on each q-q plot. If the two distributions have the same shape but different variances, then their quantiles will plot on some other straight line.

Of the four point estimation methods, the polygonal method produces estimates whose distribution is closest to the distribution of true values. Its q-q plot is roughly linear over the range from 0 to 1,300 ppm, showing that the polygonal estimates have a distribution whose shape is similar to the true one. All of the other methods produce estimated distributions that are noticeably different from the true distribution.

All of the q-q plots pass through the origin, indicating that the minimum estimated value and the minimum true value are both 0 ppm. The polygonal method is the only one whose q-q plot does not deviate immediately to the right of the ideal dashed line. This deviation is evidence of the fact that there are far fewer very low values among the estimates than among the true values. This is another effect of the smoothing we remarked on earlier. The reduced variability of smoothed estimates causes the estimated distribution to contain fewer extreme values than the true distribution. The summary statistics in Table 11.5

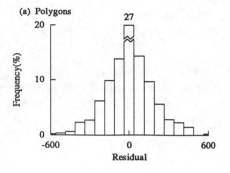

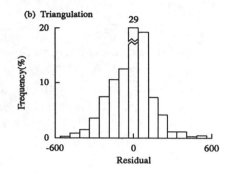

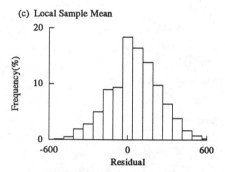

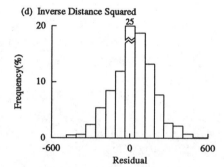

Figure 11.14 Histograms showing the distribution of the residuals for each of the four point estimation methods.

confirm that the polygonal estimates are the only ones that contain extreme values similar to the extremes found in the true distribution.

The distributions of the residuals from each of the four methods are shown in Figure 11.14 and are summarized with the statistics given in Table 11.6. The differences we noted earlier between the means of the estimated values and the true mean are reflected again in the means of the residuals. In this particular study, triangulation method has a tendency to underestimate while the inverse distance and local sample mean methods have a tendency to overestimate. The median error is closer to 0 for the polygonal and triangulation methods than for the local sample mean and the inverse distance squared approaches.

The error distributions appear quite symmetric; in all cases the difference between the mean and the median is quite small compared to the overall spread of the distribution.

If we use the spread of the error distribution as our yardstick, the triangulation estimates appear best; they have the lowest standard deviation and interquartile range.

The methods that use few of the nearby samples produce more errors close to 0 while the methods that give some weight to all nearby samples produce fewer large errors. This is confirmed by the statistical summary in Table 11.6 and by the histograms in Figure 11.14. The polygonal method and triangulation both have more errors in the class centered on 0 than either of the other methods. The polygonal method, however, produces the worst overestimation, 595 ppm, and the worst underestimation, -651 ppm. At the other end of the spectrum, the worst overestimation for the local sample mean method is only 537 ppm and its worst underestimation is only -618 ppm; its interquartile range, however, is larger than for any other method.

All of these results make it clear that the method which is "best" depends on the yardstick we choose. If we want our estimates to have the lowest standard deviation of errors and the smallest MAE, then the triangulation method works better, in this particular case, than any other method we have considered so far. If we prefer our estimates to be as unsmoothed as possible, then the polygonal approach wins out. The inverse distance squared method would be favored if we needed to minimize the largest errors.

In the next chapter we will introduce another estimation method that specifically aims at reducing the standard deviation of our residuals. Although this is a very traditional approach in statistics, the results of Table 11.6 should be kept in mind: different methods may be better for different estimation criteria.

The scatterplots of the true and estimated values from each of the four methods are given in Figure 11.15. Although the four plots look quite similar, a close examination reveals the cloud of points in Figures 11.15b and 11.15d form a tighter cluster about the 45-degree line. This observation is confirmed by the correlation coefficients: the scatterplot for the polygonal estimates (Figure 11.15a) has a correlation coefficient of 0.69; that of the triangulation estimates (Figure 11.15b) has a correlation coefficient of 0.80; that of the local sample mean estimates (Figure 11.15c) has a correlation coefficient of 0.67; and that of the inverse distance squared estimates (Figure 11.15d) has a correlation coefficient of 0.79.

Through each cloud of points we show the ideal 45-degree diago-

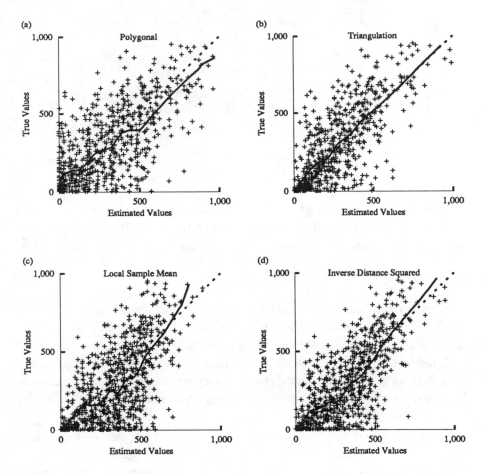

Figure 11.15 Scatterplots of true and estimated point values of V from each of four estimation methods. The correlation coefficients for the four scatterplots are 0.69, 0.80, 0.67, and 0.78, respectively.

nal as the dashed line and the conditional expectation curve as the solid line. We can judge the conditional bias by the deviations of the conditional expectation curve from the dashed line. The triangulation estimates appear to be closest to the 45-degree line and thus the least conditionally biased of all four methods, although there is some under-estimation of low values. The underestimation of the low values and overestimation of high values is quite marked for the polygonal method. Both the local sample mean and inverse distance squared plots show

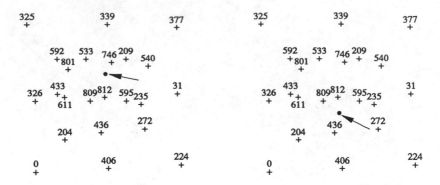

Figure 11.16 The data configuration about the point to be estimated in (a) is more or less unclustered; while the data configuration about the estimation point in (b) is clustered since the samples are not distributed uniformly around the point.

overestimation for intermediate values and underestimation for high values. Though we have specifically aimed at global unbiasedness by making the weights sum to one, this condition does not guarantee conditional unbiasedness. If we are most interested in one particular range of values then the conditional bias is a more important consideration in choosing an estimation procedure than the global bias revealed by the mean residual.

The polygonal and triangulation methods differ from the local sample mean and inverse distance squared methods in that they use very few of the nearby samples. The effect of this difference will be most noticeable when there are many nearby samples. To shed more light on the performance of the four point estimation methods we will now look only at the estimates for those points where there are at least 10 nearby samples. Since the local sample mean and the inverse distance squared methods will give some weight to all of the nearby samples, while the other two methods will use very few of them, this comparison will help us understand the effect of incorporating more sample values in the estimate.

There are at least 10 nearby samples for 345 of the points at which we calculated estimates. In some of these cases, such as the one shown in Figure 11.16a, the nearby samples are relatively unclustered, covering the local neighborhood quite evenly. In other cases, such as the one shown in Figure 11.16b, the available samples are clustered in one part

Table 11.7 Comparison of the four estimation methods for the 50 least clustered sample data configurations that contained at least 10 samples.

		Polygonal	Triangulation	Local Sample Mean	Inverse Distance Squared
	n	50	50	50	50
	m	−22.6	−49.5	−56.3	−30.1
Error	σ	187	142	212	152
Distribution	IQR	227	201	327	221
	MAE	151	127	178	123
	MSE	34,750	22,140	47,194	23,562
Correlation	ρ	0.845	0.907	0.834	0.921

Table 11.8 Comparison of the four estimation methods for the 50 most clustered sample data configurations that contained at least 10 samples.

		Polygonal	Triangulation	Local Sample Mean	Inverse Distance Squared
	n	50	43	50	50
	m	17.3	−60.5	103.7	58.0
Error	σ	182	138	221	161
Distribution	IQR	213	202	330	248
	MAE	138	106	198	142
	MSE	32,586	22,416	58,415	28,779
Correlation	ρ	0.674	0.794	0.260	0.774

of the local neighborhood. In the comparisons that follow, we look at the estimates for the 50 points whose nearby samples were least clustered and for the 50 points whose nearby samples were most clustered [4].

Table 11.7 compares the four methods for the least clustered sample data configurations; the same comparison for the most clustered sample data configurations is given in Table 11.8. These tables provide a few of the summary statistics of the error distributions along with the correlation coefficient of the true and estimated values. From the results in Table 11.7 it appears that incorporating additional nearby samples does improve our estimates. The inverse distance squared estimates correlate best with the true values and they also have the lowest mean absolute error. Table 11.8, however, shows that if the additional nearby samples are clustered, the local sample mean and inverse distance methods suffer. With either of these methods all measures of spread of the error distribution are larger for the clustered configurations in Table 11.8 than for the unclustered ones in Table 11.7. On the other hand, with the polygonal or triangulation methods, the measures of spread change little or improve slightly from the undersampled configurations to the clustered ones.

Though our estimates can often be improved by incorporating more nearby samples, they can also be adversely affected if the nearby samples are strongly clustered. Ideally, we would like an estimation method that uses all of the nearby samples and also accounts for the possibility of clustering in the sample data configuration.

Notes

[1] With samples located on a regular grid, there is a possibility that more than three polygons of influence will meet at some point. On a perfectly rectangular grid, for example, the polygons of influence will all be rectangles and four polygons will meet at each vertex.

[2] For the curious, the true value at 65E,137N is 824.2 ppm, making the polygon estimate the best for this case. One meter to the south and east, at 64E,136N, none of our estimates would change much, but here our triangulation estimate is best since the true value drops to 518.5 ppm

[3] The moment of inertia about the 45-degree diagonal line, which we used previously as a summary statistic of a scatterplot, could also be used to summarize the spread of the scatterplot. As it turns out, the moment of inertia is equal to half the MSE we defined in Equation 11.12.

[4] These most and least clustered sample data configurations were selected by dividing the available samples into quadrants and using

the variance of the number of samples in the four quadrants as an index of clustering.

Further Reading

Mueller, E. , "Comparing and validating computer models of ore bodies," in *Twelfth International Symposium of Computer Applications in the Minerals Industry*, (Johnson, T. and Gentry, D. , eds.), pp. H25–H39, Colorado School of Mines, 1974.

12

ORDINARY KRIGING

In the previous chapter we compared several point estimation methods and saw that different methods were "best" according to different estimation criteria. In this chapter we will look at ordinary kriging, a method that is often associated with the acronym B.L.U.E. for "best linear unbiased estimator." Ordinary kriging is "linear" because its estimates are weighted linear combinations of the available data; it is "unbiased" since it tries to have m_R, the mean residual or error, equal to 0; it is "best" because it aims at minimizing σ_R^2, the variance of the errors. All of the other estimation methods we have seen so far are also linear and, as we have already seen, are also theoretically unbiased. The distinguishing feature of ordinary kriging, therefore, is its aim of minimizing the error variance.

The goals of ordinary kriging are ambitious ones and, in a practical sense, unattainable since m_R and σ_R^2 are always unknown. In the previous chapter, our calculations of the mean error and the error variance were possible only because we had access to the exhaustive data set. In practical situations we never know the true answers or the actual errors before we attempt our estimation. The importance of this for ordinary kriging is that we never know m_R and therefore cannot guarantee that it is exactly 0. Nor do we know σ_R^2; therefore, we cannot minimize it. The best we can do is to build a model of the data we are studying and work with the average error and the error variance for the model. In ordinary kriging, we use a probability model in which the bias and the error variance can both be calculated and then choose weights for the nearby samples that ensure that the average error for our model, $\tilde{m}_R$, is exactly 0 and that our modeled error variance, $\tilde{\sigma}_R^2$, is minimized. We will use the same convention as in Chapter 9 where the symbol ˜ is used to denote a parameter of a model, and distinguish it from a statistic of the data.

We will be using a random function model since this type of model enables us to express the error, its mean value, and its variance. In this chapter we will begin by reviewing the approach we took earlier in Chapter 9 when we first encountered random functions and tackled the problem of unbiased estimates. After we have developed an expression for the error we will apply an earlier result, also from Chapter 9, that allowed us to express the variance of a weighted linear combination of random variables. We will then be able to develop the ordinary kriging system by using introductory calculus to minimize the error variance. Following a detailed example that illustrates how the ordinary kriging weights are calculated, we will look at how the choice of a model of spatial continuity affects the ordinary kriging weights. Finally, we will extend the point estimation case study of the previous chapter to include ordinary kriging.

The Random Function Model and Unbiasedness

In Chapter 9, we introduced the concept of a random function model and showed how it could help us in deciding how to weight the nearby samples so that our estimates are unbiased. At every point where we do not have a sample, we will estimate the unknown true value using a weighted linear combination of the available samples:

$$\hat{v} = \sum_{j=1}^{n} w_j \cdot v$$

The set of weights is allowed to change as we estimate unknown values at different locations.

If we define the error, r, of any particular estimated value to be the difference between the estimated value and the true value at that same location:

$$\textit{Error of i-th estimate} = r_i = \hat{v}_i - v_i \qquad (12.1)$$

then the average error of a set of k estimates is

$$Average\ error = m_r = \frac{1}{k}\sum_{i=1}^{k} r_i = \frac{1}{k}\sum_{i=1}^{k} \hat{v}_i - v_i \qquad (12.2)$$

Unfortunately, we are unable to make much use of this equation since it involves quantities that we do not know, namely the true values $v_1, \ldots, v_k$.

The probabilistic solution to this problem consists of conceptualizing the unknown values as the outcome of a random process and solving the problem for our conceptual model. For any point at which we attempt to estimate the unknown value, our model is a stationary random function that consists of several random variables, one for the value at each of the seven sample locations, $V(x_1), \ldots, V(x_n)$, and one for the unknown value at the point we are trying to estimate, $V(x_0)$. Each of these random variables has the same probability law; at all locations, the expected value of the random variable is $E\{V\}$. Any pair of random variables has a joint distribution that depends only on the separation between the two points and not on their locations. The covariance between pairs of random variables separated by a particular distance, h, is $\tilde{C}_V(h)$.

Every value in this model is seen as the outcome of a random variable; the samples are outcomes of random variables, as is the unknown true value. Our estimate is also a random variable since it is a weighted linear combination on the random variables at the available sample locations:

$$\hat{V}(x_0) = \sum_{i=1}^{n} w_i \cdot V(x_i)$$

Similarly, the estimation error, defined as the difference between the estimate and the random variable modeling the true value, is also a random variable:

$$R(x_0) = \hat{V}(x_0) - V(x_0)$$

By substituting the previous equation which expressed $\hat{V}(x_0)$ in terms of other random variables, we can express $R(x_0)$ solely in terms of the original $n+1$ random variables in our random function model:

$$R(x_0) = \sum_{i=1}^{n} w_i \cdot V(x_i) \quad - \quad V(x_0) \qquad (12.3)$$

The error that we make when we estimate the unknown value at x_0 is an outcome of the random variable $R(x_0)$.

We can ensure that the error at any particular location has an expected value of 0 by applying the formula for the expected value of a linear combination to Equation 12.3:

$$E\{R(x_0)\} \quad = \quad E\{\sum_{i=1}^{n} w_i \cdot V(x_i) \quad - \quad V(x_0)\}$$

$$= \quad \sum_{i=1}^{n} w_i E\{V(x_i)\} \quad - \quad E\{V(x_0)\}$$

We have already assumed that the random function is stationary, which allows us to express all of the expected values on the right-hand side as $E\{V\}$:

$$E\{R(x_0)\} = \sum_{i=1}^{n} w_i E\{V\} \quad - \quad E\{V\}$$

The expected value of the error at any particular location, $E\{R(x_0)\}$ is often referred to as the bias. Setting this expected value to 0 to ensure unbiasedness results in the following conclusion:

$$E\{R(x_0)\} = 0 \quad = \quad E\{V\} \sum_{i=1}^{n} w_i \quad - \quad E\{V\}$$

$$E\{V\} \sum_{i=1}^{n} w_i \quad = \quad E\{V\}$$

$$\sum_{i=1}^{n} w_i \quad = \quad 1$$

As we noted when we first arrived at this conclusion, all of the common estimation procedures we used in our case studies in the previous chapter all make use of this unbiasedness condition.

The Random Function Model and Error Variance

As an estimation methodology, ordinary kriging distinguishes itself by its attempt to produce a set of estimates for which the variance of the errors is minimum. The error variance, σ_R^2, of a set of k estimates can be written as

$$\sigma_R^2 \quad = \quad \frac{1}{k} \sum_{i=1}^{k} (r_i - m_R)^2$$

$$= \frac{1}{k}\sum_{i=1}^{k}[\hat{v}_i - v_i - \frac{1}{k}\sum_{i=1}^{k}(\hat{v}_i - v_i)]^2$$

$v_1,\dots,v_n$ are the true values and $\hat{v}_1,\dots,\hat{v}_n$ are the corresponding estimates. If we are willing to assume that we have a mean error of 0, we can simplify this equation somewhat:

$$\sigma_R^2 = \frac{1}{k}\sum_{i=1}^{k}(r_i - 0)^2$$

$$= \frac{1}{k}\sum_{i=1}^{k}[\hat{v}_i - v_i]^2$$

As with Equation 12.2, which provided an expression for the mean error, we cannot get very far with this equation for the error variance because it calls for knowledge of the true values.

To get out of this unfortunate dead end, we will again turn to random function models. As in the previous section, we begin with $n+1$ random variables, n of which model the behavior of the phenomenon at the nearby sample locations and one of which models its behavior at the location whose value we are trying to estimate. The available samples will be combined in a weighted linear combination to form our estimate:

$$\hat{V}(x_0) = \sum_{i=1}^{n} w_i V(x_i) \qquad (12.4)$$

The difference between the true value and the corresponding estimate will be our error or residual:

$$R(x_0) = \hat{V}(x_0) - V(x_0) \qquad (12.5)$$

As we did with the unbiasedness problem, we will transfer the original problem into the corresponding model problem. Though we cannot minimize the variance of our actual errors, we can minimize the variance of our modeled error $R(x_0)$. This minimization will be accomplished by finding an expression for the modeled error variance, $\tilde{\sigma}_R^2$, and setting to 0 the various partial derivatives of this expression.

Our first task, then, is to find an expression for the variance of the error. This error is a random variable, since it is a weighted linear combination of other random variables. In Chapter 9, when we introduced random function models, we gave a formula for the variance of a weighted linear combination:

$$Var\{\sum_{i=1}^{n} w_i \cdot V_i\} = \sum_{i=1}^{n}\sum_{j=1}^{n} w_i \cdot w_j \cdot Cov\{V_i V_j\} \qquad (12.6)$$

Using this formula with Equation 12.5, we can express the variance of the error as:

$$\begin{aligned} Var\{R(x_0)\} &= Cov\{\hat{V}(x_0)\hat{V}(x_0)\} - Cov\{\hat{V}(x_0)V(x_0)\} \\ &\quad - Cov\{V(x_0)\hat{V}(x_0)\} + Cov\{V(x_0)V(x_0)\} \\ &= Cov\{\hat{V}(x_0)\hat{V}(x_0)\} - 2Cov\{\hat{V}(x_0)V(x_0)\} \\ &\quad + Cov\{V(x_0)V(x_0)\} \end{aligned}$$

$$(12.7)$$

The first term $Cov\{\hat{V}(x_0)\hat{V}(x_0)\}$ is the covariance of $\hat{V}(x_0)$ with itself, which is equal to the variance of $\hat{V}(x_0)$, itself a linear combination $\sum_{i=1}^{n} w_i V(x_i)$ of other random variables:

$$Var\{\hat{V}(x_0)\hat{V}(x_0)\} = Var\{\sum_{i=1}^{n} w_i \cdot V_i\} = \sum_{i=1}^{n}\sum_{j=1}^{n} w_i w_j \tilde{C}_{ij}$$

The third term in Equation 12.6, $Cov\{V(x_0)V(x_0)\}$, is the covariance of the random variable $V(x_0)$ with itself and is equal to the variance of $V(x_0)$. If we assume that all of our random variables have the same variance, $\tilde{\sigma}^2$, then this third term can be expressed as

$$Cov\{V(x_0)V(x_0)\} = \tilde{\sigma}^2$$

The second term in Equation 12.6 can be written as

$$\begin{aligned} 2Cov\{\hat{V}(x_0)\hat{V}(x_0)\} &= 2Cov\{(\sum_{i=1}^{n} w_i V_i)V_0\} \\ &= 2E\{\sum_{i=1}^{n} w_i V_i \cdot V_0\} - 2E\{\sum_{i=1}^{n} w_i V_i\} \cdot E\{V_0\} \\ &= 2\sum_{i=1}^{n} w_i \cdot E\{V_i, V_0\} - 2\sum_{i=1}^{n} w_i \cdot E\{V_i\} \cdot E\{V_0\} \\ &= 2\sum_{i=1}^{n} w_i \cdot Cov\{V_i V_0\} \\ &= 2\sum_{i=1}^{n} w_i \tilde{C}_{i0} \end{aligned}$$

Combining these three terms again, we now have the following expression for the error variance:

$$\tilde{\sigma}_R^2 \;=\; \tilde{\sigma}^2 + \sum_{i=1}^{n}\sum_{j=1}^{n} w_i w_j \tilde{C}_{ij} - 2\sum_{i=1}^{n} w_i \tilde{C}_{i0} \qquad (12.8)$$

Once we have chosen our random function model parameters, specifically the variance $\tilde{\sigma}^2$ and all the covariances $\tilde{C}_{ij}$, Equation 12.8 gives us an expression for the error variance as a function of n variables, namely the weights $w_1, \ldots, w_n$.

The minimization of a function of n variables is usually accomplished by setting the n partial first derivatives to 0. This produces a system of n equations and n unknowns that can be solved by any one of several methods for solving systems of simultaneous linear equations. Unfortunately, this procedure is not quite correct for the minimization of $\tilde{\sigma}_R^2$ since we have a constraint on our solution. Earlier, we decided to use the unbiasedness condition; this means that we cannot accept any set of n weights as a solution, but must restrict possible solutions to those sets of weights that sum to 1. Such problems of constrained optimization can be solved by the technique of Lagrange parameters described in the next section.

The Lagrange Parameter

The technique of Lagrange parameters is a procedure for converting a constrained minimization problem into an unconstrained one [1]. If we try to tackle the minimization of $\tilde{\sigma}_R^2$, as expressed in Equation 12.8, as an unconstrained problem, we run into difficulties. Setting the n partial first derivatives of $\tilde{\sigma}_R^2$ to 0 will produce n equations and n unknowns. The unbiasedness condition will add another equation without adding any more unknowns. This leaves us with a system of $n + 1$ equations and only n unknowns, the solution of which is not straightforward.

To avoid this awkward problem, we introduce another unknown into our equation for $\tilde{\sigma}_R^2$. This new variable is called μ, the Lagrange parameter, and is introduced into Equation 12.8 in the following way:

$$\tilde{\sigma}_R^2 = \tilde{\sigma}^2 + \sum_{i=1}^{n}\sum_{j=1}^{n} w_i w_j \tilde{C}_{ij} - 2\sum_{i=1}^{n} w_i \tilde{C}_{i0} + 2\mu(\underbrace{\sum_{i=1}^{n} w_i - 1}_{0}) \qquad (12.9)$$

Adding variables to an equation is a tricky business; we have to be sure we do not upset the equality. The way we have chosen to do it in Equation 12.9 is safe because the term we are adding at the end is 0 due to the unbiasedness condition:

$$\sum_{i=1}^{n} w_i = 1$$

$$\sum_{i=1}^{n} w_i - 1 = 0$$

$$2\mu(\sum_{i=1}^{n} w_i - 1) = 0$$

The addition of this new term, which does not affect the equality, is all we need to convert our constrained minimization problem into an unconstrained one. The error variance for the model, as expressed in Equation 12.9, is now a function of $n + 1$ variables, the n weights and the one Lagrange parameter. By setting the $n + 1$ partial first derivatives to 0 with respect to each of these variables, we will have a system of $n + 1$ equations and $n + 1$ unknowns. Furthermore, setting the partial first derivative to 0 with respect to μ will produce our unbiasedness condition. The first three terms in Equation 12.9 do not contain μ, and do not affect the partial first derivative with respect to μ:

$$\frac{\partial(\tilde{\sigma}_R^2)}{\partial\mu} = \frac{\partial(2\mu(\sum_{i=1}^{n} w_i - 1))}{\partial\mu}$$

$$= 2\sum_{i=1}^{n} w_i - 2$$

Setting this quantity to 0 produces the unbiasedness condition:

$$\sum_{i=1}^{n} w_i = 1$$

Since the unbiasedness condition is already included in the $n + 1$ equations that result from the differentiation of $\tilde{\sigma}_R^2$, the solution of those $n + 1$ equations will produce the set of weights that minimizes $\tilde{\sigma}_R^2$ under the constraint that the weights sum to 1. This solution will also provide a value for μ that, as we will see later, is useful for calculating the resulting minimized error variance.

Minimization of the Error Variance

We will now minimize the error variance by calculating the $n+1$ partial first derivatives of Equation 12.9 and setting each one to 0. The differentiation with respect to w_1 is given in detail; the partial first derivatives with respect to the other weights can be calculated in a similar manner.

The first term on the right-hand side of Equation 12.9 does not depend on w_1, and therefore does not affect the derivative with respect to w_1. Expanding the double summation in the second term and dropping all terms that do not include w_1 gives us

$$\frac{\partial(\sum_{i=1}^{n} \sum_{j=1}^{n} w_i w_j \tilde{C}_{ij})}{\partial w_1} = \frac{\partial(w_1^2 \tilde{C}_{11} + 2w_1 \sum_{j=2}^{n} w_j \tilde{C}_{1j})}{\partial w_1}$$

$$= 2w_1 \tilde{C}_{11} + 2 \sum_{j=2}^{n} w_j \tilde{C}_{1j}$$

$$= 2 \sum_{j=1}^{n} w_j \tilde{C}_{1j}$$

The third term on the right-hand side of Equation 12.9 contains only one term that involves w_1:

$$\frac{\partial(\sum_{i=1}^{n} w_i \tilde{C}_{i0})}{\partial w_1} = \frac{\partial(w_1 \tilde{C}_{10})}{\partial w_1} = \tilde{C}_{10}$$

The last term on the right-hand side of Equation 12.9 also contains only one term that involves w_1:

$$\frac{\partial(\mu(\sum_{i=1}^{n} w_i - 1))}{\partial w_1} = \frac{\partial(\mu w_1)}{\partial w_1} = \mu$$

The first derivative of $\tilde{\sigma}_R^2$ with respect to w_1 can now be written as

$$\frac{\partial(\tilde{\sigma}_R^2)}{\partial w_1} = 2 \sum_{j=1}^{n} w_j \tilde{C}_{1j} - 2\tilde{C}_{10} + 2\mu$$

Setting this to 0 produces the following equation:

$$2 \sum_{j=1}^{n} w_j \tilde{C}_{1j} - 2\tilde{C}_{10} + 2\mu = 0$$

$$\sum_{j=1}^{n} w_j \tilde{C}_{1j} + \mu = \tilde{C}_{10} \qquad (12.10)$$

The differentiation with respect to the other weights produces similar equations:

$$\frac{\partial(\tilde{\sigma}_R^2)}{\partial w_1} = 2\sum_{j=1}^n w_j \tilde{C}_{1j} - 2\tilde{C}_{10} + 2\mu = 0 \quad \Rightarrow \quad \sum_{j=1}^n w_j \tilde{C}_{1j} + \mu = \tilde{C}_{10}$$

$$\vdots \qquad\qquad\qquad\qquad\qquad\qquad\qquad \vdots$$

$$\frac{\partial(\tilde{\sigma}_R^2)}{\partial w_i} = 2\sum_{j=1}^n w_j \tilde{C}_{ij} - 2\tilde{C}_{i0} + 2\mu = 0 \quad \Rightarrow \quad \sum_{j=1}^n w_j \tilde{C}_{ij} + \mu = \tilde{C}_{i0}$$

$$\vdots \qquad\qquad\qquad\qquad\qquad\qquad\qquad \vdots$$

$$\frac{\partial(\tilde{\sigma}_R^2)}{\partial w_n} = 2\sum_{j=1}^n w_j \tilde{C}_{nj} - 2\tilde{C}_{n0} + 2\mu = 0 \quad \Rightarrow \quad \sum_{j=1}^n w_j \tilde{C}_{nj} + \mu = \tilde{C}_{n0}$$

As we noted in the last section, the setting of the partial first derivative to 0 with respect to μ produces the unbiasedness condition.

The set of weights that minimize the error variance under the constraint that they sum to 1 therefore satisfies the following $n+1$ equations:

$$\sum_{j=1}^n w_j \tilde{C}_{ij} + \mu = \tilde{C}_{i0} \quad \forall \, i = 1, \ldots, n \qquad (12.11)$$

$$\sum_{i=1}^n w_i = 1 \qquad (12.12)$$

This system of equations, often referred to as the *ordinary kriging system*, can be written in matrix notation as

$$\mathbf{C} \qquad\qquad \cdot \qquad \mathbf{w} \quad = \quad \mathbf{D}$$

$$\underbrace{\begin{bmatrix} \tilde{C}_{11} & \cdots & \tilde{C}_{1n} & 1 \\ \vdots & \ddots & \vdots & \vdots \\ \tilde{C}_{n1} & \cdots & \tilde{C}_{nn} & 1 \\ 1 & \cdots & 1 & 0 \end{bmatrix}}_{(n+1)\times(n+1)} \cdot \underbrace{\begin{bmatrix} w_1 \\ \vdots \\ w_n \\ \mu \end{bmatrix}}_{(n+1)\times 1} = \underbrace{\begin{bmatrix} \tilde{C}_{10} \\ \vdots \\ \tilde{C}_{n0} \\ 1 \end{bmatrix}}_{(n+1)\times 1} \qquad (12.13)$$

To solve for the weights, we multiply Equation 12.13 on both sides by $\mathbf{C}^{-1}$, the inverse of the left-hand side covariance matrix:

$$\begin{aligned} \mathbf{C} \cdot \mathbf{w} &= \mathbf{D} \\ \mathbf{C}^{-1} \cdot \mathbf{C} \cdot \mathbf{w} &= \mathbf{C}^{-1} \cdot \mathbf{D} \\ \mathbf{I} \cdot \mathbf{w} &= \mathbf{C}^{-1} \cdot \mathbf{D} \\ \mathbf{w} &= \mathbf{C}^{-1} \cdot \mathbf{D} \qquad (12.14) \end{aligned}$$

After a considerable amount of mathematics, we have finally arrived at a solution. To minimize the modeled error variance, we first need to choose the $(n+1)^2$ covariances that will describe the spatial continuity in our random function model. In practice this is typically done by choosing a function $\tilde{C}(\mathbf{h})$, and calculating all of the required covariances from this function. Once the $(n+1)^2$ covariances have been chosen, the $\mathbf{C}$ and $\mathbf{D}$ matrices can be built. The set of weights that will produce unbiased estimates with the minimum error variance for our random function model is given by Equation 12.14.

Having gone to considerable trouble to minimize the error variance, we may be interested in knowing this minimum value. We could substitute the weights we have obtained into Equation 12.8 to find the actual value of the minimized error variance. There is also a quicker way that avoids the n^2 terms in the double summation. Multiplying each of the n equations given in Equation 12.11 by w_i produces the following result:

$$w_i(\sum_{j=1}^{n} w_j \tilde{C}_{ij} + \mu) = w_i \tilde{C}_{i0} \quad \forall\, i = 1,\ldots,n$$

Summing these n equations leads to an expression for the double summation:

$$\sum_{i=1}^{n} w_i \sum_{j=1}^{n} w_j \tilde{C}_{ij} + \sum_{i=1}^{n} w_i \mu = \sum_{i=1}^{n} w_i \tilde{C}_{i0}$$

$$\sum_{i=1}^{n} w_i \sum_{j=1}^{n} w_j \tilde{C}_{ij} = \sum_{i=1}^{n} w_i \tilde{C}_{i0} - \sum_{i=1}^{n} w_i \mu$$

Since the weights sum to 1, the last term is simply μ, which gives us

$$\sum_{i=1}^{n} \sum_{j=1}^{n} w_i w_j \tilde{C}_{ij} = \sum_{i=1}^{n} w_i \tilde{C}_{i0} - \mu$$

Substituting this into Equation 12.8 allows us to express the minimized error variance as

$$\tilde{\sigma}_R^2 = \tilde{\sigma}^2 + \sum_{i=1}^{n} w_i \tilde{C}_{i0} - \mu - 2\sum_{i=1}^{n} w_i \tilde{C}_{i0}$$

$$= \tilde{\sigma}^2 - (\sum_{i=1}^{n} w_i \tilde{C}_{i0} + \mu) \tag{12.15}$$

Or, in terms of the matrices we defined earlier,

$$\tilde{\sigma}_R^2 = \tilde{\sigma}^2 - \mathbf{w} \cdot \mathbf{D} \tag{12.16}$$

This minimized error variance is usually referred to as the *ordinary kriging variance*, for which we will use the notation σ_{OK}^2; though the tilde has been dropped from the notation, the OK subscript should serve as a reminder that this error variance was calculated from a model.

Ordinary Kriging Using γ or ρ

When we derived the expression for the error variance, we assumed that the random variables in our random function model all had the same mean and variance. These two assumptions also allow us to develop the following relationship between the model variogram and the model covariance:

$$
\begin{aligned}
\gamma_{ij} &= \frac{1}{2} E\{[V_i - V_j]^2\} \\
&= \frac{1}{2} E\{V_i^2\} + \frac{1}{2} E\{V_i^2\} - E\{V_i \cdot V_j\} \\
&= E\{V^2\} - E\{V_i \cdot V_j\} \\
&= E\{V^2\} - \tilde{m}^2 - [E\{V_i \cdot V_j\} - \tilde{m}^2] \\
&= \tilde{\sigma}^2 - \tilde{C}_{ij}
\end{aligned} \tag{12.17}
$$

There is also a relationship between the model correlogram and the model covariance:

$$\tilde{\rho}_{ij} = \frac{\tilde{C}_{ij}}{\tilde{\sigma}^2} \tag{12.18}$$

These relationships are valid for a random function model in which we have made the assumptions that the random variables all have the same mean and variance. This does not entail that the same relationships exist between the variogram, covariance, and correlation functions of an actual data set. Nevertheless, the validity of these relationships for our random function model allows us to express the ordinary kriging equations in terms of the variogram or the correlogram.

In terms of the variogram, the ordinary kriging system can be written as

$$\sum_{j=1}^{n} w_j \tilde{\gamma}_{ij} - \mu = \tilde{\gamma}_{i0} \quad \forall \, i = 1, \ldots, n \tag{12.19}$$

$$\sum_{i=1}^{n} w_i = 1$$

with the modeled error variance given by

$$\tilde{\sigma}_R^2 = \sum_{i=1}^{n} w_i \tilde{\gamma}_{i0} + \mu \qquad (12.20)$$

In terms of the correlogram, the ordinary kriging system can be written as

$$\sum_{j=1}^{n} w_j \tilde{\rho}_{ij} + \mu = \tilde{\rho}_{i0} \quad \forall \ i = 1, \dots, n \qquad (12.21)$$

$$\sum_{i=1}^{n} w_i = 1$$

with the modeled error variance given by

$$\tilde{\sigma}_R^2 = \tilde{\sigma}^2 (1 - (\sum_{i=1}^{n} w_i \tilde{\rho}_{i0} + \mu)) \qquad (12.22)$$

The common practice in geostatistics is to calculate modeled variogram values then, for reasons of computational efficiency, to subtract them from some constant, usually $\tilde{\sigma}^2$. The net result is that although geostatisticians eventually resort to solving the ordinary kriging equations in terms of covariances, most of the initial calculations are done in terms of variograms [2].

An Example of Ordinary Kriging

Once the ordinary kriging method has been developed, several small examples will be given to demonstrate how the various model parameters affect the estimates. Finally, to allow a comparison of ordinary kriging with the other point estimation methods we have seen earlier, the case study from the previous chapter will be extended to include estimates calculated by ordinary kriging.

Let us return to the seven sample data configuration we used earlier to see a specific example of how ordinary kriging is done. The data configuration is shown again in Figure 12.1; we have labeled the point we are estimating as location 0, and the sample locations as 1 through

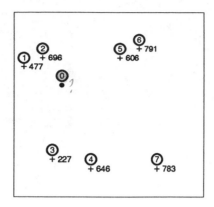

Figure 12.1 An example of a data configuration to illustrate the kriging estimator. This configuration was given earlier in Figure 11.1 where it was used to illustrate other estimation techniques discussed in Chapter 11. The sample value is given immediately to the right of the plus sign.

Table 12.1 Coordinates and sample values for the data shown in Figure 12.1.

Sample No.	X	Y	V	Distance from 65E,137N
1	225	61	139	477	4.5
2	437	63	140	696	3.6
3	367	64	129	227	8.1
4	52	68	128	646	9.5
5	259	71	140	606	6.7
6	436	73	141	791	8.9
7	366	75	128	783	13.5

7. The coordinates of these eight points are given in Table 12.1, along with the available sample values.

To calculate the ordinary kriging weights, we must first decide what pattern of spatial continuity we want our random function model to have. To keep this example relatively simple, we will calculate all of

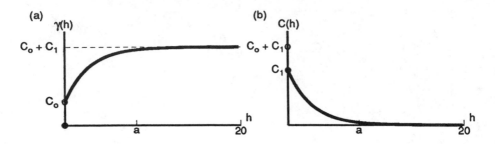

Figure 12.2 An example of an exponential variogram model (a) and an exponential covariance function (b).

our covariances from the following function:

$$\tilde{C}(\mathbf{h}) = \begin{cases} C_0 + C_1 & \text{if } |\mathbf{h}| = 0 \\ C_1 exp(\frac{-3|\mathbf{h}|}{a}) & \text{if } |\mathbf{h}| > 0 \end{cases} \qquad (12.23)$$

Using Equation 12.17, this covariance function corresponds to the following variogram:

$$\tilde{\gamma}(\mathbf{h}) = \begin{cases} 0 & \text{if } |\mathbf{h}| = 0 \\ C_0 + C_1(1 - exp(\frac{-3|\mathbf{h}|}{a})) & \text{if } |\mathbf{h}| > 0 \end{cases} \qquad (12.24)$$

Both of these functions, shown in Figure 12.2, can be described by the following parameters:

- C_0, commonly called the *nugget effect*, which provides a discontinuity at the origin.

- a, commonly called the *range*, which provides a distance beyond which the variogram or covariance value remains essentially constant.

- $C_0 + C_1$, commonly called the *sill* [3], which is the variogram value for very large distances, $\gamma(\infty)$. It is also the covariance value for $|\mathbf{h}| = 0$, and the variance of our random variables, $\tilde{\sigma}^2$.

Geostatisticians normally define the spatial continuity in their random function model through the variogram and solve the ordinary

Table 12.2 A table of distances, from Figure 12.1, between all possible pairs of the seven data locations.

Location	distance							
	0	1	2	3	4	5	6	7
0	0.00	4.47	3.61	8.06	9.49	6.71	8.94	13.45
1	4.47	0.00	2.24	10.44	13.04	10.05	12.17	17.80
2	3.61	2.24	0.00	11.05	13.00	8.00	10.05	16.97
3	8.06	10.04	11.05	0.00	4.12	13.04	15.00	11.05
4	9.49	13.04	13.00	4.12	0.00	12.37	13.93	7.00
5	6.71	10.05	8.00	13.04	12.37	0.00	2.24	12.65
6	8.94	12.17	10.05	15.00	13.93	2.24	0.00	13.15
7	13.45	17.80	16.97	11.05	7.00	12.65	13.15	0.00

kriging system using the covariance. In this example, we will use the covariance function throughout.

By using the covariance function given in Equation 12.23, we have chosen to ignore the possibility of anisotropy for the moment; the covariance between the data values at any two locations will depend only on the distance between them and not on the direction. Later, when we examine the effect of the various parameters, we will also study the important effect of anisotropy.

To demonstrate how ordinary kriging works, we will use the following parameters for the function given in Equation 12.23:

$$C_0 = 0, \quad a = 10, \quad C_1 = 10$$

These are not necessarily good choices, but they will make the details of the ordinary kriging procedure easier to follow since our covariance model now has a quite simple expression:

$$\tilde{C}(\mathbf{h}) = 10e^{-0.3|\mathbf{h}|} \qquad (12.25)$$

Having chosen a covariance function from which we can calculate all the covariances required for our random function model, we can now build the **C** and **D** matrices. Using Table 12.2, which provides the distances between every pair of locations, and Equation 12.25 above, the **C** matrix is

$$
\mathbf{C} = \begin{bmatrix}
\tilde{C}_{11} & \tilde{C}_{12} & \tilde{C}_{13} & \tilde{C}_{14} & \tilde{C}_{15} & \tilde{C}_{16} & \tilde{C}_{17} & 1 \\
\tilde{C}_{21} & \tilde{C}_{22} & \tilde{C}_{23} & \tilde{C}_{24} & \tilde{C}_{25} & \tilde{C}_{26} & \tilde{C}_{27} & 1 \\
\tilde{C}_{31} & \tilde{C}_{32} & \tilde{C}_{33} & \tilde{C}_{34} & \tilde{C}_{35} & \tilde{C}_{36} & \tilde{C}_{37} & 1 \\
\tilde{C}_{41} & \tilde{C}_{42} & \tilde{C}_{43} & \tilde{C}_{44} & \tilde{C}_{45} & \tilde{C}_{46} & \tilde{C}_{47} & 1 \\
\tilde{C}_{51} & \tilde{C}_{52} & \tilde{C}_{53} & \tilde{C}_{54} & \tilde{C}_{55} & \tilde{C}_{56} & \tilde{C}_{57} & 1 \\
\tilde{C}_{61} & \tilde{C}_{62} & \tilde{C}_{63} & \tilde{C}_{64} & \tilde{C}_{65} & \tilde{C}_{66} & \tilde{C}_{67} & 1 \\
\tilde{C}_{71} & \tilde{C}_{72} & \tilde{C}_{73} & \tilde{C}_{74} & \tilde{C}_{75} & \tilde{C}_{76} & \tilde{C}_{77} & 1 \\
1 & 1 & 1 & 1 & 1 & 1 & 1 & 0
\end{bmatrix}
$$

$$
= \begin{bmatrix}
10.00 & 5.11 & 0.44 & 0.20 & 0.49 & 0.26 & 0.05 & 1.00 \\
5.11 & 10.00 & 0.36 & 0.20 & 0.91 & 0.49 & 0.06 & 1.00 \\
0.44 & 0.36 & 10.00 & 2.90 & 0.20 & 0.11 & 0.36 & 1.00 \\
0.20 & 0.20 & 2.90 & 10.00 & 0.24 & 0.15 & 1.22 & 1.00 \\
0.49 & 0.91 & 0.20 & 0.24 & 10.00 & 5.11 & 0.22 & 1.00 \\
0.26 & 0.49 & 0.11 & 0.15 & 5.11 & 10.00 & 0.19 & 1.00 \\
0.05 & 0.06 & 0.36 & 1.22 & 0.22 & 0.19 & 10.00 & 1.00 \\
1.00 & 1.00 & 1.00 & 1.00 & 1.00 & 1.00 & 1.00 & 0.00
\end{bmatrix}
$$

The **D** matrix is

$$
\mathbf{D} = \begin{bmatrix}
\tilde{C}_{10} \\
\tilde{C}_{20} \\
\tilde{C}_{30} \\
\tilde{C}_{40} \\
\tilde{C}_{50} \\
\tilde{C}_{60} \\
\tilde{C}_{70} \\
1
\end{bmatrix} = \begin{bmatrix}
2.61 \\
3.39 \\
0.89 \\
0.58 \\
1.34 \\
0.68 \\
0.18 \\
1.00
\end{bmatrix}
$$

The inverse of **C** is

$$
\mathbf{C}^{-1} = \begin{bmatrix}
0.127 & -0.077 & -0.013 & -0.009 & -0.008 & -0.009 & -0.012 & 0.136 \\
-0.077 & 0.129 & -0.010 & -0.008 & -0.015 & -0.008 & -0.011 & 0.121 \\
-0.013 & -0.010 & 0.098 & -0.042 & -0.010 & -0.010 & -0.014 & 0.156 \\
-0.009 & -0.008 & -0.042 & 0.102 & -0.009 & -0.009 & -0.024 & 0.139 \\
-0.008 & -0.015 & -0.010 & -0.009 & 0.130 & -0.077 & -0.012 & 0.118 \\
-0.009 & -0.008 & -0.010 & -0.009 & -0.077 & 0.126 & -0.013 & 0.141 \\
-0.012 & -0.011 & -0.014 & -0.024 & -0.012 & -0.013 & 0.085 & 0.188 \\
0.136 & 0.121 & 0.156 & 0.139 & 0.118 & 0.141 & 0.188 & -2.180
\end{bmatrix}
$$

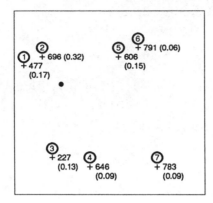

Figure 12.3 The ordinary kriging weights for the seven samples using the isotropic exponential covariance model given in Equation 12.25. The sample value is given immediately to the right of the plus sign while the kriging weights are shown in parenthesis.

The set of weights that will provide unbiased estimates with a minimum estimation variance is calculated by multiplying $\mathbf{C}^{-1}$ by $\mathbf{D}$:

$$
\mathbf{w} = \begin{bmatrix} w_1 \\ w_2 \\ w_3 \\ w_4 \\ w_5 \\ w_6 \\ w_7 \\ \mu \end{bmatrix} = \mathbf{C}^{-1} \cdot \mathbf{D} = \begin{bmatrix} 0.173 \\ 0.318 \\ 0.129 \\ 0.086 \\ 0.151 \\ 0.057 \\ 0.086 \\ 0.907 \end{bmatrix}
$$

Figure 12.3 shows the sample values along with their corresponding weights. The resulting estimate is

$$
\begin{aligned}
\hat{v}_0 &= \sum_{i=1}^{n} w_i v_i \\
&= (0.173)(477) + (0.318)(696) + (0.129)(227) + (0.086)(646) + \\
&\quad (0.151)(606) + (0.057)(791) + (0.086)(783) \\
&= 592.7 \ \text{ppm}
\end{aligned}
$$

The minimized estimation variance is

$$
\begin{aligned}
\tilde{\sigma}_R^2 &= \tilde{\sigma}^2 - \sum_{i=1}^{n} w_i \tilde{C}_{i0} + \mu \\
&= 10 - (0.173)(2.61) - (0.318)(3.39) - (0.129)(0.89) - \\
&\quad (0.086)(0.58) - (0.151)(1.34) - (0.057)(0.68) - \\
&\quad (0.086)(0.18) + 0.907 \\
&= 8.96 \text{ ppm}^2
\end{aligned}
$$

Ordinary Kriging and the Model of Spatial Continuity

Earlier, when we tackled the problem of unbiasedness, the unbiasedness condition did not require us to specify any parameters of our random function model. Though we had to assume that the mean of the random variables was the same, we did not have to specify its actual value; the condition that the weights sum to one does not involve the parameter $\tilde{m}$. In the minimization of $\tilde{\sigma}_R^2$, however, our solution does involve model parameters; the ordinary kriging weights and the resulting minimized error variance directly depend on our choice of the covariances for the $\mathbf{C}$ and $\mathbf{D}$ matrices.

The choice of a covariance model (or, if one prefers, a variogram model or a correlogram model) is a prerequisite for ordinary kriging. Though this makes ordinary kriging more time consuming than the estimation procedures we looked at in the previous chapter, it also makes it more flexible. We saw earlier that the exponent for inverse distance estimation gave us an ability to modulate the estimation procedure from a polygonal estimation to a moving average estimation. The covariance model in ordinary kriging provides a similar but much more powerful ability to customize the ordinary kriging estimation procedure. In addition to allowing us to modulate between polygonal estimates and moving average estimates, the covariance model also provides a vehicle for incorporating valuable qualitative insights such as the pattern of anisotropy.

In practice, the pattern of spatial continuity chosen for the random function model is usually taken from the spatial continuity evident in the sample data set. Once the sample variogram has been calculated, a function is fit to it; Chapter 16 discusses this procedure in detail. There are two reasons why the sample variogram cannot be used directly in the ordinary kriging system.

First, the **D** matrix may call for variogram values for distances that are not available from the sample data. There are often situations in which the distance from the point being estimated to a particular sample is smaller than the distance between any pair of available samples. Since the sample data set cannot provide any pairs for these small distances, we must rely on a function that provides variogram values for all distances and directions, even those that are not available from sample data.

Second, the use of the sample variogram does not guarantee the existence and uniqueness of the solution to the ordinary kriging system. The system of $n + 1$ equations and $n + 1$ unknowns described by Equation 12.13 does not necessarily have a unique solution. Certain choices of the covariances in the **C** and **D** matrices may cause the system to have no solution; other choices may cause the system to have several solutions. To be guaranteed of having one and only one solution, we must ensure that our system has a property known as *positive definiteness*. Even if the sample data are regularly gridded and all of the distances for which the **D** matrix requires values are available from sample data, the use of the sample variogram, unfortunately, does not guarantee positive definiteness. There are many ways of checking for positive definiteness [4]; in practice, however, we guarantee the existence and uniqueness of our solution by fitting the sample variogram with functions that are known to be positive definite.

Though fitting a function to the sample variogram is the most common approach to choosing the pattern of spatial continuity for the random function model, it is not the only one nor is it necessarily the best one. There are many situations in which it is better to base the choice of a pattern of spatial continuity on a more qualitative interpretation. Experience with similar data sets may often be a better guide than pattern of spatial continuity shown by too few available samples.

Frequently, the sample data set does not show any clear pattern of spatial continuity. The lack of evident structure in the available samples does not justify using a spatially uncorrelated random function model. In earth science data sets there is nearly always some pattern of spatial continuity. It may not be evident from the available samples due to their insufficient number, sampling error, erratic values, or possible outlier values.

Even in situations where the sample data set does exhibit a clear pattern of spatial continuity, the decision to use the sample spatial continuity for the random function model should still be considered carefully. If the samples are clustered in particular areas, one should consider how appropriate the sample variogram is for the points that will be estimated. In the Walker Lake sample data set, for example, the preferential clustering of the samples in the Wassuk Range anomaly causes the sample variogram to be more representative of that particular region than of the entire area. If we intend to perform estimation only at locations within this anomalous area, then the sample variogram may be appropriate. If we intend to calculate estimates throughout the entire Walker Lake area, however, then the use of the sample variogram for our random function model is questionable.

The decision to use the sample spatial continuity should be carefully considered even for sample data sets in which clustering is not a significant problem. For example, the anisotropy may not be adequately captured by an analysis of the sample data set. As we will see shortly, anisotropy is an important element of the pattern of spatial continuity in the random function model. The analysis of spatial continuity in a sample data set usually involves the calculation and summary of h-scatterplots for particular directions. As explained in Chapter 7, the use of a tolerance on the direction is necessary in practice; unfortunately, the use of a directional tolerance may cause the anisotropy evident from sample variograms to be weaker than that which would be observed if exhaustive information was available. This is clear from our analysis of the spatial continuity in the exhaustive data set in Chapter 5 and of the sample data set in Chapter 7. The sample variograms and covariance functions show less anisotropy than do the exhaustive ones.

In Chapter 16 we will discuss the practical details of fitting functions to sample variograms and deriving a mathematical expression that provides variogram values for any separation vector **h**. Though the fitting of functions to sample variograms is certainly the most common approach to choosing the pattern of spatial continuity for the random function model, it should not be viewed as the only correct approach. In every study that uses geostatistical estimation methods, the geostatistician must choose the pattern of spatial continuity. The use of the most common approach does not remove the responsibility of making this choice wisely, nor does it remove the responsibility of understanding the effect of one's chosen model on the estimation procedure.

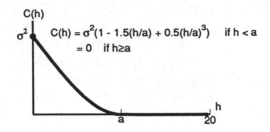

Figure 12.4   The spherical model of a covariance function.

An Intuitive Look at Ordinary Kriging

To understand how parameters such as the nugget effect, the range and the sill affect our estimates, it will help to have a more intuitive understanding of the role of the **C** and **D** matrices used in the ordinary kriging system. To many practitioners, the mathematical development of the ordinary kriging system presented earlier is tedious and virtually incomprehensible. The following explanation is not rigorous and may provide a more intuitive feel for what the ordinary kriging procedure is doing. While the earlier development provided a rationale for the procedure, the following one probably provides a better understanding of its practical success. Understanding the role of the **C** and **D** matrices in intuitive terms also allows the practitioner to make ad hoc adjustments that, despite their lack of apparent theoretical rigor, may actually improve the estimation procedure.

Taken by itself, the **D** matrix on the right-hand side of Equation 12.13 provides a weighting scheme similar to that of the inverse distance methods. Like an inverse distance weight, the covariance between any particular sample and the point being estimated generally decreases as the sample gets farther away. This can be seen in the example used in the previous section: sample 2 is closest to the point being estimated and $\tilde{C}_{20}$ is the largest covariance in **D**; sample 7 is the farthest, and $\tilde{C}_{70}$ is the smallest covariance in **D**. Unlike inverse distance weights, which are limited to the form $|\mathbf{h}|^{-p}$, the covariances calculated for our model can come from a much larger family of functions. For example, the covariance function shown in Figure 12.4 provides decreasing weights up to some distance, a, and provides a weight of 0 for distances greater than a.

The **D** matrix therefore contains a type of inverse distance weighting in which the "distance" is not the geometric distance to the sample but rather its statistical distance. What really distinguishes ordinary kriging from inverse distance methods, however, is not the use of statistical distance instead of geometric distance, but rather the role played by the **C** matrix. From our earlier example, it is clear that the multiplication of **D** by $\mathbf{C}^{-1}$ does considerably more than rescale the covariances in **D** so that they sum to one. For example, sample 4 is farther from the point being estimated than is sample 6; this is recorded in the **D** matrix by the fact that $\tilde{C}_{40}$ is smaller than $\tilde{C}_{60}$. The ordinary kriging weight for sample 4, however, is larger than that for sample 6.

The **C** matrix records distances between each sample and every other sample, providing the ordinary kriging system with information on the clustering of the available sample data. If two samples are very close to each other, this will be recorded by a large value in the **C** matrix; if two samples are far apart, this will be recorded by a low value. The multiplication of **D** by $\mathbf{C}^{-1}$ adjusts the raw inverse statistical distance weights in **D** to account for possible redundancies between the samples.

In our earlier example, though sample 6 was the closer than sample 4 to the point we were trying to estimate, its usefulness was reduced by its proximity to sample 5. When **D** is multiplied by the inverse of **C**, the net result is that some of the weight that was allocated to sample 6 because of its closeness to the point being estimated is redistributed to other samples that are farther away yet less redundant.

Like the **D** matrix, the **C** matrix records the distance in terms of statistical distance rather than geometric distance. The possible redundancy between samples depends not simply on the distance between them but also on the spatial continuity. For example, two measurements of the elevation of the water table taken 10 m apart are likely to be much more redundant than two measurements of the gold concentration in a vein-type gold deposit also taken 10 m apart. The fact that the elevation of the water table is much more spatially continuous than the concentration of gold would be captured by the covariance functions of the two data sets. It makes good sense, therefore, that clustering information be recorded in terms of the statistical distance, using some measure of spatial continuity such as the covariance or the variogram.

The ordinary kriging system therefore takes into account two of the

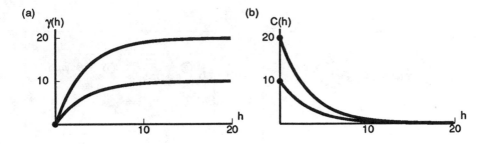

Figure 12.5   Two variograms and their corresponding covariance functions that differ only by their scale.

important aspects of estimation problems, distance, and clustering:

$$\mathbf{w} = \underbrace{\mathbf{C}^{-1}}_{\text{Clustering}} \cdot \underbrace{\mathbf{D}}_{\text{Distance}}$$

The information on the distances to the various samples and the clustering between the samples is all recorded in terms of a statistical distance, thereby customizing the estimation procedure to a particular pattern of spatial continuity.

Variogram Model Parameters

We will now look at how the various parameters of the covariance or variogram model affect the ordinary kriging weights. Much of the terminology for these parameters has evolved from the traditional use of the variogram, and we will refer to variogram models throughout this section. The same remarks apply to covariance functions and correlograms; in the following examples we will show the covariance functions that correspond to the variograms we discuss so that it is clear how the changes in various parameters manifest themselves on the covariance function.

The following observations serve as additional support for the argument that even with regularly gridded and well-behaved sample data, the exercise of fitting a function to the sample variogram model involves important choices on the part of the practitioner. As we noted earlier, the sample variogram does not provide any information for distance shorter than the minimum spacing between the sample data. Unless

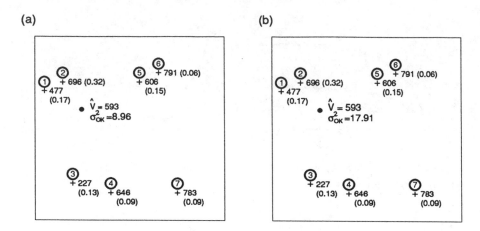

Figure 12.6 Ordinary kriging results using two different covariance functions that differ only in their scale. (a) shows the kriging weights for the variogram with a sill of 20 while (b) shows the weights for a sill of 10. The two covariance functions are given in Figure 12.5.

the sampling includes duplicates at the same location, the nugget effect and the behavior of the variogram near the origin can not be determined from the sample variogram. Yet the following examples demonstrate that these two parameters have the biggest effect on the ordinary kriging weights and on the resulting estimate.

The Effect of Scale. Figure 12.5 shows two variogram models that differ only in their scale. $\gamma_1(h)$ is the variogram that corresponds to the covariance function we chose for our earlier detailed example of ordinary kriging; $\gamma_2(h)$ has the same shape as $\gamma_1(h)$, but it is exactly twice as big:

$$\gamma_1(h) = 10(1 - e^{-.3|h|})$$
$$\gamma_2(h) = 20(1 - e^{-.3|h|}) = 2\gamma_1(h)$$

The results of using these two covariance models for ordinary kriging are shown in Figure 12.6. Rescaling the variogram values has not affected the ordinary kriging weights or the ordinary kriging estimate; however, it has affected the ordinary kriging variance. These effects will be observed with any rescaling; while the estimate itself is unchanged,

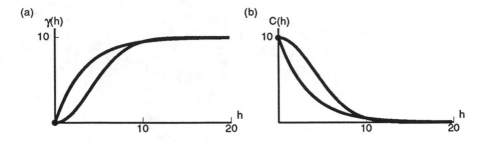

Figure 12.7 Two variograms and their corresponding covariance functions that differ only in their shape. The model that has a parabolic behavior near the origin is the Gaussian model, while the other is the exponential model.

the estimation variance increases by the same factor that was used to scale the variogram [5].

The Effect of Shape. Figure 12.7 shows two variogram models that reach the same sill but have different shapes. $\gamma_1(h)$ is the same as the first variogram model from the previous example:

$$\gamma_1(h) = 10(1 - exp(-3\frac{|h|}{10}))$$

$\gamma_2(h)$ has a similar expression, but the square in the exponent causes it to behave more like a parabola near the origin:

$$\gamma_2(h) = 10(1 - exp(-3(\frac{|h|}{10})^2))$$

The results of using these two variogram models for ordinary kriging are shown in Figure 12.8. With the second model, more weight is given to the three values that surround the point being estimated (the same three that were used for the triangulation estimate in Chapter 11); the remaining points all receive less weight, with most of them actually receiving a negative weight. A parabolic behavior near the origin is indicative of a very continuous phenomena so the estimation procedure makes much more use of the closest samples.

The appearance of negative weights is a result of an effect often referred to as the screen effect. A particular sample is said to be screened if another sample falls between it and the point being estimated. For the data configuration we are using, sample 6 is strongly screened by

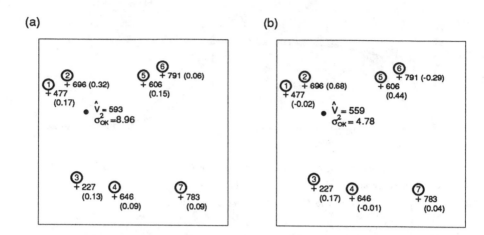

Figure 12.8 Ordinary kriging results using two different covariance functions that differ only in their shape. (a) shows the kriging weights for a exponential model while (b) shows the weights for a Gaussian model. The two covariance functions are given in Figure 12.7.

sample 5; to a lesser extent, sample 1 is partially screened by sample 2. It makes sense to reduce the weights of samples that are partially or totally screened by others; this is part of what the multiplication by C^{-1} accomplishes. The degree to which screened samples lose their influence depends on the pattern of spatial continuity. The use of a variogram with a parabolic behavior near the origin will cause the screen effect to be much more pronounced, often producing negative weights even larger than the ones we calculated in our last example.

Even with variogram models that are linear rather than parabolic near the origin, it is possible to produce negative weights for samples that are strongly screened by others. None of the other estimation procedures we looked at earlier can produce negative weights. The advantage of a procedure that can assign weights smaller than 0 or greater than 1 (but still respect the constraint that the sum of the weights is 1) is that it can yield estimates larger than the largest sample value or smaller than the smallest sample value. All procedures that restrict the weights to be between 0 and 1 can only produce estimates that are between the minimum and maximum sample values. It is unlikely that the sample data set includes the most extreme values and it is reasonable to imagine that the true values we are trying to estimate may be beyond the extremes of the available sample values. The disadvantage of negative weights is that they also create the possibility of negative estimates if a particularly high sample value is associated with a negative weight. In most earth science applications, the variable being estimated is necessarily positive. Ore grades and tonnages, porosities, permeabilities, pollutant concentrations, densities, depths to geologic horizons, and thicknesses of strata are common examples of variables that one may be interested in estimating and that are never negative. For such variables, if ordinary kriging produces estimates that are negative, one is perfectly justified in setting such estimates to 0.

For data sets in which the variable of interest is indeed extremely continuous, such as the depth to a particular horizon or the thickness of a certain zone, the sample variogram often shows a definite parabolic behavior near the origin. Even in such situations where the spatial continuity of the sample data set can correctly be extended to the points at which estimates will be required, variogram models with parabolic behavior near the origin are avoided in practice since the negative weights they may produce tend to make the estimation very erratic.

The Nugget Effect. Figure 12.9 shows two variogram models that differ only in their nugget effect. While $\gamma_1(h)$ has no nugget effect, $\gamma_2(h)$ has a nugget effect that is 50% of the sill:

$$\gamma_1(h) = 10(1 - e^{-.3|h|})$$

$$\gamma_2(h) = \begin{cases} 0 & \text{if } h = 0 \\ 5 + 5(1 - e^{-.3|h|}) & \text{if } h > 0 \end{cases}$$

The results of using these two variogram models for ordinary kriging are shown in Figure 12.10. The weights calculated using $\gamma_2(h)$ are more similar to one another than are those calculated using $\gamma_1(h)$. With $\gamma_2(h)$, the smallest weight is 0.125 and the largest weight is 0.178; with $\gamma_1(h)$, the smallest weight is 0.057 and the largest weight is 0.318. The more equal distribution of weight causes the estimated value to be somewhat higher. The other noticeable result of using a higher nugget effect is that the ordinary kriging variance is higher.

Increasing the nugget effect makes the estimation procedure become more like a simple averaging of the available data. If the vari-

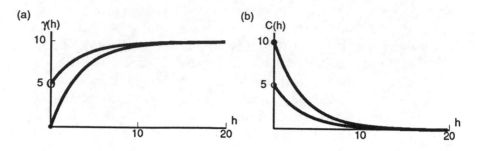

Figure 12.9 Two variograms and their corresponding covariance functions that differ only in their nugget effect.

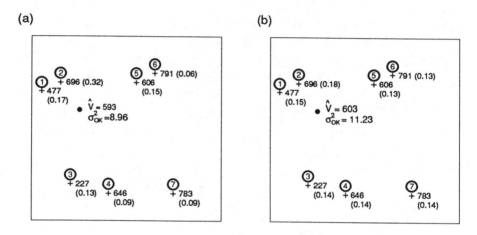

Figure 12.10 Ordinary kriging results using two different covariance functions that differ only in their nugget effect. (a) shows the kriging weights for no nugget effect while (b) shows the weights for a relative nugget of one-half. The two covariance functions are given in Figure 12.9.

ogram model is a pure nugget effect:

$$\gamma(h) = \begin{cases} 0 & \text{if } h = 0 \\ C_0 & \text{if } h > 0 \end{cases}$$

there is no redundancy between any of the samples and, in terms of statistical distance, none of the samples is any closer to the point being estimated than any other. The result is that for ordinary kriging with a pure nugget effect model of spatial continuity, all weights are equal

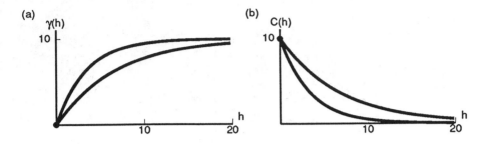

Figure 12.11 Two variograms and their corresponding covariance functions that differ only in their range.

to $\frac{1}{n}$. A pure nugget effect model entails a complete lack of spatial correlation; the data value at any particular location bears no similarity even to very nearby data values. While this produces a simple solution to the ordinary kriging system, it is not a desirable situation in terms of the ordinary kriging variance. The only use of additional samples is to reduce the uncertainty about the unknown mean of the random variables in our random function model. The ordinary kriging variance for spatially uncorrelated phenomena is the variance we have assumed for our random variables, plus the variance of the unknown mean:

$$\sigma_{OK}^2 = \underbrace{\tilde{\sigma}^2}_{\substack{\text{Variance of} \\ \text{random variables}}} + \underbrace{\frac{\tilde{\sigma}^2}{n}}_{\substack{\text{Variance of} \\ \text{unknown mean}}}$$

The Effect of the Range. Figure 12.11 show two variogram models that differ only in their ranges. $\gamma_2(h)$ has a range twice that of $\gamma_1(h)$:

$$\gamma_1(h) = 10(1 - e^{-.3|h|})$$
$$\gamma_2(h) = 10(1 - e^{-.15|h|}) = \gamma_1(\tfrac{1}{2}h)$$

The change of the range has a relatively minor effect on the ordinary kriging weights; none of the weights changes by more than 0.06. Even so, these relatively small adjustments in the weights do cause a noticeable change in the estimate. The ordinary kriging variance is lower since the effect of doubling the range in $\gamma_2(h)$ is to make the

(a)   (b)

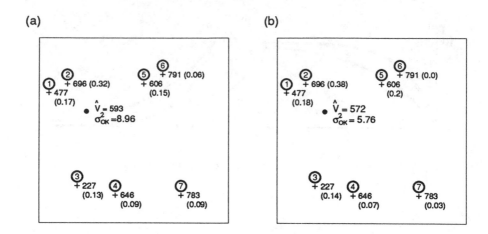

Figure 12.12   Ordinary kriging results using two different covariance functions that differ only in their range. (a) shows the kriging weights for a range of 10 while (b) shows the weights for a range of 20. The two covariance functions are given in Figure 12.11.

samples appear to be twice as close, in terms of statistical distance, as they originally were with $\gamma_1(h)$. If the range becomes very small, then all samples appear to be equally far away from the point being estimated and from each other, with the result being similar to that of a pure nugget effect model: the weights all become $\frac{1}{n}$ and the estimation procedure becomes a simple average of the available sample data.

The Effect of Anisotropy. In all of the examples of ordinary kriging we have looked at so far, we have used only the magnitude of the vector **h**, thus ignoring the influence of direction. All of our variogram models have been isotropic; a contour map of the variogram or covariance surface, such as the one shown in Figure 12.13, would show circular contour lines. In many data sets the data values are more continuous along certain directions than along others. The covariance surface contoured in Figure 12.14 rises more rapidly in the N45°E direction than in the N45°W direction. The directional covariance functions and variograms along these axes of maximum and minimum continuity are shown in Figure 12.15; the anisotropy ratio is 2:1. In Chapter 16 we will show how two variogram models for perpendicular directions can be combined into a single function that

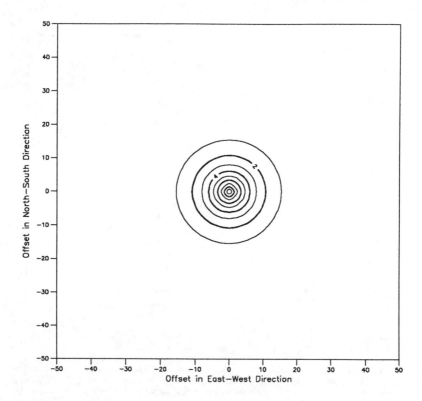

Figure 12.13 A contour map of an isotropic covariance surface. The contour map of the corresponding variogram surface appears identical except that the contours would show a hole rather than a peak.

describes the spatial continuity for all distances and directions. For the moment, however, let us concentrate on the effect of anisotropy and not worry about the precise mathematical description of the variogram model shown in Figure 12.14.

The results of using the isotropic variogram model shown in Figure 12.13 and the anisotropic variogram model shown in Figure 12.14 are shown in Figure 12.16. With the anisotropic model, more of the weight is given to samples 1 and 2, which lie in the direction of maximum continuity and considerably less is given to sample 5, which lies in the direction of minimum continuity.

If we rotate the axes of the anisotropy, as shown in Figure 12.17, so that the N45°E direction is now the major direction of continuity,

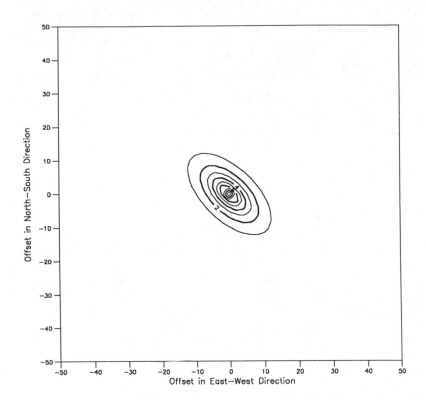

Figure 12.14 A contour map of an anisotropic covariance surface whose major axis of continuity is oriented along N45°W. The contour map of the corresponding variogram surface appears identical except that the contours would show a hole rather than a peak.

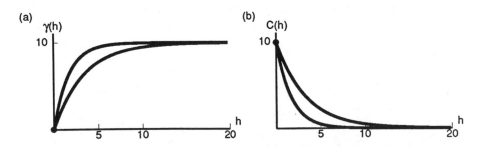

Figure 12.15 Directional variograms of the surface shown in Figure 12.14 along the axes of maximum and minimum continuity.

(a)   (b)

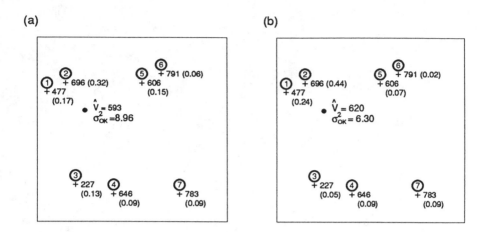

Figure 12.16 Ordinary kriging results using two different covariance functions that differ only in their anisotropy. (a) shows the kriging weights for the isotropic model shown in Figure 12.13 and (b) for the anisotropic model shown in Figure 12.14.

the ordinary kriging weights will reflect this new choice. Figure 12.18 shows the result of using the variogram model shown in Figure 12.17. The choice of a pattern of spatial continuity, which identifies N45°E as the direction of maximum continuity, causes sample 5 to receive the largest weight despite the fact that sample 2 is much closer in terms of geometric distance to the point being estimated.

Finally, the anisotropy ratio plays as important a role as the direction of anisotropy. Figure 12.19 shows a variogram surface that, like the one shown in Figure 12.14, has N45°W as the direction of maximum continuity. The variogram model shown in Figure 12.19, however, has a much higher anisotropy ratio of 10:1. The ordinary kriging weights, which are calculated using this model, are shown in Figure 12.20. Nearly all of the weight is now given to three samples: samples 1 and 2 to the northwest of the point we are estimating and sample 7 far to the southeast. In terms of geometric distance, sample 7 is the farthest from the point we are estimating, but since it lies in the direction of maximum continuity and since we have chosen a high anisotropy ratio, it becomes one of the most influential samples.

The possibility of choosing strongly anisotropic patterns of spatial continuity for our random function model gives us a powerful ability

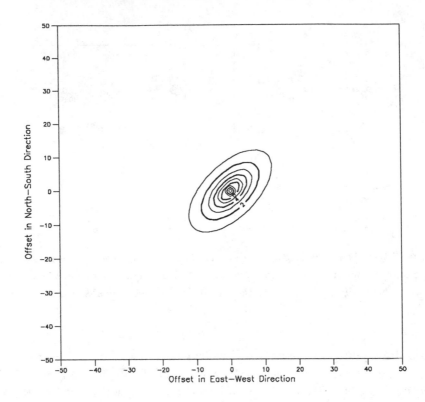

Figure 12.17 A contour map of an anisotropic covariance surface. The direction of maximum continuity is N45°E. The contour map of the corresponding variogram surface appears identical except that the contours would show a hole rather than a peak.

to customize the estimation procedure. Qualitative information such as a geologic interpretation for an ore deposit, a knowledge of the prevailing wind direction in a study of airborne pollution, or tracer tests that reveal preferred flow directions in a reservoir, can be incorporated through the anisotropy of the variogram model. In many data sets, the direction of maximum continuity is not the same throughout the area of interest; there may be considerable local fluctuations in the direction and the degree of the anisotropy. In such situations, the sample variograms may appear isotropic only because we are unable to sort out the undulating character of the anisotropy. If qualitative information offers a way to identify the direction and the degree of the anisotropy,

(a)   (b)

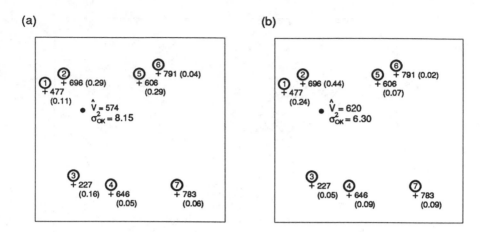

Figure 12.18 Ordinary kriging results using two different covariance functions that differ only in their anisotropy. (a) shows the kriging weights for the anisotropic model shown in Figure 12.17 and (b) for the anisotropic model shown in Figure 12.14.

then the estimation procedure will benefit greatly from a decision to base the choice of the spatial continuity model on qualitative evidence rather than on the quantitative evidence of the sample variogram.

Comparison of Ordinary Kriging to Other Estimation Methods

To compare ordinary kriging to the other estimation methods we looked at in the last chapter, we have repeated the exercise of estimating the V value at points located on a regular square grid. As before, the origin of this grid is 5E,5N and the spacing between points is 10 m in both the north-south and the east-west directions.

For the spatial continuity of our random function model, we have chosen the traditional method of fitting a function to our sample variogram. The directional sample variograms calculated in Chapter 7 as part of the analysis of the spatial continuity in the sample data set are shown again in Figure 12.22. These sample variograms show an anisotropy that is preserved in the complete variogram model shown in Figure 12.21. The variogram models for the directions of maximum and minimum continuity shown in Figure 12.22 are: Direction of

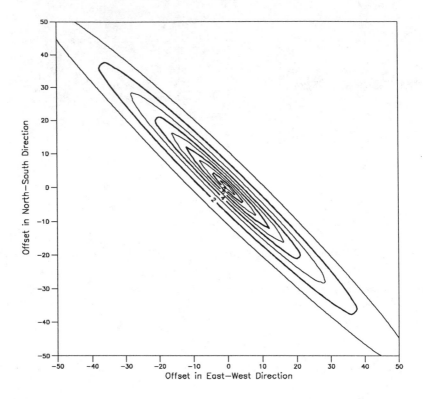

Figure 12.19 A contour map of a severe anisotropic covariance surface. The direction of maximum continuity is N45°W and the anisotropy ratio is 10:1. The contour map of the corresponding variogram surface appears identical except that the contours would show a hole rather than a peak.

maximum continuity (N14°W):

$$\gamma_{max}(h) = \begin{cases} 0 & \text{if } h = 0 \\ 22,000 + 40,000\text{Sph}_{30}(h) + 45,000\text{Sph}_{150}(h) & \text{if } h > 0 \end{cases}$$

$$(12.26)$$

Direction of minimum continuity (N76°E):

$$\gamma_{min}(h) = \begin{cases} 0 & \text{if } h = 0 \\ 22,000 + 40,000\text{Sph}_{25}(h) + 45,000\text{Sph}_{50}(h) & \text{if } h > 0 \end{cases}$$

$$(12.27)$$

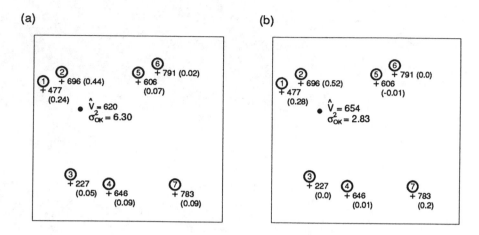

Figure 12.20 Ordinary kriging results using two different covariance functions that differ only in their anisotropy ratio. (a) shows the kriging weights for the anisotropic model shown in Figure 12.14 and (b) for the anisotropic model shown in Figure 12.19.

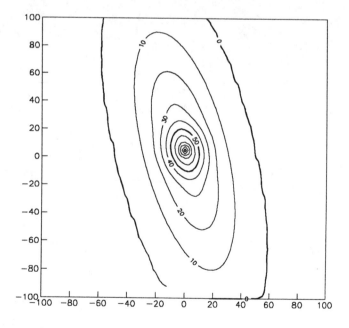

Figure 12.21 A contour map of the anisotropic model fitted to the sample variogram of V.

(a) $\gamma_V(h'_x) = 22{,}000 + 40{,}000 \, Sph_{25}(h'_x) + 45{,}000 \, Sph_{50}(h'_x)$

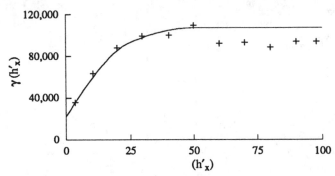

(b) $\gamma_V(h'_y) = 22{,}000 + 40{,}000 \, Sph_{30}(h'_y) + 45{,}000 \, Sph_{150}(h'_y)$

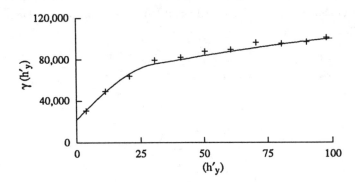

(c)

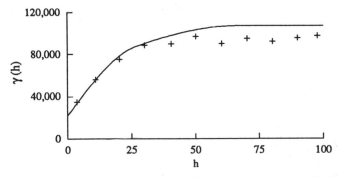

Figure 12.22 The directional sample variograms of V with their fitted model.

Table 12.3 Univariate statistics of ordinary kriged estimates compared to the true values.

	True	OK
n	780	780
m	283	283
σ	251	202
min	0	0
Q_1	70	127
M	219	251
Q_3	446	392
max	1323	1227

Table 12.4 Summary statistics for the error distributions from OK and each of the previous four point estimation methods.

	Polygonal	Triangulation	Local Sample Mean	Inverse Distance Squared	OK
n	780	672	780	780	780
m	1.1	−16.8	51.1	26.8	-0.2
σ	175.2	153.7	187.5	156.0	144.2
min	−651	−546	−618	−630	−472
Q_1	−97	−58	−66	−65	−86.8
M	0	3.9	48.9	24.4	9.1
Q_3	95	105	175	121	79.9
max	595	524	537	473	657
MAE	128	111	154	121	108
MSE	30,642	23,885	37,741	25,000	20,769
$\rho_{\dot{b}v}$	0.69	0.80	0.67	0.78	0.82

$\text{Sph}_a(h)$ is a positive definite function that is commonly used in practice for fitting sample variograms and is defined as follows:

$$\text{Sph}_a(h) = \begin{cases} 0 & \text{if } h = 0 \\ 1.5\frac{h}{a} - 0.5(\frac{h}{a})^3 & \text{if } 0 < h < a \\ 1 & \text{if } h \geq a \end{cases} \qquad (12.28)$$

The details of how these two directional variogram models are combined to make the complete model shown in Figure 12.21 will be deferred to Chapter 16.

Like the case studies presented in the last chapter, the ordinary

Table 12.5 Comparison of OK to the previous four estimation methods for the 50 least clustered sample data configurations that contained at least 10 samples.

		Polygonal	Triangulation	Local Sample Mean	Inverse Distance Squared	OK
	n	50	50	50	50	50
	m	-22.6	-49.5	-56.3	-30.1	-31.7
Error	σ	187	142	212	152	126
Distribution	IQR	227	201	327	221	203
	MAE	151	127	178	123	109
	MSE	34,750	22,140	47,194	23,562	23,562
Correlation	ρ	0.845	0.907	0.834	0.921	0.930

kriging exercise used all samples that fell within 25 m of the point being estimated.

Table 12.3 presents a comparison of the univariate distributions of the estimates and the true values. The results of the estimation studies in the previous chapter have been repeated in Tables 12.4, 12.5, and 12.6 so that we can compare the ordinary kriging results with the results of those earlier methods. The mean of the ordinary kriged estimates is the same as the true mean; this nearly exact match between the estimated and true means is quite fortuitous and one should not expect such a close agreement in all situations. The ordinary kriging estimates are less variable than the true values; their standard deviation and interquartile range are both lower than those of the true values, and the maximum estimated value is lower than the maximum true value. The degree of smoothing of the ordinary kriging estimates is more severe than that of the polygonal and triangulation estimates, and similar in magnitude to the degree of smoothing of the inverse distance squared estimates. The q-q plot of the estimated and true distributions shown in Figure 12.23 further reveals the effects of the smoothing. There are fewer extremely low estimates than there are extremely low true value; very high values are also more numerous in the true values than in the estimates.

Table 12.4, which summarizes the error distributions for the various methods, shows that the standard deviation of the errors is lower for the ordinary kriging estimates than for any of the other techniques.

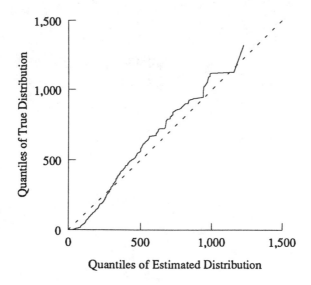

Figure 12.23 A q-q plot of the distribution of OK estimates and the distribution of their true values.

Table 12.6 Comparison of OK to the previous four estimation methods for the 50 most clustered sample data configurations that contained at least 10 samples.

		Polygonal	Triangulation	Local Sample Mean	Inverse Distance Squared	OK
	n	50	43	50	50	50
	m	17.3	−60.5	103.7	58.0	-29.6
Error	σ	182	138	221	161	125
Distribution	IQR	213	202	330	248	177
	MAE	138	106	198	142	102
	MSE	32,586	22,416	58,415	28,779	16,099
Correlation	ρ	0.674	0.794	0.260	0.774	0.853

Though the specific aim of ordinary kriging was only to minimize the error variance (or, equivalently, the error standard deviation), we can see that the ordinary kriging estimates are also very good according to many other criteria. They have the lowest mean absolute error and also the lowest mean squared error. The distribution of the errors

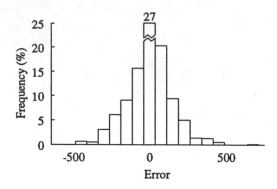

Figure 12.24 A histogram of the 780 OK estimation errors.

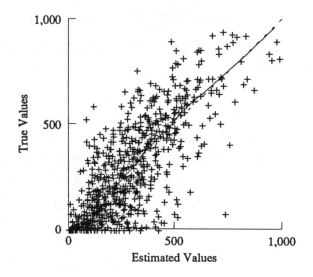

Figure 12.25 A scatterplot of the 780 OK estimates and their true values. The conditional expectation line is also included in the figure.

shown by the histogram in Figure 12.24 is fairly symmetric and does not have long tails. The correlation coefficient between the true values and the estimates is higher for ordinary kriging than for any of the other method we have tried.

The scatterplot of the 780 estimated and true values shown in Figure 12.25 along with the conditional expectation curve shows that there is very little conditional bias for the entire range of estimated values.

In the previous chapter we saw that all of the estimation methods suffered from the effects of clustering. We saw that our estimates were generally poorer for the 50 most clustered data configurations than for the 50 least clustered configurations. In Tables 12.5 and 12.6 this comparison is extended to include the ordinary kriging exercise. Though the ordinary kriging estimates are definitely adversely affected by clustering, they do not deteriorate as much as the estimates for other techniques. For example, the triangulation estimates for the least clustered configurations correlate almost as well with the true values as do the ordinary kriging estimates; both have a correlation coefficient of about 0.9. For the most clustered configurations, however, the correlation coefficient for the triangulation estimates drops to 0.78 while the correlation coefficient for the ordinary kriging estimates drops slightly to 0.85.

By trying to account for the possibility of redundancy in the sample data set through the covariances in the C matrix, ordinary kriging handles the adverse effects of clustering much better than other methods. Though it is certainly a more time-consuming procedure, it does generally produce better estimates.

The success of ordinary kriging is due to its use of a customized statistical distance rather than a geometric distance and to its attempt to decluster the available sample data. Its use of a spatial continuity model that describes the statistical distance between points gives it considerable flexibility and an important ability to customize the estimation procedure to qualitative information.

Notes

[1] A good presentation of Lagrange multipliers and constrained maximization-minimization problems with theorems and proofs is given in:
Edwards, C. and Penney, D. , *Calculus and Analytical Geometry.* Englewood Cliffs, N.J.: Prentice-Hall, 1982.

[2] The covariance is used by many algorithms in solving the ordinary kriging matrices for the sake of convenience. By using the covariance, the largest elements of the covariance matrix will be located on the diagonal. Thus for a system solver based on Gaussian elim-

ination, for example, there is no need for a pivot search and the exchange of rows.

[3] The name *sill* comes from the variogram, which typically reaches a plateau; although it makes less sense when discussing a covariance function, we will still refer to $C_0 + C_1$ as the sill.

[4] Each of the following tests is a necessary and sufficient condition for a real symmetric matrix $\mathbf{C}$ to be positive definite.

$\mathbf{w}^T \mathbf{C} \mathbf{w} > 0$ for all nonzero vectors $\mathbf{w}$.

All the eigenvalues of $\mathbf{C}$ are greater than 0.

All the submatrices of $\mathbf{C}$ have positive determinants.

All the pivots (without row exchanges) are greater than 0.

[5] The fact that the variogram can be rescaled by any constant without changing the estimate enables one to use the relative variogram without fear of altering the estimate. In the case where local relative variograms differ one from another by only a rescaling factor, only the kriging variance will be affected. Each one of the local relative variograms will provide an identical kriging estimate.

Further Reading

Brooker, P. , "Kriging," *Engineering and Mining Journal,* vol. 180, no. 9, pp. 148–153, 1979.

Buxton, B. , *Coal Reserve Assessment: A Geostatistical Case Study.* Master's thesis, Stanford University, 1982.

David, M. , *Geostatistical Ore Reserve Estimation.* Amsterdam: Elsevier, 1977.

Journel, A. G. and Huijbregts, C. J. , *Mining Geostatistics.* London: Academic Press, 1978.

Royle, A. , "Why geostatistics?," *Engineering and Mining Journal,* vol. 180, no. 5, pp. 92–102, 1979.

13

BLOCK KRIGING

In the last two chapters, we have concentrated on the goal of point estimation. Often, however, we require a block estimate or, more precisely, an estimate of the average value of a variable within a prescribed local area [1].

One method for obtaining such an estimate is to discretize the local area into many points and then average the individual point estimates all together to get the average over the area. Though conceptually simple, this procedure may be computationally expensive. For example, in mining applications hundreds of thousands of block estimates may be required. If each block is discretized into 100 points, and each point estimate is made by ordinary kriging, there will be several million kriging systems to solve. In this chapter we will see how the number of computations can be significantly reduced by constructing and solving only one kriging system for each block estimate.

We begin this chapter with the development of the block kriging system and follow with an example demonstrating the equivalence between block kriging and the averaging of kriged point estimates within the block. We then use the sample V data to explore how the discretization of the block affects the estimates. The chapter concludes with a case study that compares block estimates calculated by kriging and by inverse distance squared to the true values.

The Block Kriging System

The block kriging system is similar to the point kriging system given in the previous chapter:

$$
\begin{array}{ccc}
\mathbf{C} & \cdot & \mathbf{w} & = & \mathbf{D}
\end{array}
$$

$$
\underbrace{\begin{bmatrix}
\tilde{C}_{11} & \cdots & \tilde{C}_{1n} & 1 \\
\vdots & \ddots & \vdots & \vdots \\
\tilde{C}_{n1} & \cdots & \tilde{C}_{nn} & 1 \\
1 & \cdots & 1 & 0
\end{bmatrix}}_{(n+1)\times(n+1)}
\cdot
\underbrace{\begin{bmatrix}
w_1 \\
\vdots \\
w_n \\
\mu
\end{bmatrix}}_{(n+1)\times 1}
=
\underbrace{\begin{bmatrix}
\tilde{C}_{10} \\
\vdots \\
\tilde{C}_{n0} \\
1
\end{bmatrix}}_{(n+1)\times 1}
\tag{13.1}
$$

The matrix $\mathbf{C}$ consists of the covariance values $\tilde{C}_{ij}$ between the random variables V_i and V_j at the sample locations. The vector $\mathbf{D}$ consists of the covariance values $\tilde{C}_{i0}$ between the random variables V_i at the sample locations and the random variable V_0 at the location where we need an estimate. The vector $\mathbf{w}$ consists of the kriging weights $w_1, \ldots, w_n$ and the Lagrange parameter μ. It should be remembered that the random variables V_i, V_j, and V_0 are models of the phenomenon under study and the tilde above the C reminds us that these are parameters of a random function model.

However, suppose we wish to estimate the mean value of some phenomenon over a local area, rather than at a point location. Within the framework of the random function model that we used earlier, the mean value of a random function over a local area is simply the average (a linear combination) of all the point random variables contained within the local area. Recall from Chapter 9 that a linear combination of random variables is also a random variable, thus the mean value over a local area can be described as follows:

$$
V_A = \frac{1}{|A|} \sum_{j|j \in A} V_j \tag{13.2}
$$

where V_A is a random variable corresponding to the mean value over an area A and V_j are random variables corresponding to point values within A.

If we examine the ordinary point kriging system given in Equation 13.1 with a view towards modifying it for block estimation, we will soon see that the location of the point or block that we are estimating has absolutely nothing to do with the construction of the has absolutely nothing to do with the construction of the covariance matrix **C**. This matrix is independent of the location at which the estimate is required and so we can correctly conclude that the matrix **C** does not require any modifications for block kriging. However, the covariance vector **D** consists of covariance values between the random variables at the sample locations and the random variable at the location that we are trying to estimate. For point estimation, these covariances are point-to-point covariances. By analogy, for block estimation, the covariance values required for the covariance vector **D** are the point-to-block covariances. In fact, by making this single alteration, we can convert the ordinary point kriging system to a ordinary block kriging system.

The point-to-block covariances that are required for block kriging can be developed as follows:

$$
\begin{aligned}
\tilde{C}_{iA} &= Cov\{V_A V_i\} \\
&= E\{V_A V_i\} - E\{V_A\}E\{V_i\} \\
&= E\{\frac{1}{|A|} \sum_{j|j \in A} V_j V_i\} - E\{\frac{1}{|A|} \sum_{j|j \in A} V_j\}E\{V_i\} \\
&= \frac{1}{|A|} \sum_{j|j \in A} E\{V_j V_i\} - \frac{1}{|A|} \sum_{j|j \in A} E\{V_j\}E\{V_i\} \\
&= \frac{1}{|A|} \sum_{j|j \in A} [E\{V_j V_i\} - E\{V_j\}E\{V_i\}] \\
&= \frac{1}{|A|} \sum_{j|j \in A} Cov\{V_j V_i\}
\end{aligned}
$$

The covariance between the random variable at the ith sample location and the random variable V_A representing the average value of the phenomenon over the area A is the same as the average of the point-to-point covariances between V_i and the random variables at all the points within A. The block kriging system can therefore be written as

$$
\mathbf{C} \qquad \cdot \qquad \mathbf{w} \qquad = \qquad \mathbf{D}
$$

$$
\underbrace{\begin{bmatrix} \tilde{C}_{11} & \cdots & \tilde{C}_{1n} & 1 \\ \vdots & \ddots & \vdots & \vdots \\ \tilde{C}_{n1} & \cdots & \tilde{C}_{nn} & 1 \\ 1 & \cdots & 1 & 0 \end{bmatrix}}_{(n+1)\times(n+1)} \cdot \underbrace{\begin{bmatrix} w_1 \\ \vdots \\ w_n \\ \mu \end{bmatrix}}_{(n+1)\times 1} = \underbrace{\begin{bmatrix} \bar{\tilde{C}}_{1A} \\ \vdots \\ \bar{\tilde{C}}_{nA} \\ 1 \end{bmatrix}}_{(n+1)\times 1} \qquad (13.3)
$$

The bar above the covariances on the right-hand side reminds us that the covariance is no longer a point-to-point covariance, but the average covariance between a particular sample location and all of the points within A:

$$
\bar{\tilde{C}}_{iA} = \frac{1}{|A|} \sum_{j|j\in A} \tilde{C}_{ij} \qquad (13.4)
$$

Later in this chapter we will determine the number of discretizing points that are needed within A to give us an adequate approximation of $\bar{\tilde{C}}_{iA}$.

The block kriging variance is given by:

$$
\tilde{\sigma}_{OK}^2 = \bar{\tilde{C}}_{AA} - \left(\sum_{i=1}^{n} w_i \bar{\tilde{C}}_{iA} + \mu \right) \qquad (13.5)
$$

The value $\bar{\tilde{C}}_{AA}$ in this equation is the average covariance between pairs of locations within A:

$$
\bar{\tilde{C}}_{AA} = \frac{1}{|A|^2} \sum_{i|i\in A} \sum_{j|j\in A} \tilde{C}_{ij} \qquad (13.6)
$$

In practice, this average block-to-block covariance is approximated by discretizing the area A into several points. It is important to use the same discretization for the calculations of the point-to-block covariances in $\mathbf{D}$ and for the calculation of the block-to-block covariance in Equation 13.5 [2].

The advantage of using the block kriging system given in Equation 13.3 is that it produces an estimate of the block average with the solution of only one kriging system. The disadvantage is that the calculation of the average covariances involves slightly more computation than the calculation of the point-to-point covariances in the point kriging system. However, the computational savings in requiring only one set of simultaneous linear equations far outweighs the additional cost of calculating average point-to-block covariances.

The convenience of estimating block values directly is a feature of the ordinary kriging system that is not shared by other estimation methods. Though some methods can be adapted in a similar manner, the results are not consistent. For example, inverse distance methods can be adapted so that the weight is proportional to the average distance between a nearby sample and the block. Unfortunately, the estimate calculated by this procedure is not the same as the one calculated by averaging individual point estimates within the block.

In the next section, we will give an example that demonstrates that the block kriging system given in Equation 13.3 produces an estimate identical to that obtained by averaging the point estimates produce by Equation 13.1.

Block Estimates Versus the Averaging of Point Estimates

An example of block kriging is shown in Figure 13.1a. Five sampled locations are marked by plus signs; the sample value is given immediately to the right of each sign with the corresponding block kriging weight enclosed in parentheses. The variogram model used to build the kriging matrices was the one used earlier in the case study in Chapter 12 and developed in detail in Chapter 16. The block whose average value we wish to estimate is shown as a shaded square. For the purposes of calculating the various average covariances, this block has been discretized by four points shown as black dots in Figure 13.1a. The block estimate is $\hat{V}_A = 337$ ppm. Figures 13.1b through 13.1e show the kriging weights and the point estimates for each one of the four point locations within the shaded square. As shown in Table 13.1, the average of the four point estimates is the same as the direct block estimate and the average of the point kriging weights for a particular sample is the same as the block kriging weight for that sample.

Varying the Grid of Point Locations Within a Block

When using the block kriging approach, one has to decide how to discretize the local area or block being estimated. The grid of discretizing points should always be regular; the spacing between points, however,

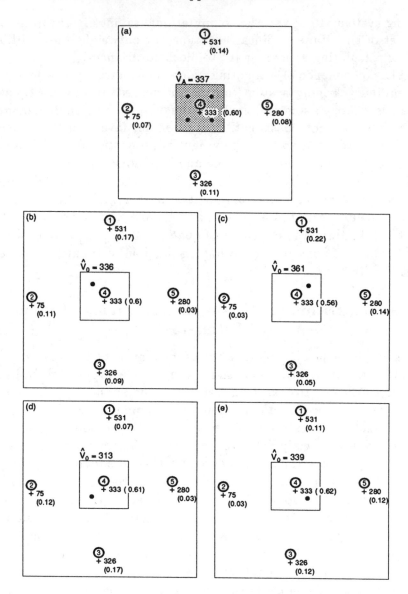

Figure 13.1 In (a) the shaded block is estimated directly using block kriging, with the block being approximated by the four points shown as dots. The nearby sample locations are marked with a plus sign. The value posted immediately to the right of the plus sign is the sample value and the value enclosed in parentheses is the corresponding kriging weight. Figures (b) to (e) show the results for the point kriging of each one of the four points within the shaded square. The average of the point estimates in (b) to (e) is identical to the block estimate $\hat{V}_A{=}337$ in (a).

Table 13.1 The point estimates and weights shown in Figure 13.1(b) - (e) are tabulated and averaged, demonstrating their equivalence to the direct block estimate shown in Figure 13.1(a).

Figure	Estimate	Kriging weights for samples				
		1	2	3	4	5
13.1(b)	336	0.17	0.11	0.09	0.60	0.03
13.1(c)	361	0.22	0.03	0.05	0.56	0.14
13.1(d)	313	0.07	0.12	0.17	0.61	0.03
13.1(e)	339	0.11	0.03	0.12	0.62	0.12
Average	337	0.14	0.07	0.11	0.60	0.08
13.1(a)	337	0.14	0.07	0.11	0.60	0.08

may be larger in one direction than the other if the spatial continuity is anisotropic. An example of such an anisotropic grid is given in Figure 13.2. The shaded block is approximated by six points located on a regular 2 x 3 grid. The closer spacing of the points in the north-south direction reflects a belief that there is less continuity in this direction than in the east-west direction. Despite the differences in the east-west and north-south spacing, the regularity of the grid ensures that each discretizing point accounts for the same area, as shown by the dashed lines.

If one chooses to use fewer discretizing points, less computer memory is required and the computations are faster. This computational efficiency must be weighed against the desire for accuracy, which calls for as many points as possible.

Table 13.2 provides several examples of the effect of the number of discretizing points on the final estimate. The table shows estimates of the average V value within 10 x 10 m^2 blocks using ordinary block kriging to calculate weights for the nearby samples. In these examples, the search strategy included all samples within a 25 m radius of the center of the block; the variogram model used to calculate the various covariances is the same one used in earlier studies and given in Equations 12.25–12.27. The only parameter changed from one kriging to the next is the number of points used to discretize a 10 x 10 m^2 block.

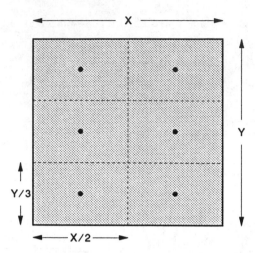

Figure 13.2 An example showing the design of a regular 2 x 3 grid of point locations within a block. The points are located within the square so that each discretizing point accounts for the same area, as shown by the dashed lines.

The entries in Table 13.2 show significant differences in the estimates using grids containing fewer than 16 discretizing points; with more than 16 points, however, the estimates are all very similar. For a two-dimensional block, sufficient accuracy can usually be obtained with a 4 x 4 grid containing 16 points. For a three dimensional block, more points are usually required; a 4 x 4 x 4 grid containing 64 points is usually sufficient.

A Case Study

Next we present a case study that compares ordinary block kriging estimates of the average value of V within 10 x 10 m^2 blocks to block estimates using the inverse distance squared method.

For this study, we have used a very fine discretization, with each 10 x 10 m^2 block being discretized by 100 points. Ordinary block kriging estimates were calculated using all samples within 25 m. The variogram model used to calculate the various covariances required by Equation 13.3 was the same one that we used for the point kriging case study in Chapter 12.

The inverse distance squared estimates were calculated using the

Table 13.2 Examples of estimates of the average value of V within 10 x 10 m^2 blocks using ordinary block kriging and various discretizing grids within the block.

Block center		Grid size within the block				
E	N	1x1	2x2	4x4	6x6	10x10
80	80	584.67	576.41	574.30	573.98	573.81
100	80	408.53	418.29	419.19	419.38	419.47
80	90	538.36	519.89	520.58	520.53	520.47
100	90	460.13	479.73	480.35	480.52	480.61
80	100	497.66	547.87	549.40	550.13	550.51
100	100	530.37	513.32	513.56	513.47	513.42
80	110	781.17	737.04	732.29	731.06	730.41
100	110	591.13	580.73	578.75	578.74	578.72

same search strategy as that used for kriging. For each block, 100 inverse distance point estimations were averaged together to obtain the block estimate.

Summary statistics for the estimates are given in Table 13.3. The average of the kriged block estimates is closer to the true mean than the inverse distance squared estimates. The correlation of the kriged estimates with the true values is also larger. In this particular case, the kriged block estimates have a standard deviation closer to the true one and are therefore not as smoothed as the inverse distance estimates.

The summary statistics of the estimation error are given in Table 13.4. The errors from ordinary block kriging have a mean closer to 0 than those from inverse distance squared. Their spread, as measured by the standard deviation or the interquartile range, is also lower. The MAE and MSE, which provide measures of the combination of bias and spread, both favor the ordinary block kriging estimates.

Figures 13.3 and 13.4 provide a comparison of the two sets of errors using grayscale maps. A plus symbol denotes a positive estimation error while a minus symbol denotes a negative estimation error. The relative magnitude of the error corresponds to the degree of shading indicated by the grey scale at the top of the figure.

The most striking feature of these maps is the consistency of the overestimations in the area of Walker Lake itself and the corresponding

Table 13.3 Summary statistics for the ordinary kriged block estimates, the inverse distance squared block estimates, and the true block values of V.

	True Values	Ordinary Block Kriging	Inverse Distance Squared
n	780	780	780
m	278	284	319
σ	216	194	186
CV	0.77	0.68	0.58
min	0	5	7
Q_1	103	136	178
M	239	258	310
Q_3	404	389	431
max	1,247	1,182	1,112
$\rho_{\hat{V}V}$		0.90	0.87

Table 13.4 Summary statistics for the error distributions from the ordinary block kriging and the inverse distance squared method.

	Ordinary Block Kriging	Inverse Distance Squared
n	780	780
m	6.2	40.1
σ	92.9	105.4
min	-405.6	-336.5
Q_1	-45.5	-21.4
M	17.6	42.6
Q_3	65.3	116.8
max	286.8	392.1
MAE	71.9	90.2
MSE	8,674	12,763

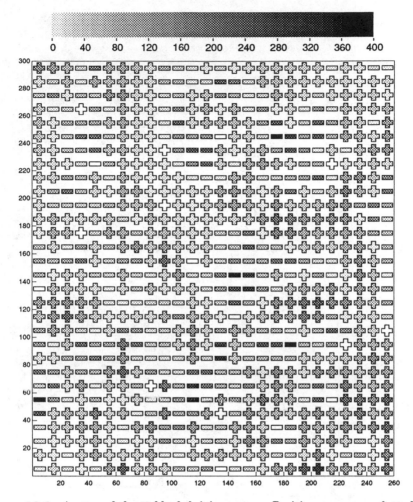

Figure 13.3 A map of the *V* block kriging errors. Positive errors are plotted as plus symbols while negative errors are plotted as minus symbols. The magnitude of the error is given by the grey scale at the top of the figure.

underestimations in the Wassuk range area. This pattern of the over- and underestimations illustrates the smoothing inherent in both the inverse distance squared method and in kriging; high values tend to be underestimated while low values tend to be overestimated.

Though the pattern of over and underestimation is the same in both maps, it is clear that the inverse distance squared method produces

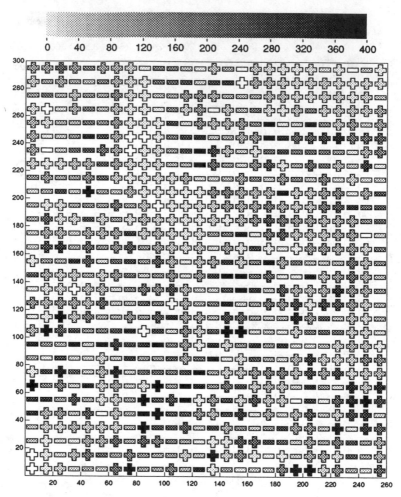

Figure 13.4 A plot of the inverse distance squared block estimation errors. A comparison of these errors to the block kriging errors in Figure 13.3 reveals that they are, in general, larger.

more large errors. In particular, there are several quite large overestimations (dark plus signs) in the vicinity of (80E,180N). In these areas, the relatively sparse sampling of the low-valued areas meets the much denser sampling in the Wassuk range. The inverse distance squared method did not correctly handle the clustered samples, giving too much weight to the additional samples in the high-valued areas.

Table 13.5 Summary statistics of the distribution of point estimation errors and block estimation errors for V.

	Block Kriging	Point Kriging
n	780	780
m	6.2	-0.2
σ	92.9	144.2
min	-405.6	-472.0
Q_1	-45.5	-86.8
M	17.6	9.1
Q_3	65.3	79.9
max	286.8	657.0
MAE	71.9	108.
MSE	8,674	20,769
$\rho_{\hat{V}V}$	0.90	0.82

The map of the ordinary block kriging errors shows that it has fewer large overestimations but still has several quite large underestimations. This is due to a combination of factors. The smoothing effect that we discussed earlier causes the distribution of block estimates to have a lower spread and also to be generally less skewed than the corresponding distribution of true block values. This, combined with the positive skewness of the distribution of true block values, creates more chances for a severe underestimation than for a severe overestimation. In the case of extreme smoothing, where all estimates are close to the global mean, there would be fewer underestimates but their magnitudes would generally be much greater than those for the more numerous overestimates.

The results shown in Table 13.4 also allow us to make an important remark about the accuracy of block estimation versus point estimation. In Chapter 12 we presented point kriging for the same variable V and gave the summary statistics of the estimation error in Table 12.4. These statistics are given again in Table 13.5, where they are compared to the summary statistics of the error distribution for block estimation. Though the point estimates have a better mean error, this is not a

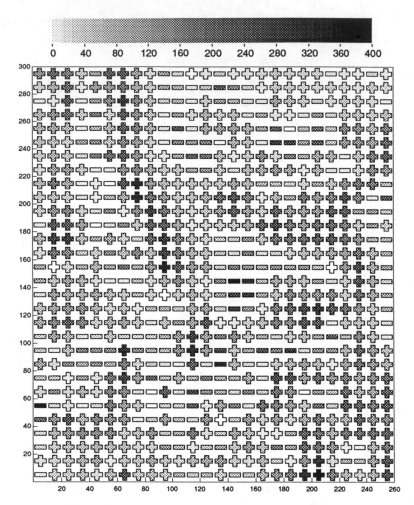

Figure 13.5 A map of the V point kriging errors. The darker shading of the symbols in this figure shows that the magnitude of the point estimation errors are larger than the block estimation errors shown in Figure 13.3.

general rule. By any other statistical criterion, the block estimates are better. They have a lower spread and correlate better with the true values.

Figure 13.5 is a map of the errors from the point kriging case study. The overall pattern of estimation errors is similar to the pattern of block kriging errors; the point estimation errors, however, tend to be larger. The symbols in Figure 13.5 tend to be darker than those in Figure 13.3, which lends further evidence to the fact that block estimation is generally more accurate than point estimation.

Notes

[1] For the sake of convenience, we use the phrase "block estimation" to describe the estimation of the mean value of a spatial variable within a prescribed local area; in three dimensions "local area" should be interpreted as a local volume.

[2] If one uses different discretizations for the two calculations, there is a risk of getting negative error variances from Equation 13.5.

14

SEARCH STRATEGY

We have not yet confronted the issue of what counts as a "nearby" sample. In our earlier examples of the various point estimation methods in Chapter 11, we simply accepted the seven samples that appeared on our diagram showing the local sample data configuration (Figure 11.1). Later, when we compared the estimates for the entire Walker Lake area, we accepted all samples that fell within 25 m of the point being estimated. The choice of a search strategy that controls the samples that are included in the estimation procedure is an important consideration in any approach to local estimation.

In this chapter we will use the following four questions to guide our discussion of the search strategy:

- Are there enough nearby samples?

- Are there too many samples?

- Are there nearby samples that are redundant?

- Are the nearby samples relevant?

The first three questions are important for estimation techniques such as kriging and inverse distance methods that can handle any number of samples. For geometric techniques, such as the polygonal method or triangulation, the samples used in the weighted linear combination are uniquely defined by the configuration of the sample data set. For methods that can handle any number of samples, the most common search strategy is to use all samples that fall within a certain search neighborhood or window. After presenting the basic idea of a search neighborhood in the next section, we will examine each of the first three questions above and see how they affect the definition of the search neighborhood.

The most important of these questions, both theoretically and practically, is the last one, which raises the issue of whether or not the nearby samples belong in the same group or population as the point being estimated. This question is important for any approach to estimation, including geometric techniques such as polygons and triangulation. After discussing the importance of carefully considering the relevance of nearby samples, we will conclude the chapter with a discussion of how this last question relates to the use of stationary random function models.

Search Neighborhoods

For estimation methods that can handle any number of nearby samples, the most common approach to choosing the samples that contribute to the estimation is to define a search neighborhood within which all available samples will be used. The search neighborhood is usually an ellipse centered on the point being estimated. The orientation of this ellipse is dictated by the anisotropy in the pattern of spatial continuity. If sample values are much more continuous in one direction than in another, then the ellipse is oriented with its major axis parallel to the direction of maximum continuity. The anisotropy of the ellipse is usually determined from the anisotropy evident on some measure of the spatial continuity, typically the sample variogram. If there is no evident anisotropy the search ellipse becomes a circle and the question of orientation is no longer relevant.

In the case studies of the various estimation methods we looked at in the previous chapters, we adopted a very simple search strategy, using all samples within a circular search neighborhood with a radius of 25 m. Given the earlier evidence in Chapter 7 of a distinct anisotropy in the pattern of spatial continuity of the V values, we might prefer to use an anisotropic search neighborhood. The ellipse that appears in Figure 7.9 has a minor axis of about 18 m and a major axis of about 30 m, with its major axis in the N10°W direction. The effects of using this ellipse as a search window rather than the simple 25 m circle

Table 14.1 A comparison of the error distributions for the inverse distance squared and ordinary kriging methods with different isotropic and anisotropic search neighborhoods.

	ISOTROPIC		ANISOTROPIC	
	Inverse Distance Squared	Ordinary Kriging	Inverse Distance Squared	Ordinary Kriging
Error Distribution				
n	780	780	780	780
m	26.8	-0.2	22.1	0.4
σ	156.0	144.2	153.5	143.9
IQR	186	167	185	167
MAE	121	108	118	108
MSE	25,000	20,769	24,008	20,682
Correlation				
ρ	0.78	0.82	0.79	0.82

were checked by repeating the inverse distance and ordinary kriging estimation using this new search strategy. The results are summarized in Table 14.1.

The use of an anisotropic search window in this example produces a small but consistent improvement for both the inverse distance and the ordinary kriging techniques. Part of the reason that the improvement is not very pronounced here is that while the N10°W direction is the clear direction of maximum continuity for the large Wassuk Range anomaly, it may not be appropriate elsewhere in the data set. In other parts of the Walker Lake area, the directions of maximum and minimum continuity may be quite different from this single predominant direction in the Wassuk Range anomaly. For example, in the region of the Excelsior Mountains in the southern part of the Walker Lake region, the axes of maximum and minimum continuity may be reversed, with N70°E being the direction of greatest spatial continuity. Were we to take the time to customize the orientation and anisotropy ratio of the search neighborhood for each point being estimated, we would expect to see larger improvements than the rather slight ones seen in Table 14.1.

Are there enough nearby samples? Having chosen an orientation and an anisotropy ratio for our search ellipse, we still have to decide how big to make it. The simple answer is that it must be big enough to include some samples; this is determined by the geometry of the sample data set. If the data are on a pseudo regular grid, one can calculate how big the search ellipse must be in order to include at least the closest four samples. In practice one typically tries to have at least 12 samples. For irregularly gridded data, the search neighborhood should be slightly larger than the average spacing between the sample data, which can be crudely calculated with the following formula:

$$Average\ spacing\ between\ data \approx \sqrt{\frac{Total\ area\ covered\ by\ samples}{Number\ of\ samples}}$$

Are there too many nearby samples? The question of how big to make the search ellipse is only partially answered. We have a minimum size determined largely by the geometry of the sample data set, but we still have to decide how much bigger than this it should be. There are two factors that limit the size of the search window in practice. First, using more samples increases the amount of computation required. Second, as samples come from farther and farther away, the appropriateness of a stationary random function model becomes more doubtful [1].

The concern about computation is a major issue for ordinary kriging. The number of computations required to solve for the ordinary kriging weights is proportional to the cube of the number of samples retained; if we double the number of samples, we increase the number of calculations eightfold.

The number of calculations can be reduced by combining several of the farthest samples into a single composite sample. An example of this is shown in Figure 14.1. Rather than view the 50 samples shown in Figure 14.1a as 50 samples, we can view them as 16 samples, with the 12 closest samples being treated as individual samples and the remaining 38 being combined into four composite samples as shown in Figure 14.1b.

The ordinary kriging system is easily adapted to handle such composite samples. Earlier, when we looked at ordinary kriging of average

(a)   (b)

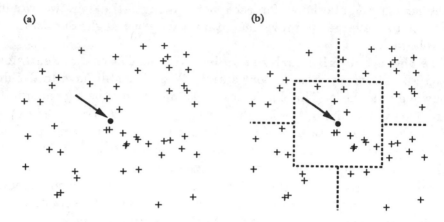

Figure 14.1 The combination of several samples into a composite sample. In (b), the farthest samples shown in (a) have been combined into four groups to reduce the number of calculations required for estimation. The weight assigned to a composite sample will be divided evenly among the individual samples that were grouped together.

block values, we combined several points into a single block and filled our right-hand side matrix, the **D** matrix, with average covariances between the points in the block and the sample data locations. In a similar way, the grouping of several samples into a composite sample can easily be accommodated in the ordinary kriging matrices simply by calculating the average covariance between any two samples. If the two samples are both individual point samples, then the average covariance between them is the point-to-point covariance we used when we first developed the ordinary kriging algorithm. If one of them is an individual point sample and the other is a composite of n samples, then the average covariance between them is the average of the n point-to-point covariances between the individual point sample and the n samples which form the composite sample. If both are composite samples, one containing n samples and the other containing m samples, then the average covariance between them is the average of the nm point-to-point covariances.

The weight assigned to a composite sample is equally distributed among the individual samples of which it is composed. In the example shown in Figure 14.1b, the weight assigned to sample 16, the composite sample in the upper right corner, would be equally distributed among the 13 samples in that group while the weight given to sample 15, the composite sample in the upper left corner, would be equally distributed among the 6 samples in that group.

The other concern about limiting the number of samples is related to the question of whether far away samples belong in the same group or population as the point being estimated. This is discussed in more detail in the final sections of this chapter. For the moment, we will simply remark that the use of a stationary model does not justify the use of extremely large search windows. There is an unfortunate misperception in practice that once the assumption of stationarity is made, one does not go any further out on a limb by using a very large search neighborhood. The assumption of stationarity pertains to the model and has nothing to do with the reality of the situation. From the perspective of the model, additional samples will always improve the estimation regardless of their distance from the point being estimated. This does not mean that the same is true from the perspective of reality.

The appropriateness of conceptualizing the sample data within the search window as an outcome of a stationary random function model often becomes more questionable as the search window gets larger. Even as the model becomes less appropriate, the estimation algorithm can still be applied, but with an ever-increasing departure between the theoretical statistical properties predicted by the model and their real counterparts. As the model becomes less realistic, the actual estimates may not share such properties as unbiasedness and minimum estimation variance with their model counterparts. By restricting the samples to a much smaller neighborhood, the stationary random function conceptualization becomes more plausible and the differences between the actual statistical properties and those of the model are less severe.

Another common misconception in practice is that the search radius should not extend beyond the range of the variogram. Experience has shown that if there are few samples within the range, then the addition of samples beyond the range often improves local estimation.

Are the nearby samples redundant? After having chosen the orientation, shape and size of our search neighborhood, the final issue we should consider is the possibility that some of the nearby samples are redundant. This is less of a concern for ordinary kriging, which accounts for possible redundancies through the left-hand side **C** matrix. For inverse distance techniques, however, a search strategy that accounts for the possibility of clustering will usually yield noticeable improvements over the more naive approach of taking all samples that fall within the search window. In addition to reducing the adverse effects of clustering, the removal of redundant samples also has the advantage of reducing the number of calculations involved. There are several procedures for reducing possible redundancies in the samples that fall within the search window, the most common of these, the quadrant search, is discussed in the next section.

Quadrant Search

Figure 14.2a shows a particular sample data configuration. The dashed lines in this figure divide the search neighborhood into quadrants. We notice that the two northern quadrants each contain more samples than either of the two southern ones. By limiting the number of nearby samples in each quadrant, we may reduce the adverse effects of clustering.

When we use a quadrant search in our choice of nearby samples we typically specify a maximum number of samples in any particular quadrant. If a particular quadrant has fewer samples than the maximum allowable, then we keep all the samples it contains; however, if a quadrant contains too many samples then we keep only the closest ones. Figure 14.2b shows the samples that would be kept from the original set of nearby samples shown in Figure 14.2a if we limited ourselves to a maximum of three samples per quadrant. Figure 14.2c shows the remaining samples if we limit ourselves to two samples per quadrant.

We will now use the exhaustive data set to examine the effect of a quadrant search. We have repeated the point estimation studies from Chapters 11 and 12 using the same 25 m search radius, but limiting ourselves to a maximum of four samples per quadrant. The effect of this quadrant search on the inverse distance squared and ordinary kriging estimates is presented in Table 14.2. The first two columns present the summary statistics of the error distributions from the studies given in the earlier chapters, where no quadrant search was used. The last two columns show how these statistics change when the search strategy limits the number of samples in each quadrant to a maximum of four.

The use of a quadrant search improves the inverse distance squared estimates, and has virtually no influence on the ordinary kriging es-

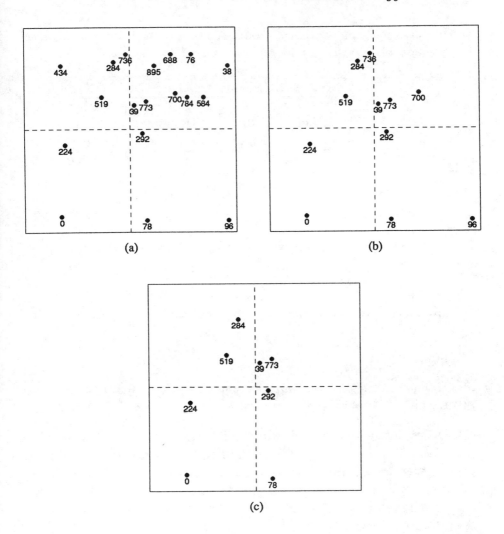

Figure 14.2 The selection of samples using the quadrant search. The samples in (a) are clustered in the two northern quadrants; in (b) the contents of each quadrant are limited to the nearest three samples, in (c), they are limited to the nearest two samples.

timates. The spread of the inverse distance squared errors, whether measured by the standard deviation or the interquartile range, is reduced slightly. The mean absolute error and the mean squared error also decrease while the correlation coefficient shows a small improve-

Table 14.2 A comparison of the error distributions for the inverse distance squared and ordinary kriging methods with and without a quadrant search.

| | NO MAXIMUM | | MAXIMUM OF 4 | |
	Inverse Distance Squared	Ordinary Kriging	Inverse Distance Squared	Ordinary Kriging
n	780	780	780	780
m	26.8	-0.2	20.5	0.4
σ	156.0	144.2	151.6	144.3
IQR	186	167	175	167
MAE	121	108	117	108
MSE	25,000	20,769	23,383	20,796
Correlation				
ρ	0.78	0.82	0.80	0.82

ment. The most significant improvement is in the global bias. For the inverse distance squared estimates the bias is reduced by about 25%.

As we noted earlier, the inverse distance squared method makes no attempt to account for the possibility of clustering; the ordinary kriging method, on the other hand, does account for this through the left-hand covariance matrix. A quadrant search accomplishes some declustering and the effect of this is more noticeable on the method that does not decluster by itself.

Once we have decided to use a quadrant search, the advantage of ordinary kriging over inverse distance squared becomes slight. Though we would likely judge the ordinary kriging estimates better, based on the statistics shown in Table 14.2, their overall advantage is reduced once the inverse distance estimates have the benefit of the quadrant search. With estimation methods that do not account for clustering, relatively simple procedures such as the quadrant search may adequately compensate for the adverse effects of clustering. In fact its generally a good idea to screen all data using a quadrant search for not only the inverse distance squared method but also before kriging.

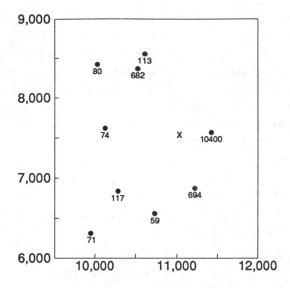

Figure 14.3 An example of a sample data configuration taken from a study of lead concentration in the soil surrounding a smelter. The nearby samples are shown by the dots, with the sample value (in ppm) given beneath the dot. The X marks the point at which we would like to estimate the unknown lead concentration.

Are the Nearby Samples Relevant?

Every estimate involves the following assumption that is easily over-looked or treated casually: that the sample values used in the weighted linear combination are somehow relevant and that they belong to the same group or population as the point being estimated. Unfortunately, even when this assumption is carefully considered it cannot be verified; it is necessarily a subjective decision that benefits most from qualitative information. To further complicate matters, the validity of this assumption may also depend on the goals of the study.

Figure 14.3 provides a specific example that demonstrates the importance of carefully considering the relevance of the nearby samples. The sample values shown in the figure are measurements of the lead concentration in the soil from an area surrounding a lead smelter. The goal of the study was to provide local estimates of the lead concentration due to the smelter. For the estimation of the lead concentration at the location marked X, Figure 14.3 shows the nearby samples that might be used.

With the naive search strategies we used earlier, all estimation methods, including polygons and triangulation, would make use of the $10,400$ ppm sample value. The logs that describe the sampling campaign make it clear that this particular sample was collected near a local dump; rather than being related to the smelter contamination, the $10,400$ ppm value likely reflects contamination due to leakage from discarded car batteries [2]. Despite the fact that it is the nearest, this $10,400$ ppm sample is not relevant for the estimation of contamination due to the smelter. However, if the goal of the study were to estimate the lead concentration in the soil from all possible sources, the use of this sample would be appropriate.

Similar problems arise in most practical geostatistical studies. For example, in structurally complex ore deposits good estimates may not be obtainable with any estimation method until a sound geologic interpretation dictates which samples are relevant for the estimation at each particular location. In studies of petroleum reservoirs or groundwater aquifers the existence of faulting may require that samples be divided into separate groups since the overall pattern of spatial continuity is unlikely to persist across a fault.

Deciding which samples are relevant for the estimation of a particular point may be more important than the choice of an estimation method. The initial definition of a boundary within which estimation will occur is an appropriate first step in any estimation exercise. In addition to this, it is also worthwhile to examine the configuration of nearby samples for each point being estimated and to decide which samples should be used. Though it is a common practice to define a single search strategy for an entire area under study, this is not always a good approach. What works in certain areas of a particular data set may not work in others. For example, in the center of a data set, it may be possible to choose relevant samples simply by taking all samples within a certain radius; near the fringes of the same data set, this same strategy may be unwise. "Black-box" computer programs make it easy for one to avoid a tedious case by case determination of relevant samples. Rather than yield to this temptation, one should consider using computer programs that permit more interaction. If such programs are not available, a cross-validation study (see Chapter 15) may be helpful in deciding if certain areas require special treatment.

Relevance of Nearby Samples and Stationary Models

When we introduced random function models in Chapter 9, we discussed the notion of stationarity. A random function model was said to be first order stationary if the mean of the probability distribution of each random variable is the same. Later, because we used this assumption to develop the unbiasedness condition, our conclusion that unbiasedness is guaranteed when the weights sum to one is limited to first order stationary models.

The decision to view a particular sample data configuration as an outcome of a stationary random function model is strongly linked to the decision that these samples can be grouped together. Both of these are decisions that cannot be checked quantitatively; they are neither right nor wrong and no proof of their validity is possible. However, they can be judged as appropriate or inappropriate. Such a judgment must take into account the goals of the study and will benefit considerably from qualitative information about the data set.

The cost of choosing to use an inappropriate model is that statistical properties of the actual estimates may be very different from their model counterparts. As we have already noted, the use of weighted linear combinations whose weights sum to one does not guarantee that the actual bias is 0. The actual bias will depend on several factors, including the appropriateness of conceptualizing each sample data configuration as an outcome of a stationary random function.

It should be emphasized that all of the estimation methods we looked at in Chapters 11 and 12 implicitly assume a first order stationary model through their use of the unbiasedness condition. It is naive, therefore, to dismiss ordinary kriging with the argument that it requires first order stationarity; the same argument criticism applies to *all* of the methods we have studied. If estimation is performed blindly, with no thought given to the relevance of nearby samples, the methods that make use of more samples may produce worse results than the methods that make use of few nearby samples. If one does not have the time or curiosity required for good estimation, then the use of polygonal method or triangulation may limit the damage done by a poor search strategy.

Notes

[1] Some practitioners make maps of estimates in typical areas using

various search strategies. Estimates are first made using a large number of samples. Then the search strategy is changed to reduce the number of samples and the corresponding estimates mapped. The search strategy is deemed appropriate just before the mapped estimates begin to show noticeable differences with less samples.

[2] A complete case study of the Dallas data set is presented in: Isaaks, E. H. , *Risk qualified mappings for hazardous waste sites: A case study in distribution free geostatistics*. Master's thesis, Stanford University, 1985.

15

CROSS VALIDATION

In our case studies comparing different point estimation methods we relied on our access to the true values in the exhaustive data set. Such comparisons between true and estimated values are useful in helping us to understand the different approaches to estimation. In many practical situations, we would like to check the results of different approaches and choose the one that works best. Unfortunately, we never have an exhaustive data set in practice and are unable to conduct the kind of comparisons we showed in earlier chapters.

In this chapter we will look at cross validation, a technique that allows us to compare estimated and true values using only the information available in our sample data set. Like the comparisons based on the exhaustive data set that we showed earlier, a cross validation study may help us to choose between different weighting procedures, between different search strategies, or between different variogram models. Regrettably, cross validation results are most commonly used simply to compare the distributions of the estimation errors or residuals from different estimation procedures. Such a comparison, especially if similar techniques are being compared, typically falls far short of clearly indicating which alternative is best. Cross validated residuals have important spatial information, and a careful study of the spatial distribution of cross validated residuals, with a specific focus on the final goals of the estimation exercise can provide insights into where an estimation procedure may run into trouble. Since such insights may lead to case-specific improvements in the estimation procedure, cross validation is a useful preliminary step before final estimates are calculated.

Cross Validation

In a cross validation exercise, the estimation method is tested at the locations of existing samples. The sample value at a particular location is temporarily discarded from the sample data set; the value at the same location is then estimated using the remaining samples. This procedure, shown in Figure 15.1, can be seen as an experiment in which we mimic the estimation process by pretending that we had never sampled a certain location. Once the estimate is calculated we can compare it to the true sample value that was initially removed from the sample data set. This procedure is repeated for all available samples. The resulting true and estimated values can then be compared using the same statistical and visual tools we have been using in Chapters 11 and 12 for comparing estimates to actual values from the exhaustive data set.

Cross Validation as a Quantitative Tool

We will now use cross validation to evaluate the differences between estimates calculated by ordinary kriging and those calculated by the polygonal method. This is similar to an earlier study in which we compared estimates at 780 regularly gridded locations with the corresponding true values, except we now rely only on the information available in the sample data set and make no use of the information in the exhaustive data set.

At each of the 470 locations where we have a sample we discard the known V value, pretending that we had never sampled that particular location, and we estimate the V value using the other nearby samples. The polygonal estimate is simply the value of the nearest sample. The ordinary kriging estimate was calculated using the model fit to the sample variogram in Chapter 16, Equation 16.34, Figure 16.10; the search strategy consisted of using a circular search neighborhood with a radius of 25 m and retaining only the closest four samples in each quadrant. This procedure gives us 470 polygonal estimates and 470 ordinary kriging estimates that we can compare with our 470 true sample values. All of the statistical tools we used earlier to compare sets of estimates—summary statistics of estimated and true values, q-q

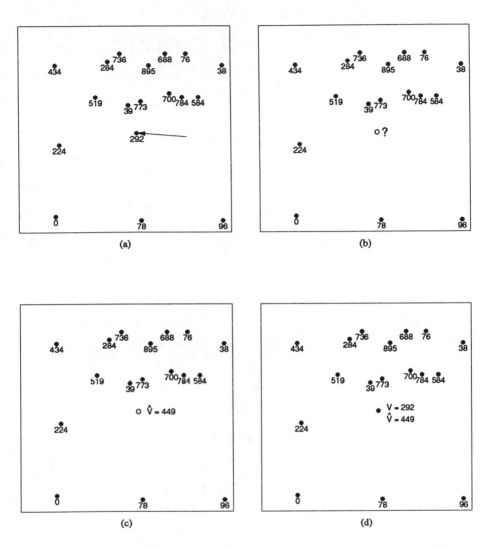

Figure 15.1 An example of the cross validation procedure. The sample at the location highlighted by the arrow in (a) is removed, leaving us with the 17 samples shown in (b). Using only these 17 samples, the value at the location marked o is estimated; in (c), the estimate is calculated using inverse distance squared weighting. This estimate can now be compared with the actual value which was removed earlier, giving us one pair of estimated and true values.

Table 15.1 A comparison of the estimates calculated by the polygonal method and by ordinary kriging to the true sample values.

	True	Polygons	Ordinary Kriging
n	470	470	470
m	436.5	488.0	444.5
σ	300.2	291.0	229.4
CV	0.69	0.60	0.52
min	0	0	5.8
Q_1	184.8	240.3	277.0
M	425.3	515.9	433.4
Q_3	645.4	673.4	593.7
max	1,528.1	1,528.1	1,170.4

Table 15.2 A comparison of the residuals from polygonal estimation and from ordinary kriging.

	Polygons	Ordinary Kriging
n	470	470
m	51.5	8.0
σ	251.4	177.8
IQR	310.1	243.0
MAE	200.4	140.5
MSE	65,729	31,614

plots, summary statistics of residuals, conditional expectation curves—we can now use again to compare these sets of estimated and true values that were generated by cross validation.

The results of this cross validation study are summarized in Tables 15.1 and 15.2 as well as in Figures 15.2 and 15.3. Table 15.1 compares the distributions of the two sets of estimates to the distribution of the true values. This comparison agrees with many of the

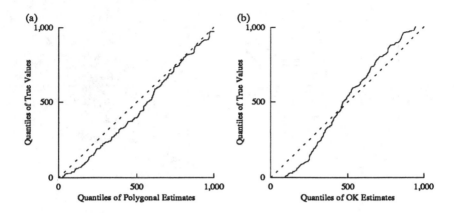

Figure 15.2 Quantile-quantile plots of the distribution of estimates versus the distribution of true values. The distribution of the polygonal estimates is compared to the true distribution in (a); the distribution of ordinary kriging estimates is compared to the true distribution in (b).

observations we were able to make earlier when we used the exhaustive data set. With the exception of its mean, the distribution of polygonal estimates is more similar to the distribution of the true values than is the distribution of ordinary kriging estimates. This is supported by the q-q plots in Figure 15.2. While the quantiles of the distribution of polygonal estimates are consistently larger than the corresponding quantiles of the true sample values, the slope of the solid line in Figure 15.2a is close to one, which shows that the two distributions have very similar shapes. In Figure 15.2b, however, the q-q plot of the ordinary kriging estimates and the true sample values shows that the two are less similar. The distribution of ordinary kriging estimates has less spread than the distribution of the true sample values. These observations provide further evidence of the smoothing effect of kriging. A procedure that uses few nearby sample values, such as the polygonal method, will produce estimates that are less smoothed than one that combines many nearby values, such as ordinary kriging.

Table 15.2 compares the distributions of the residuals (estimated minus true values) from the two estimation procedures. Like our earlier comparison using the exhaustive data set, this cross validation exercise

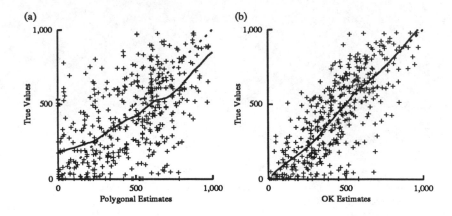

Figure 15.3 Conditional bias of the polygonal estimates and the ordinary kriging estimates. The conditional expectation curve is shown as a solid line through the cloud of points; the estimates are conditionally unbiased if their conditional expectation curve falls on the 45-degree line shown as a dashed line.

shows that the estimates obtained using ordinary kriging are better according to the various statistical criteria we adopted earlier. The estimation errors from ordinary kriging have a mean closer to 0 and have less spread.

Figure 15.3 shows scatterplots of the estimated and true values from the two procedures, along with their conditional expectation curves and their correlation coefficients.

In addition to having a lower overall bias, the ordinary kriging estimates also have less conditional bias. For the broad range of values shown in Figure 15.3 the conditional expectation curve of the ordinary kriging estimates lies closer to the 45-degree line than does the conditional expectation curve of the polygonal estimates. Furthermore, the correlation coefficient between true and estimated values is higher for the ordinary kriging estimates than for the polygonal ones.

Despite the fact that this cross validation study agrees in many ways with the earlier study based on the exhaustive data set, there are some troublesome discrepancies that highlight the weaknesses of cross validation as a purely quantitative tool. Though the cross validation concept is a clever trick for producing pairs of true and estimated values in practical situations, its results can be misleading.

The summary statistics of the residual distribution given in Table 15.2 led us to believe that the polygonal method produces quite biased results. With a true mean value of 437 ppm, and an average residual of 51.5 ppm, the cross validation study suggests that the polygonal method will overestimate by more than 10%. Our earlier comparison using the exhaustive data set in Chapter 10 showed that the polygonal method actually does a very good job in terms of global bias for our Walker Lake example. Similarly, the conditional bias as revealed by Figure 15.3a is much worse than the actual conditional bias as revealed by Figure 11.15a in Chapter 11.

One of the factors that limits the conclusions that can legitimately be drawn from a cross validation exercise is the recurring problem of clustering. If our original sample data set is spatially clustered, then so, too, are our cross validated residuals. Ideally we would like to have residuals that adequately represent the entire area of interest and the complete range of values. In practice, our residuals may be more representative of only certain regions or particular ranges of values. The fact that our Walker Lake samples are preferentially located in areas with anomalously high V values causes our cross validated residuals to be more representative of these particular anomalies than of the whole area. Since we are interested in estimating unknown values not only in these high areas but throughout the Walker Lake map area, the inevitable spatial clustering of our cross validated residuals reduces their usefulness. Some conclusions drawn from the cross validated residuals may be applicable to the entire map area, others may not.

We can try to account for the effect of clustering by calculating a declustered mean of the residuals using one of the global declustering approaches we discussed in Chapter 10. Unfortunately, this does not help much in this particular case. Using cell declustering weights calculated with a 20 x 20 m^2 cell, the declustered mean residual for the polygonal approach is 61.5 ppm, and for the ordinary kriging approach is 29.4 ppm. Even after the effect of clustering has been taken into account, the mean cross validated residual is simply misleading in this particular example.

Another approach to accounting for the clustering is to not perform cross validation at every sample location, but only at a selected subset of locations that is representative of the entire area. In the Walker Lake example, the initial 195 samples on a pseudo regular 20 x 20 m^2 grid provide a good coverage of the entire area. If we limit our attention to these, however, the mean residual for the polygonal estimation is 53.1 ppm, still not an accurate reflection of the actual performance of the polygonal method.

There are other limitations of cross validation that should be kept in mind when analyzing the results of a cross validation study. For example, it can generate pairs of true and estimated values only at sample locations. Its results usually do not accurately reflect the actual performance of an estimation method because estimation at sample locations is typically not representative of estimation at all of the unsampled locations.

In our Walker Lake example, the distance to nearby samples is generally greater in the cross validation study than it was for the earlier estimation of the 780 points on a 10 x 10 m² grid. For many of the cross validated locations the closest samples are 20 m away; in the actual estimation of any point in the Walker Lake map area, the nearest sample is never farther than 15 m away. The result of this is that the spread of the errors calculated in our cross validation study is higher than the spread of the actual errors. This is confirmed by comparing the standard deviations of the cross validated residuals to the standard deviations of the actual errors given in Table 12.4. The cross validated residuals have a standard deviation of 251.4 ppm for the polygonal estimates and 177.8 ppm for the ordinary kriging estimates; the actual standard deviations calculated earlier were 175.2 ppm for the polygonal method and 144.2 ppm for ordinary kriging.

In other practical situations, particularly three-dimensional data sets where the samples are located very close to one another vertically but not horizontally, cross validation may produce very optimistic results. Discarding a single sample from a drill hole and estimating the value using other samples from the same drill hole will produce results that make any estimation procedure appear to perform much better than it will in actual use. The idea of cross validation is to produce sample data configurations that mimic the conditions under which the estimation procedure will actually be used. If very close nearby samples will not be available in the actual estimation, it makes little sense to include them in cross validation. In such situations, it is common to discard more than the single sample at the location we are trying to estimate; it may, for example, be wiser to discard all of the samples from the same drill hole. This, however, puts us back in the situation of producing cross validated results which are probably a bit too pessimistic.

Cross Validation as a Qualitative Tool

Cross validation also offers qualitative insights into how any estimation method performs. The spatial features of the residuals are often overlooked when cross validation is used as a purely quantitative tool for choosing between estimation procedures. The tools we presented for spatial description in Chapter 4 can be used to reveal problems with an estimation procedure; an analysis of the spatial arrangement of the residuals often suggests further improvements to the estimation method.

Just as we prefer estimates to be conditionally unbiased with respect to any range of values, we also prefer them to be conditionally unbiased with respect to their location. Within any region we hope that the center, the spread, and the skewness of the residuals will all be as close to 0 as possible. A contour map of the residuals can reveal areas where the estimates are consistently biased; maps of moving window statistics can be used to show how the spread of the residuals varies throughout the entire area.

A posting of the residuals shows all of the local detail but often contains too much detail to be an effective visual display. An alternative is simply to record where underestimation occurs and where overestimation occurs. Figure 15.4 shows such a map of the ordinary kriging residuals from the cross validation study. Rather than post the value of each residual, we use a + symbol to show where overestimation occurs and a − symbol for underestimation. The shading of the symbol is used to reflect the magnitude of the residual, with the darkest symbols corresponding to the largest errors. On this type of display we hope to see the + and − symbols well mixed, with no obvious regions of consistent overestimation or underestimation. If such regions do exist, the reasons for the local bias should be investigated.

In Figure 15.4 there is a fairly large patch of positive residuals around 110E,180N. The sample values in this area are all quite low and are easily overestimated in a cross validation study due to their closeness to some of the extremely high values in the adjacent Wassuk Range anomaly. Most of the samples in this area are type 1 samples, and this observation prompts us to consider how the ordinary kriging

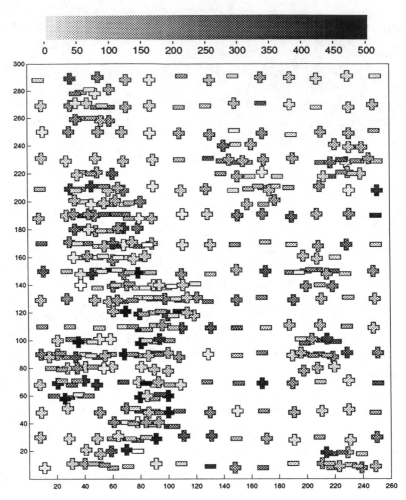

Figure 15.4 A posting of the residuals from the cross validation of the 470 *V* sample values using ordinary kriging. The search strategy consisted of using a circular search neighborhood with a radius of 25 m, retaining only the the nearest four samples per quadrant for estimation. A "+" symbol indicates an overestimation, and a "-" symbol shows an underestimation. The darkness of the shading of the symbol is proportional to the magnitude of the residual.

approach performs for the other type 1 samples. In Figure 15.5 only the residuals at the 45 type 1 locations are shown. From this map it is clear that we have a major problem with the estimation at type

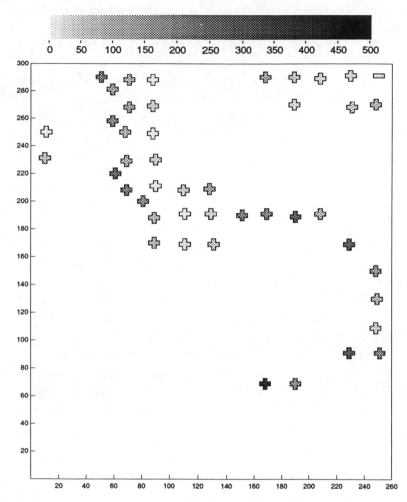

Figure 15.5 Cross validated residuals from ordinary kriging for type 1 sample locations

1 locations. Of the 45 locations shown on Figure 15.5, 44 of them show positive residuals; virtually all of the type 1 sample values were overestimated in the cross validation exercise. The mean of the 45 type 1 sample values is 40 ppm; the mean of the cross validated estimates calculated using ordinary kriging is 148 ppm —an error of more than 250%!

Having observed this problem, we can consider how we might improve our estimation procedure. The problem is largely with the search strategy; type 2 samples should not be combined with type 1 samples if the goal is the accurate estimation of type 1 values. A fairly simple refinement to our search strategy is to separate the two types; if we are estimating the value at a type 1 location, we will use only type 1 samples. With this change to our search strategy, and expanding the circular search neighborhood to 30 m (since one of the type 1 samples is more than 25 m away from its nearest type 1 neighbors), the cross validation exercise was repeated. The residuals from this exercise are shown in Figure 15.6. We now have a more reassuring mixture of overestimates and underestimates; there is still a persistent overestimation in the area of Walker Lake itself, but the magnitude of the errors is smaller than it was when type 2 samples were included.

Though we have been able to improve our cross validation results for type 1 estimates, the improvement of our final estimates is a more difficult problem. In the estimation of values at unsampled locations we will face the additional problem of determining the appropriate type for each of the points we will be estimating. In the cross validation this was not a problem since all of our estimates were at sample locations where we knew the correct type. In the actual estimation at unsampled locations, the type is unknown. If we want to separate the estimation of type 1 values from the estimation of type 2 values, we will have to find some way of figuring out the type before we do the actual estimation of V. While it provides satisfaction for our curiosity about how our estimation method might perform, cross validation can also bring considerable frustration since it often reveals problems that do not have straightforward solutions.

There are many other aspects of our cross validated residuals that we could examine with the tools presented for spatial description in Chapter 4. We could explore the possible correlation between errors at different locations using a correlogram of the residuals. A posting of the largest residuals would reveal that the largest estimation errors are occurring in the areas with the highest values; we could better document this relationship between the magnitude of the residuals and their variability with a scatterplot of local means versus local standard deviations. Each cross validated data set will suggest other informative displays as one tries to understand its various peculiarities. Reduc-

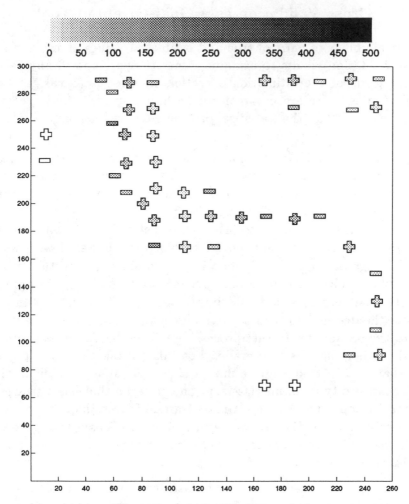

Figure 15.6 Cross validated residuals for type 1 sample locations using ordinary kriging and a search strategy that uses only those nearby samples that are also type 1.

ing cross validated results to a few summary statistics, often only a global mean and variance, is a wasteful (but common) practice; there is much to be learned from a thorough spatial analysis of the residuals.

Cross Validation as a Goal-Oriented Tool

The availability of pairs of true and estimated values allows us to use cross validation results to examine issues that are specifically related to the goals of the estimation exercise. In most practical geostatistical studies there are economic or technical criteria that are much more relevant than the statistical criteria we have been using so far for evaluating a set of estimates.

As an example, we can imagine for the moment that the Walker Lake data set is an ore deposit, with the V values representing the concentration of some metal. Not all of the material will be treated as ore; below some cutoff value, the quantity of metal contained in the rock will be too small to justify the cost of its extraction. Let us suppose that this economic cutoff is 300 ppm; material with an ore grade of greater than 300 ppm will be classified as ore that will be stockpiled and eventually processed, material less than 300 ppm will be classified as waste that will simply be stockpiled. In such an operation, the decision to send a particular load of rock to the waste pile or to the ore pile for further processing will be based on its estimated grade rather than its true grade. For this reason, there is a possibility that certain loads will be misclassified. As shown on Figure 15.7, there are two types of misclassification: material that has a true grade above 300 ppm but whose estimated grade is less than 300 ppm will be incorrectly classified as waste; material that has a true grade less than 300 ppm but whose estimated grade is greater than 300 ppm will be incorrectly classified as ore.

Though we have discussed misclassification in terms of a mining example, the same problem occurs in many other applications. For example in the assessment of hazardous waste sites if the pollutant concentration in a certain region is above a specified threshold, then that region of the site will have to be cleaned up. The decision to clean up or not to clean up is based on estimates of the pollutant concentration and this gives rise to the possibility of misclassification. In such applications, the misclassification is often referred to as *false positive* or *false negative* error. A false positive error is the type of misclassification in which estimates above the critical threshold are assigned to locations whose true value is actually below the critical threshold; the other type, where estimates below the critical threshold are assigned to locations whose true value is actually above the critical

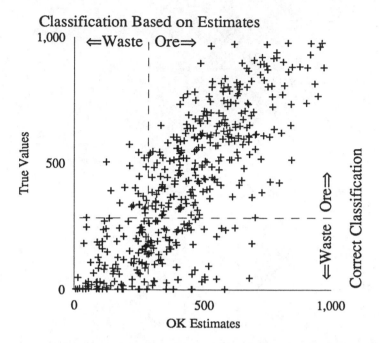

Figure 15.7 Misclassification due to selection based on estimated rather than true values. The points that are misclassified are those in the upper left and lower right quadrants. The points in the upper left quadrant are those for which the true value is above the cutoff while the estimated value is below the cutoff; the points in the lower right quadrant are those for which the true value is below the cutoff while the estimated value is above cutoff. The points in the other two quadrants are correctly classified despite the fact that the estimated value is not exactly the same as the true value.

threshold, is called a false negative error. Areas in which false negative errors have occurred will be left uncleaned when they should, in fact, have been cleaned up; areas in which false positive errors occur will be cleaned unnecessarily.

For applications in which the problem of misclassification has important consequences, the minimization of the misclassification may be a much more relevant criterion for judging the goodness of sets of estimates than are the various statistical criteria we are accustomed to using. The magnitude of overestimation or underestimation may be unimportant as long as the classification remains correct. In our

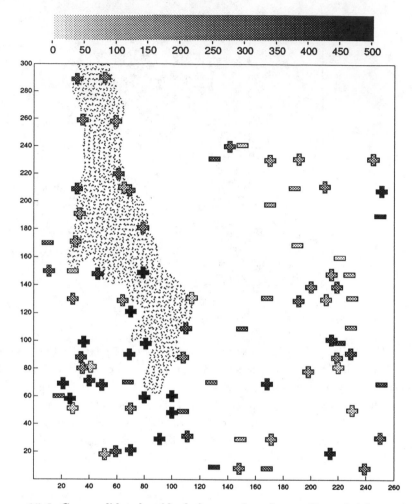

Figure 15.8 Cross validated residuals for samples whose ordinary kriging estimate would result in misclassification at the 300 ppm cutoff.

mining example with the Walker Lake data set, most of the type 1 locations have values much lower than 300 ppm and will be discarded as waste. The overestimation of these type 1 locations does not matter unless the overestimate is so large that waste material is incorrectly classified as ore. Similarly, the underestimation of some of the very rich areas in the Wassuk Range anomaly is not important unless the underestimation is so large that material that should have been treated as ore is actually classified as waste.

Unfortunately, we do not yet have estimation procedures that aim specifically at minimizing misclassification in the same way that ordinary kriging aims specifically at minimizing the error variance. Even so, we should still be interested in the misclassification problem and cross validation can help us address this important concern.

From the set of ordinary kriging residuals shown earlier in Figure 15.4, only the residuals for those samples that are incorrectly classified at a 300 ppm cutoff are shown on Figure 15.8. Also shown on this figure is the approximate extent of the Wassuk Range anomaly. This figure provides us with several interesting insights into how we might run into trouble making ore/waste decisions based on our ordinary kriging estimates.

In the heart of the Wassuk Range anomaly there is only one sample that is incorrectly classified, so despite the many large overestimates and underestimates in this region, ordinary kriging does not lead to any wrong decisions. Along the edges of the anomaly, however, there are several misclassified samples; all of these are waste samples that would be incorrectly classified as ore based on their ordinary kriging estimate. This forewarns us that our perception of the size of the Wassuk Range anomaly based on ordinary kriging estimates might be a bit larger than the actual extent of the economically viable anomaly.

Figure 15.8 also shows a few other areas in which we may get waste where we expected to find ore. A check of Figure 5.9f, which showed an indicator map of the exhaustive data set for a cutoff very close to 300 ppm, shows that one of these areas, the group of five overestimations near (210E,140N), coincides with a small zone of waste. The group of six overestimations near (40E,80N), however, appears to be a case of bad luck in our sampling campaign; the values in the region around those six overestimations is, in reality, predominantly above the 300 ppm cutoff, with occasional patches of waste material. Our sampling seems to have hit most of these waste patches. In an actual study, we would not have access to an exhaustive indicator map that tells us which of the misclassified areas from the cross validation study we should ignore and which ones we should worry about. The potential problem areas identified by the cross validation study may warrant additional sampling if the misclassification of waste as ore has major economic consequences. Since a goal-oriented cross validation study focuses specifically on the errors that have the most serious practical impact (and these may not be the largest errors), it may prompt practically important changes in the estimation methodology.

16

MODELING THE SAMPLE
VARIOGRAM

In Chapter 7 we discussed the computation of sample variograms that describe how the spatial continuity changes with distance and direction. Although a set of directional sample variograms provides an excellent descriptive summary of the spatial continuity, it most likely will not provide all of the variogram values needed by the the kriging system. When the ordinary kriging system was presented in Chapter 12, we saw that, not only is the variogram (or covariance) between all pairs of sample locations called for, but variogram values are also called for between all sample locations and the locations where we wish to make an estimate [1]. The separation vector between the sample locations and the locations where we require estimates conceivably could involve a direction and distance for which we do not have a sample variogram value. Therefore, in order to ensure that we can build the ordinary kriging matrices, we require a model that will enable us to compute a variogram value for any possible separation vector. In this chapter we will see how to construct such a model.

We will begin by examining the constraints that a model must respect. We cannot use any arbitrary function, but must choose functions that obey certain rules. After examining these constraints we will present several "basic" variogram models that respect all constraints and that will form the building blocks for more complex models. Then we will show how these basic models can be combined to build a more general model of the sample variogram in one direction. Then we will show how a geometric and zonal anisotropy in two or three dimensions can be modeled using different combinations of the basic models. Explicit examples will also be given showing how a mixture of the geometric and zonal anisotropies in two or three dimensions can be modeled.

In the presence of an anisotropy, it is easiest to build the variogram model with reference to the axes of the anisotropy. However, these anisotropy axes usually will not coincide with the axes of the data coordinate system. Thus we may end up requiring variogram values between locations that are referenced in the data coordinate system from a variogram model that is referenced by a second set of axes that are not coincident with the first. To get around this problem we will provide a simple method that will enable us to go from the coordinate system of one set of axes to the other. Finally, we will provide a detailed example and show how the sample variograms of V, U, and V crossed with U given in Chapter 7 can be modeled.

Restrictions on the Variogram Model

Our need for a model comes from the fact that we may need a variogram value for some distance or direction for which we do not have a sample variogram value. For example, the ordinary kriging system presented earlier in Chapter 12 will undoubtedly require variogram values at distances and directions for which we have not calculated sample variogram values.

Naïvely we might simply consider interpolating between the known values of the directional sample variograms. Although this method will provide numbers, it will also provide a major problem: the solution of the ordinary kriging equations derived using these numbers may not exist or if it does exist, it may not be unique. This is because the kriging matrices built using such variogram values are not likely to be *positive definite* [2].

If we wish the ordinary kriging equations to have one, and only

(a)

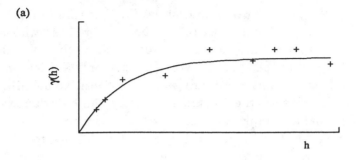

(b)

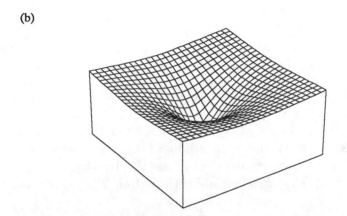

Figure 16.1 The solid line in (a) is the variogram model fit to the directional sample variogram indicated by the "+" symbols. A complete variogram surface is shown in (b). The depression or hole at the center of the surface represents the variogram at zero distance. As one moves out from the hole in any direction, the height of the meshed surface above the base of the slab is proportional to the value of the variogram.

one, stable solution, we must ensure that the left-hand matrix:

$$\mathbf{K} = \underbrace{\left[\begin{array}{cccc} \tilde{C}_{00} & \tilde{C}_{01} & \cdots & \tilde{C}_{0n} \\ \tilde{C}_{10} & \tilde{C}_{11} & \cdots & \tilde{C}_{1n} \\ \vdots & \vdots & \ddots & \vdots \\ \tilde{C}_{n0} & \tilde{C}_{n1} & \cdots & \tilde{C}_{nn} \end{array} \right]}_{(n+1)\times(n+1)} \qquad (16.1)$$

satisfies a mathematical condition known as positive definiteness. An explanation of why this condition guarantees existence and uniqueness of the solution is beyond the scope of this book. For those who wish to understand positive definiteness in greater depth, we have included references at the end of the chapter. We will make only a brief remark on how one can ensure that any kriging matrix is positive definite.

The matrix $\mathbf{K}$ given earlier includes not only the covariances between the n available sample locations, but also the covariances between the sample locations and the location at which we require an estimate. A necessary condition that guarantees that the matrix $\mathbf{K}$ is positive definite is given by

$$\mathbf{w}^t \mathbf{K} \mathbf{w} = \sum_{i=0}^{n} \sum_{j=0}^{n} w_i w_j \tilde{C}_{ij} > 0 \qquad (16.2)$$

where $\mathbf{w}$ is any vector of weights, $(w_0, w_1, \cdots, w_n)$, one of which must be nonzero [3].

It is interesting to note that Equation 16.2 is the same equation we encountered when we discussed random function models. In Chapter 9, the variance of random variable defined by a weighted linear combination of other random variables (Equation 9.14) was given as

$$Var\{\sum_{i=1}^{n} w_i \cdot V_i\} = \sum_{i=1}^{n} \sum_{j=1}^{n} w_i \cdot w_j \cdot Cov\{V_i V_j\} \qquad (16.3)$$

Thus the positive definite condition given in Equation 16.2, can also be seen as a guarantee that the variance of any random variable formed by a weighted linear combination of other random variables will be positive. One such random variable that we have already discussed is the difference between our estimate and the unknown value, which is the estimation error or residual

$$R_0 = \sum_{i=1}^{n} w_i \cdot V_i \quad - \quad V_0 \qquad (16.4)$$

Thus by adhering to the positive definiteness condition, we guarantee that the estimation error will have a positive variance.

Positive Definite Variogram Models

One way of satisfying the positive definiteness condition is to use only a few functions that are known to be positive definite. Although this may seem restrictive at first, we can combine those functions that we know are positive definite to form new functions that are also positive definite. In this section we will present some of the functions that are known to be positive definite. Although it is possible to concoct a new function and to verify its positive definiteness, it is not worth the effort. The functions we present in this section are varied enough to enable a satisfactory fit to all sample variograms likely to be encountered in practice.

It should be noted that the variogram models introduced in this section are those that we consider to be the "basic models". They are simple, isotropic models, independent of direction. The basic variogram models can be conveniently divided into two types; those that reach a plateau and those that do not. Variogram models of the first type are often referred to as *transition* models. The plateau they reach is called the *sill* and the distance at which they reach this plateau is called the *range*. Some of the transition models reach their sill asymptotically. For such models, the range is arbitrarily defined to be that distance at which 95% of the sill is reached. In this section, the sill of all transition models has been standardized to one.

Variogram models of the second type do not reach a plateau, but continue increasing as the magnitude of **h** increases. Such models are often necessary when there is a trend or drift in the data values.

Nugget Effect Model. As was discussed in Chapter 7, many sample variograms have an obvious discontinuity at the origin. While the variogram value for **h** = 0 is strictly 0, the variogram value at very small separation distances may be significantly larger than 0 giving rise to a discontinuity. We can model such a discontinuity using a discontinuous positive definite transition model that is 0 when h is equal to 0 and 1 otherwise. This is the nugget effect model and its equation is given by

$$\gamma_0(h) = \begin{cases} 0 & \text{if } h = 0 \\ 1 & \text{otherwise} \end{cases} \tag{16.5}$$

In geostatistical literature the nugget effect is not usually given explicitly as a basic model, but rather appears as a constant C_0 in the variogram equation, with the understanding that this "constant" is 0 when $h = 0$. Our notation for the nugget effect is $w_0\gamma_0(h)$, where w_0 is the height of the discontinuity at the origin and $\gamma_0(h)$ is the standardized basic model given in Equation 16.5. This notation is consistent

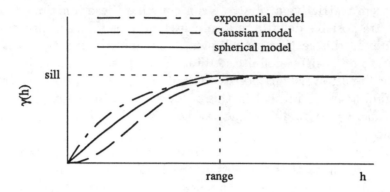

Figure 16.2 The three most commonly used transition models–the spherical, exponential, and Gaussian models–shown here with the same range and sill.

with the presentation of the basic models in this section and will become convenient later when we combine the basic models.

Spherical Model. Perhaps the most commonly used variogram model is the spherical model, whose standardized equation is

$$\gamma(h) = \begin{cases} 1.5\frac{h}{a} - 0.5(\frac{h}{a})^3 & \text{if } h \leq a \\ 1 & \text{otherwise} \end{cases} \qquad (16.6)$$

where a is the range. This model is shown in Figure 16.2 by the solid line. It has a linear behavior at small separation distances near the origin but flattens out at larger distances, and reaches the sill at a. In fitting this model to a sample variogram it is often helpful to remember that the tangent at the origin reaches the sill at about two thirds of the range.

The Exponential Model. Another commonly used transition model is the exponential model. Its standardized equation is

$$\gamma(h) = 1 - exp(-\frac{3h}{a}) \qquad (16.7)$$

This model reaches its sill asymptotically, with the practical range a defined as that distance at which the variogram value is 95% of the sill. [4]. Like the spherical model, the exponential model is linear at very short distances near the origin, however it rises more steeply and then flattens out more gradually. In Figure 16.2 it is shown by the alternating long and short dashes. In fitting this model to a sample variogram its helpful to remember that the tangent at the origin reaches the sill at about one fifth of the range.

The Gaussian Model. The Gaussian model is a transition model that is often used to model extremely continuous phenomena. Its equation is

$$\gamma(h) = 1 - exp(-\frac{3h^2}{a^2}) \qquad (16.8)$$

Like its cousin, the exponential model, the Gaussian model reaches its sill asymptotically, and the parameter a is defined as the practical range or distance at which the variogram value is 95% of the sill [4]. The distinguishing feature of the Gaussian model is its parabolic behavior near the origin; in Figure 16.2 it is shown using the long dashed line. It is the only transition model presented whose shape has an inflection point.

The Linear Model. The linear model is not a transition model since it does not reach a sill, but increases linearly with h [5]. In its standardized form it is written simply as

$$\gamma(h) = |h| \qquad (16.9)$$

Models in One Direction

At this point we have a few basic models that we can fit to a directional sample variogram. Actually, in the isotropic case the sample variogram depends only on the separation distance and not on direction and thus all the directional sample variograms will be the same. In such cases we can model the omnidirectional sample variogram as though it were a directional sample variogram and we are finished. In fact the omnidirectional sample variogram is preferred since it is usually "better behaved" and thus easier to model.

Although one can sometimes model a directional sample variogram satisfactorily using one basic model, more often a combination of basic models is required to obtain a satisfactory fit. This brings us to an important property of positive definite variogram models: any linear combination of positive definite variogram models with positive coefficients is also a positive definite model [6].

This property provides us with a very large family of positive definite models since any model of the form

$$\gamma(h) = \sum_{i=1}^{n} |w_i| \gamma_i(h) \qquad (16.10)$$

will be positive definite as long as the n individual models are all positive definite.

In geostatistical jargon, a linear combination of basic variogram models is said to form a model of "nested structures," where each one of the nested structures corresponds to a term of the linear combination in Equation 16.10.

To fit a combination of basic variogram models to a particular directional sample variogram, we must decide which of the basic model(s) best describes the overall shape. If the sample variogram has a plateau, one of the transition models will be most appropriate; if not, perhaps the linear model may be more appropriate.

Among the three transition models given in the previous section, the choice usually depends on the behavior of the sample variogram near the origin. If the underlying phenomenon is quite continuous, the sample variogram will likely show a parabolic behavior near the origin; in such situations, the Gaussian model will usually provide the best fit. If the sample variogram has a linear behavior near the origin, either the spherical or exponential model is preferable. Often one can quickly fit a straight line to the first few points on the sample variogram. If this line intersects the sill at about one fifth of the range, then an exponential model will likely fit better than a spherical. If it intersects at about two thirds of the range, then the spherical model will likely fit better.

In using Equation 16.10, we are not limited to combining models of the same shape. Often the sample variogram will require a combination of different basic models. For example, a sample variogram that does not reach a stable sill but has a parabolic behavior near the origin may require some combination of the Gaussian model and the linear model.

Beginners at the art of variogram modeling often have a tendency to overfit the sample variogram. Three or more basic models may be combined to capture each and every kink of the sample variogram points. Such complicated models usually do not lead to estimates more accurate than those provided by simpler models. If the major features of the sample variogram can be captured by a simple model, then it will provide solutions that are as accurate as those found using a more complex model. The principle of parsimony is a good guide in variogram modeling. For example, if a single exponential model fits the sample variogram as well as two nested spherical models, then the simpler exponential model is preferable.

In deciding whether or not a particular feature of the sample variogram should be modeled, it is wise to consider whether or not a physical explanation exists for the feature. If qualitative information about the genesis of the phenomenon explains or confirms a particular feature of the sample variogram, then it is worth building a model which includes that feature. If there is no explanation, however, the feature may be spurious and not worth modeling.

Once the basic models are chosen, modeling the sample variogram becomes an exercise in curve fitting in which there are several parameters with which to play. For the transition models given earlier, the range parameter a can usually be picked quite easily from the sample variogram. The nugget effect can also often be picked from the sample variogram by extrapolating the linear behavior of the first few points back to the y axis. Choosing the coefficients for the basic models ($w_1, \ldots, w_n$ in Equation 16.10) is a bit trickier. The most useful guideline for choosing these coefficients is to remember that their sum must equal the sill of the sample variogram. Within this guideline, there are many possible choices, and a satisfactory fit often requires a trial and error approach. A good interactive graphical program can be tremendously helpful.

Models of Anisotropy

Often, directional sample variograms will reveal major changes in the range or sill as the direction changes. The example in Figure 16.3a shows an isometric view of a variogram surface where the range changes with direction, while the sill remains constant. This type of anisotropy is known as the *geometric anisotropy*. In case of the *zonal anisotropy*, the sill changes with direction while the range remains constant. The example in Figure 16.3b shows both the range and sill changing with direction and is a mixture of both the geometric and zonal anisotropies.

Given a set of sample variograms which show the range and/or sill obviously changing with direction, one begins by identifying the anisotropy axes. This is usually done by experimentally determining

(a)

(b)

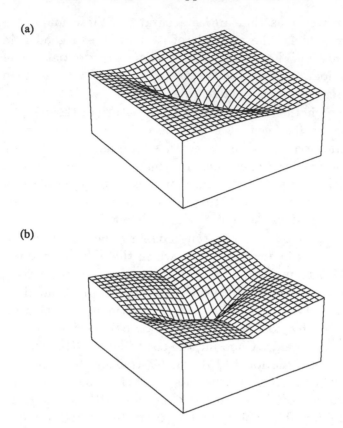

Figure 16.3 In (a), the isometric drawing of the variogram surface shows an example of a geometric anisotropy where the range changes with direction while the sill remains constant. Another anisotropy type is the zonal, where the sill changes with direction while the range remains constant. A mixture of the two anisotropy types is shown in (b).

the directions corresponding to the minimum and maximum range or the maximum and minimum sill in the case of the zonal anisotropy [7]. Contour maps of the variogram surface, such as the one shown in Figure 7.6, are tremendously useful for determining these directions. Another alternative, also shown in Chapter 7, is to plot the experimental ranges for the different directional sample variograms on a rose diagram.

Qualitative information, such as the orientation of lithologic units

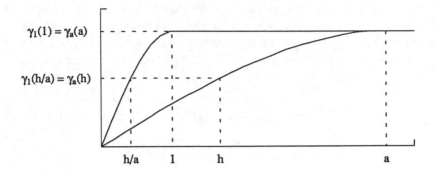

Figure 16.4 This figure shows the equivalence between a transition model with range *a* and the transition model with range 1 and equal sill values. These two models are equivalent providing they are evaluated using the vectors **h** and **h**/*a* respectively.

or bedding planes, is usually very helpful in identifying the axes of the anisotropy. A knowledge of the genesis of the phenomenon under study can also be very helpful. For example, airborne pollution is likely to be more continuous in the prevailing wind direction than in a perpendicular direction.

Having identified the axes of the anisotropy, the next step is to put together a model that describes how the variogram changes as the distance and *direction* change. For now we will work in the coordinate system defined by the axes of the anisotropy. Once we have seen how to build a complete model in this coordinate system we will look at a method that will enable the variogram model to be used in the data coordinate system.

One method for combining the various directional models into a model that is consistent in all directions is to define a transformation that reduces all directional variograms to a common model with a standardized range of 1. The trick is to transform the separation distance so that the standardized model will provide us with a variogram value that is identical to any of the directional models for that separation distance.

For example, two transitional variogram models with the same sill are shown in Figure 16.4. One has a range of 1 while the other has a range of *a*. Note that if we evaluate the model with range 1 at a distance of h/a we will get the same value that we would get by evaluating the model with range a at a distance of h. Thus we have effectively reduced the model with range a to an equivalent model with range 1 by reducing the separation distance h to h/a. We can express this equivalence as

$$\gamma_1(\frac{h}{a}) = \gamma_a(h) \quad \text{or} \quad \gamma_1(h) = \gamma_a(ah) \qquad (16.11)$$

or, if we let h_1 equal $\frac{h}{a}$ then,

$$\gamma_1(h_1) = \gamma_a(h) \qquad (16.12)$$

Thus any directional model with range a can be reduced to a standardized model with a range of 1 simply by replacing the separation distance, h, by a reduced distance $\frac{h}{a}$.

The concept of an equivalent model and reduced distance can be extended to two-dimensions. If a_x is the range in the x direction and a_y the range in the y direction, then the anisotropic variogram model can be expressed as

$$\gamma(\mathbf{h}) = \gamma(h_x, h_y) = \gamma_1(h_1) \qquad (16.13)$$

and the reduced distance h_1 is given by

$$h_1 = \sqrt{(\frac{h_x}{a_x})^2 + (\frac{h_y}{a_y})^2} \qquad (16.14)$$

where h_x is the component of $\mathbf{h}$ along the x axis and h_y is its component along the y axis.

Similarly, the anisotropic variogram model in three dimensions, with ranges a_x, a_y, and a_z can be expressed as

$$\gamma(\mathbf{h}) = \gamma(h_x, h_y, h_z) = \gamma_1(h_1) \qquad (16.15)$$

and the reduced distance h_1 is given by

$$h_1 = \sqrt{(\frac{h_x}{a_x})^2 + (\frac{h_y}{a_y})^2 + (\frac{h_z}{a_z})^2} \qquad (16.16)$$

The method of using equivalent models and reduced distances also works with models that do not reach a sill. Equations 16.12 through

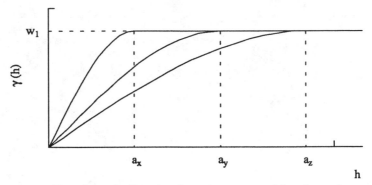

Figure 16.5 Transitional directional variogram models along the axes of a three-dimensional geometric anisotropy.

16.16 can also be applied to linear directional models with slopes a_x, a_y, and a_z. The reduced distances would be $a_x h$, $a_y h$, and $a_z h$ and the equivalent isotropic model with slope 1 is given by $\gamma_1(a_x h) = \gamma_{a_x}(h)$ etc.

In the following paragraphs we have provided several examples of different models that use the equivalent model and reduced distance trick. The equation of the final model in three dimensions is also provided for each example.

Geometric Anisotropy - One Structure. Recall that the geometric anisotropy is characterized by directional sample variograms that have approximately the same sill but different ranges (in the case of a linear variogram, the slope will vary with direction). Figure 16.5 shows three directional variogram models along the three perpendicular axes of the anisotropy. Each directional model consists of only one structure, and all three have the same sill value; however, their ranges are different. The equivalent three-dimensional variogram model for Figure 16.5 is given by

$$\gamma(\mathbf{h}) = |w_1|\gamma_1(h_1) \tag{16.17}$$

and the reduced distance h_1 is

$$h_1 = \sqrt{(\frac{h_x}{a_x})^2 + (\frac{h_y}{a_y})^2 + (\frac{h_z}{a_z})^2} \tag{16.18}$$

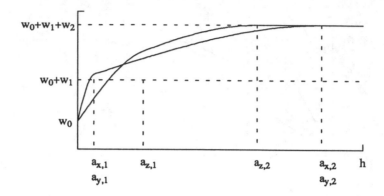

Figure 16.6 A second example of transitional directional variogram models along the axes of a three-dimensional geometric anisotropy. In this example, each directional model consists of three structures. The first is the the nugget with coefficient w_0. The second is the transition structure with ranges $a_{x,1}$, $a_{y,1}$, $a_{z,1}$ and coefficient w_1. The third is the transition structure with ranges $a_{x,2}$, $a_{y,2}$, $a_{z,2}$ and coefficient w_2.

where a_x, a_y, and a_z are the ranges of the directional variogram models along the axes of the anisotropy; h_x, h_y, and h_z are the components of **h** in the x, y, and z directions of the anisotropy axes and $\gamma_1(h_1)$ is the equivalent model with a standardized range of 1. *Note that for each nested structure, the directional models must all be the same type.* That is, the directional models must all be either spherical, exponential, or some other suitable model for each nested structure; however, the model types can differ from one nested structure to the next. For example the first nested structure for each of the directionals might consist of a spherical model while the second nested structure could be an exponential model.

Geometric Anisotropy - Nugget and Two Structures. A modeling problem that is commonly encountered in practice is shown in Figure 16.6. Each directional model consists of three structures, the nugget effect and two additional transition structures. The nugget effect is isotropic in three directions while the remaining two structures are isotropic in the x, y plane but show an anisotropy between the z direction and the x, y plane. We will build the three-dimensional model for this example by considering each nested structure in turn.

The nugget effect is isotropic and can be modeled straightforwardly. Its equation is given by

$$\gamma(\mathbf{h}) = w_0 \gamma_0(h) \tag{16.19}$$

where $\gamma_0(h)$ is defined by Equation 16.5.

The next structure is identified in Figure 16.6 by the ranges $a_{x,1}$, $a_{y,1}$, and $a_{z,1}$ and the coefficient w_1. This structure is isotropic in the x, y plane but shows an anisotropy between the x, y plane and the z direction. The equivalent isotropic model of this structure is given by

$$\gamma(\mathbf{h}) = |w_1| \gamma_1(h_1) \tag{16.20}$$

and the reduced distance h_1 is

$$h_1 = \sqrt{(\frac{h_x}{a_{x,1}})^2 + (\frac{h_y}{a_{y,1}})^2 + (\frac{h_z}{a_{z,1}})^2} \tag{16.21}$$

For the last structure, the ranges along the three principal axes of the anisotropy are given by $a_{x,2}$, $a_{y,2}$, and $a_{z,2}$ and the coefficient by w_2. This structure is also isotropic in the x, y plane and shows an anisotropy between the x, y plane and the z direction. The equivalent isotropic model is given by

$$\gamma(\mathbf{h}) = |w_2| \gamma_1(h_2) \tag{16.22}$$

and the reduced distance h_2 is

$$h_2 = \sqrt{(\frac{h_x}{a_{x,2}})^2 + (\frac{h_y}{a_{y,2}})^2 + (\frac{h_z}{a_{z,2}})^2} \tag{16.23}$$

The complete three dimensional anisotropic model is obtained by combining the three equivalent isotropic models of Equations 16.19 through 16.22 to obtain

$$\gamma(\mathbf{h}) = w_0 \gamma_0(h) + w_1 \gamma_1(h_1) + w_2 \gamma_1(h_2) \tag{16.24}$$

To summarize, the geometric anisotropy requires some foresight in modeling the directional sample variograms. All the directional variogram models must have identical sill values. Each nested structure in any particular directional variogram model must appear in all the other directional models with the same coefficient w.

(a)

(b)

(c)

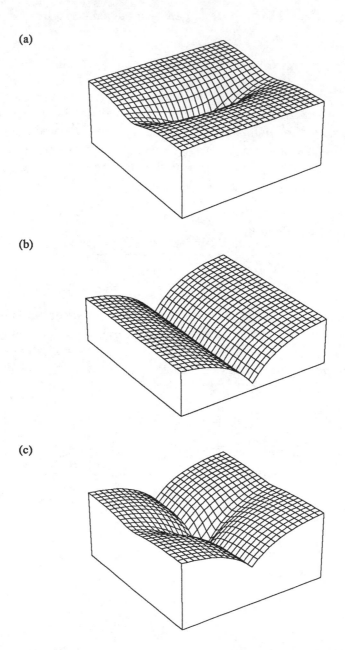

Figure 16.7 The geometric anisotropic variogram model shown in (a) has been combined with the directional (zonal) component shown in (b) and the resulting mixture is shown in (c).

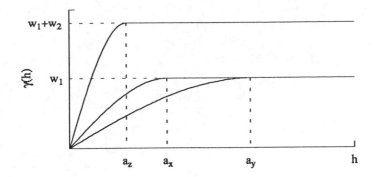

Figure 16.8 A third example of transitional directional variogram models along the axes of a three-dimensional anisotropy. The directional models along the x and y axis show the same sill, w_1, but different ranges. However, the directional model along the z axis shows a different range and sill and indicates a mixture of geometric and zonal anisotropies.

Zonal and Geometric Anisotropy. A zonal anisotropy is one in which the sill value changes with direction while the range remains constant. In practice we rarely find a pure zonal anisotropy; it is more common to find a mixture of the zonal and geometric anisotropies together.

Figure 16.7a shows a perspective view of the geometric variogram surface where the range changes with direction while the sill remains constant. Figure 16.7b shows a zonal variogram surface where the variogram only changes in one direction. Figure 16.7c, shows the resulting model obtained by combining the geometric anisotropy shown in Figure 16.7a with the zonal model shown in Figure 16.7b. Both the sill and the range change with direction in the combined model.

The example in Figure 16.8 consists of three directional variogram models along the axes of the anisotropy. Each directional model consists of one structure. The directional models along the x and y axes have the same sill, but different ranges. The directional model along the z axis has a shorter range and a larger sill than the directional models for x and y. The isotropic model for these anisotropies consists of two structures similar to the models illustrated in Figures 16.7a and b. The first structure is modeled as a geometric anisotropy, while the second is modeled as a zonal component using a directional model.

The equivalent model for the first structure will be an isotropic model with a sill of w_1 and a range of 1. It is important to note that this model is isotropic, which means it must return a value of w_1 when evaluated for the vector $(0, 0, a_z)$. Its equation is given by

$$\gamma(\mathbf{h}) = |w_1| \gamma_1(h_1) \tag{16.25}$$

and the reduced distance h_1 is

$$h_1 = \sqrt{(\frac{h_x}{a_x})^2 + (\frac{h_y}{a_y})^2 + (\frac{h_z}{a_z})^2} \tag{16.26}$$

The second structure has a sill equal to w_2 and exists only in the h_z direction. This zonal component is modeled using an equivalent directional variogram in the z direction;

$$\gamma(\mathbf{h}) = w_2 \gamma_1(h_2) \tag{16.27}$$

and the reduced distance h_2 is

$$h_2 = \frac{h_z}{a_z} \tag{16.28}$$

The complete model is given by

$$\gamma(\mathbf{h}) = w_1 \gamma_1(h_1) + w_2 \gamma_1(h_2) \tag{16.29}$$

Matrix Notation

The method for reducing directional variogram models to an isotropic model with a standardized range of 1 can be conveniently summarized using matrix notation. In general, for those final models that contain n nested structures, one reduced distance h_n will be needed for each structure. These reduced distances can be summarized by the vector $\mathbf{h}_n$ that is defined by

$$\mathbf{h}_n = \mathbf{T}\mathbf{h} \tag{16.30}$$

where the matrix $\mathbf{T}$ is given by

$$\mathbf{T} = \begin{bmatrix} \frac{1}{a_x} & 0 & 0 \\ 0 & \frac{1}{a_y} & 0 \\ 0 & 0 & \frac{1}{a_z} \end{bmatrix} \tag{16.31}$$

and a_x, a_y, and a_z are the ranges of the directional models (or inverse of the slope in the case of the linear model) along the anisotropy axes x, y, and z. For example, the vector $\mathbf{h_1}$ in Equation 16.25 can expressed as

$$\mathbf{h_1} = \begin{bmatrix} \frac{1}{a_x} & 0 & 0 \\ 0 & \frac{1}{a_y} & 0 \\ 0 & 0 & \frac{1}{a_z} \end{bmatrix} \cdot \begin{bmatrix} h_x \\ h_y \\ h_z \end{bmatrix} \qquad (16.32)$$

Similarly, the vector $\mathbf{h_2}$ in Equation 16.27 can be expressed by replacing h_x and h_y with zeros. We can think of the ranges a_x and a_y as being very large or infinite, making the corresponding elements of $\mathbf{T}$ approach 0, so that the multiplication of any vector $\mathbf{h}$ by $\mathbf{T}$ provides only the zonal component $\frac{h_z}{a_z}$

$$\mathbf{h_2} = \begin{bmatrix} 0 & 0 & 0 \\ 0 & 0 & 0 \\ 0 & 0 & \frac{1}{a_z} \end{bmatrix} \cdot \begin{bmatrix} h_x \\ h_y \\ h_z \end{bmatrix} \qquad (16.33)$$

The length of each of the vectors $\mathbf{h_1}$ and $\mathbf{h_2}$ is given by Equations 16.25 and 16.27.

The final model should always be checked by evaluating it for key distances and directions. A good starting point is to ensure that the model reaches the sill at the appropriate distance in each direction. For example, the model given in Equation 16.29 must return the variogram values w_1, w_1, and $w_1 + w_2$, respectively, when evaluated for the three vectors

$$(h_x, h_y, h_z) = (a_x, 0, 0) \qquad (16.34)$$
$$(h_x, h_y, h_z) = (0, a_y, 0) \qquad (16.35)$$
$$(h_x, h_y, h_z) = (0, 0, a_z) \qquad (16.36)$$

The model should also be evaluated for a series of distances along each anisotropy axis to ensure that it returns the correct directional variogram in each case. An additional check that is often used is to calculate values from the directional variogram model along some arbitrary direction in between the directions of the anisotropy axes and then plot these values and compare them to the directional sample variograms that have been calculated in the same direction. They should compare quite well if the anisotropy directions have been correctly identified and the modeling has been done correctly.

Coordinate Transformation by Rotation

Now that we know how to model the sample variogram within the coordinate system coincident with the anisotropy axes, we need to consider the case where the anisotropy axes do not coincide with the axes of the data coordinate system.

In general, the orientation of the anisotropy is usually controlled by some physical feature inherent in the phenomenon represented by the data, whereas the orientation of the data coordinate system is often arbitrary. There is no reason why these two axial systems should coincide with each other.

The procedures we have described in this chapter provide us with a variogram model that can be evaluated for any distance and direction expressed in the coordinate system coincident with the anisotropy axes. Given the components (h'_x, h'_y, h'_z) of any vector h in the coordinate system coincident with the anisotropy axes, we can calculate a reduced distance and use an equivalent isotropic model with a standardized range of 1. However, when the anisotropy axes are not aligned with the axes of the data coordinate system the components (h_x, h_y, h_z) of the separation vectors in the data coordinate system will have quite different values when referenced in the coordinate system coincident with the anisotropy axes. Thus before the model can be evaluated for any vector referenced in the data coordinate system, references must be established that link that vector with the coordinate system coincident with the anisotropy axes. Perhaps it is easier to think of this process as a transformation. We simply transform the vector from the data coordinate system to the coordinate system coincident with the anisotropy axes and then evaluate the anisotropic variogram model using the transformed vector.

The method presented here for making the transformation is not unique, but it is reasonably straightforward and easy to apply. We will use a matrix $\mathbf{R}$ to make the transformation as follows

$$\mathbf{h}' = \mathbf{R}\mathbf{h} \qquad (16.37)$$

where $\mathbf{h}$ is the vector in the data coordinate system and $\mathbf{h}'$ is the same vector transformed to the anisotropic coordinate system. Thus once we have defined $\mathbf{R}$, we will be able to transform any vector from the data coordinate system to the anisotropic system, and then evaluate the isotropic variogram model using the transformed vector $\mathbf{h}'$.

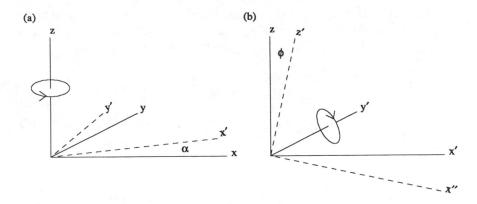

Figure 16.9 The transformation of a three dimensional coordinate system can be defined by two angles of rotation, where the first angle α, shown in (a), is defined as the clockwise rotation angle about the z axis, forming the new axes x' and y'. The second rotation angle ϕ, shown in (b), is the angle of clockwise rotation about the y' axis, forming the new axes x'' and z'.

To begin the definition of **R**, we define two angles of rotation, following the trigonometric conventions of the Cartesian coordinate system. The positive directions of the x, y, and z axes are shown in Figure 16.9a. The first rotation angle, α, is defined as the clockwise rotation about the z axes forming the new axes x' and y'. (Note that a *clockwise* direction around an axis is defined by looking in the positive direction of that axis). The second rotation angle, ϕ, shown in Figure 16.9b is defined as the clockwise rotation about the y' axes forming the new axes x'' and z'. These two rotation angles enable us to define the transformation matrix **R** as follows

$$\mathbf{R} = \begin{bmatrix} cos(\alpha)cos(\phi) & sin(\alpha)cos(\phi) & sin(\phi) \\ -sin(\alpha) & cos(\alpha) & 0 \\ -cos(\alpha)sin(\phi) & -sin(\alpha)sin(\phi) & cos(\phi) \end{bmatrix} \qquad (16.38)$$

To summarize, given that x, y, and z are the axes of the data coordinate system and x'', y', and z' are the axes of the anisotropy system, we can define a transformation matrix **R** using Equation 16.38 which will transform any vector **h** defined in the data system of coordinates to **h'**, defined in the anisotropic coordinate system. Then the anisotropic variogram model can be correctly evaluated using the vector **h'**.

The calculation of the vector $\mathbf{h}_n$ containing the reduced distances given in Equation 16.30 can be combined with the transformation of coordinate systems given by Equation 16.37 to obtain the transformed reduced vector $\mathbf{h}'_n$ as follows

$$\mathbf{h}'_n = \mathbf{TRh} \qquad (16.39)$$

Note that the order of $\mathbf{T}$ and $\mathbf{R}$ cannot be reversed. The vector $\mathbf{h}$ must be defined in the anisotropic coordinate system before a reduced vector can be calculated.

The Linear Model of Coregionalization

So far in this chapter, we have discussed only the modeling of a single variable. The linear model of coregionalization provides a method for modeling the auto- and cross-variograms of two or more variables so that the variance of any possible linear combination of these variables is always positive. Each variable is characterized by its own sample autovariogram and each pair of variables by their own sample cross-variogram. The model for each of these sample variograms may consist of one or more basic models as shown in Equation 16.10; however, the same basic model must appear in each auto- and cross-variogram. In other words, each auto- and cross-variogram model must be constructed using the same basic variogram models. For example, consider two variables, U and V. The auto- and cross-variogram models of U and V must be constructed using the same basic variogram models as follows

$$
\begin{aligned}
\gamma_U(\mathbf{h}) &= u_0\gamma_0(\mathbf{h}) + u_1\gamma_1(\mathbf{h}) + \cdots + u_m\gamma_m(\mathbf{h}) \\
\gamma_V(\mathbf{h}) &= v_0\gamma_0(\mathbf{h}) + v_1\gamma_1(\mathbf{h}) + \cdots + v_m\gamma_m(\mathbf{h}) \\
\gamma_{UV}(\mathbf{h}) &= w_0\gamma_0(\mathbf{h}) + w_1\gamma_1(\mathbf{h}) + \cdots + w_m\gamma_m(\mathbf{h})
\end{aligned}
\qquad (16.40)
$$

where $\gamma_U(\mathbf{h})$, $\gamma_V(\mathbf{h})$, and $\gamma_{UV}(\mathbf{h})$ are the auto- and cross-variogram models for U and V, respectively; the basic variogram models are given by $\gamma_0(\mathbf{h})$, $\gamma_1(\mathbf{h})$, $\cdots$, $\gamma_m(\mathbf{h})$; u, v and w are coefficients, possibly negative. We can rewrite Equation 16.40 in matrix form as combinations of each basic model as

- Combinations of the first basic model, $\gamma_0(\mathbf{h})$.

$$
\begin{bmatrix} \gamma_{U,0}(\mathbf{h}) & \gamma_{UV,0}(\mathbf{h}) \\ \gamma_{VU,0}(\mathbf{h}) & \gamma_{V,0}(\mathbf{h}) \end{bmatrix} = \begin{bmatrix} u_0 & w_0 \\ w_0 & v_0 \end{bmatrix} \bullet \begin{bmatrix} \gamma_0(\mathbf{h}) & 0 \\ 0 & \gamma_0(\mathbf{h}) \end{bmatrix}
$$

$$(16.41)$$

- Combinations of the second basic model, $\gamma_1(\mathbf{h})$.

$$\begin{bmatrix} \gamma_{U,1}(\mathbf{h}) & \gamma_{UV,1}(\mathbf{h}) \\ \gamma_{VU,1}(\mathbf{h}) & \gamma_{V,1}(\mathbf{h}) \end{bmatrix} = \begin{bmatrix} u_1 & w_1 \\ w_1 & v_1 \end{bmatrix} \bullet \begin{bmatrix} \gamma_1(\mathbf{h}) & 0 \\ 0 & \gamma_1(\mathbf{h}) \end{bmatrix}$$
(16.42)

- Combinations of the mth basic model, $\gamma_m(\mathbf{h})$.

$$\begin{bmatrix} \gamma_{U,m}(\mathbf{h}) & \gamma_{UV,m}(\mathbf{h}) \\ \gamma_{VU,m}(\mathbf{h}) & \gamma_{V,m}(\mathbf{h}) \end{bmatrix} = \begin{bmatrix} u_m & w_m \\ w_m & v_m \end{bmatrix} \bullet \begin{bmatrix} \gamma_m(\mathbf{h}) & 0 \\ 0 & \gamma_m(\mathbf{h}) \end{bmatrix}$$
(16.43)

To ensure the linear model given in Equation 16.40 is positive definite, it is sufficient to ensure that all the matrices of the coefficients u, v, and w in Equations 16.41 to 16.43 are positive definite. This implies that the coefficients must be chosen so that

$$\begin{aligned} u_j > 0 \text{ and } v_j > 0, & \quad \text{for all } j = 0, \cdots, m \\ u_j \cdot v_j > w_j \cdot w_j, & \quad \text{for all } j = 0, \cdots, m \end{aligned}$$
(16.44)

The restrictions imposed by Equations 16.44 can make the modeling of a coregionalization somewhat difficult. Often one of the auto- or cross-models may not fit its sample variogram very well, while the others fit quite well. In such situations, one must think of each individual model as a small part the total model and judge the overall fit accordingly. Equations 16.44 suggest two points that are helpful when modeling a coregionalization. First, a basic model that appears in any auto variogram model does not necessarily have to be included in the cross-variogram model. Second, any basic model that is included in the cross-variogram model must necessarily be included in all the auto variogram models.

Models for the Walker Lake Sample Variograms

In Chapter 7 we presented sample variograms for the variables V, U, and cross-variograms between U and V. In this section we will model these sample variograms using the linear model of coregionalization. The auto variogram models of V and U are required for the ordinary kriging of V and U in Chapter 12. The complete linear model of coregionalization is required for the cokriging of U in Chapter 17.

Table 16.1   A summary of the basic models appearing in each of the isotropic models for V, U, and U crossed with V.

Basic model	Range	Direction	Cross-variogram	Auto-variograms
Spherical	25	N76°E	✓	✓
Spherical	50	N76°E	✓	✓
Spherical	30	N14°W	✓	✓
Spherical	150	N14°W	✓	✓

The directional sample variograms and their models for V, U, and U crossed with V are given in Figures 16.10, 16.11, and 16.12, respectively. Each figure contains three directional variograms. The variograms along the minor axis, N76°E, of the anisotropy are given in (a) of each figure, while those along the major axis, N14°W, are shown in (b). An average of the directionals along the intermediate directions, N31°E and N59°W, is shown in (c) to verify the anisotropic model in these directions.

Recall from the previous section that any basic model appearing in the cross-variogram of the linear model of coregionalization must necessarily appear in all the auto variogram models. A summary of all the basic models used is given in Table 16.1.

The anisotropic auto-variogram model for V is given by

$$\gamma_V(\mathbf{h}) = 22{,}000 + 40{,}000\ Sph_1(\mathbf{h}_1') + 45{,}000\ Sph_1(\mathbf{h}_2') \qquad (16.45)$$

where the reduced distance vectors $\mathbf{h}_n'$ are calculated using Equation 16.39 as follows;

$$\mathbf{h}_1' = \begin{bmatrix} h_{x,1}' \\ h_{y,1}' \end{bmatrix} = \begin{bmatrix} \frac{1}{25} & 0 \\ 0 & \frac{1}{30} \end{bmatrix} \cdot \begin{bmatrix} cos(14) & sin(14) \\ -sin(14) & cos(14) \end{bmatrix} \cdot \begin{bmatrix} h_x \\ h_y \end{bmatrix} \qquad (16.46)$$

and

$$\mathbf{h}_2' = \begin{bmatrix} h_{x,2}' \\ h_{y,2}' \end{bmatrix} = \begin{bmatrix} \frac{1}{50} & 0 \\ 0 & \frac{1}{150} \end{bmatrix} \cdot \begin{bmatrix} cos(14) & sin(14) \\ -sin(14) & cos(14) \end{bmatrix} \cdot \begin{bmatrix} h_x \\ h_y \end{bmatrix} \qquad (16.47)$$

(a) $\gamma_V(h'_x) = 22{,}000 + 40{,}000\ \mathrm{Sph}_{25}(h'_x) + 45{,}000\ \mathrm{Sph}_{50}(h'_x)$

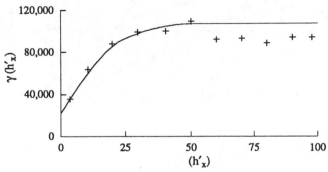

(b) $\gamma_V(h'_y) = 22{,}000 + 40{,}000\ \mathrm{Sph}_{30}(h'_y) + 45{,}000\ \mathrm{Sph}_{150}(h'_y)$

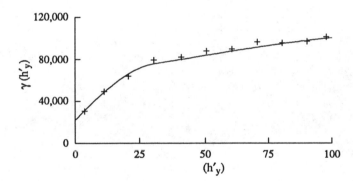

(c)

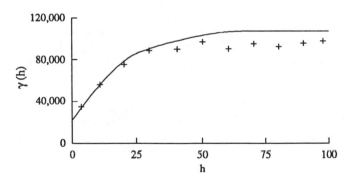

Figure 16.10 The variogram model for V is shown along the minor anisotropy axis, N76°E, in (a), along the major axis, N14°W, in (b). The average of the intermediate directions, N31°E and N59°W, is shown in (c).

(a) $\gamma_U(h'_x) = 440{,}000 + 70{,}000\ \mathrm{Sph}_{25}(h'_x) + 95{,}000\ \mathrm{Sph}_{50}(h'_x)$

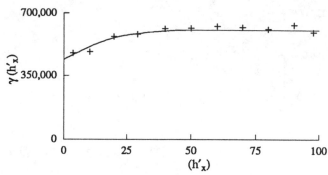

(b) $\gamma_U(h'_y) = 440{,}000 + 70{,}000\ \mathrm{Sph}_{30}(h'_y) + 95{,}000\ \mathrm{Sph}_{150}(h'_y)$

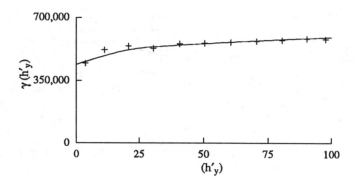

(c)

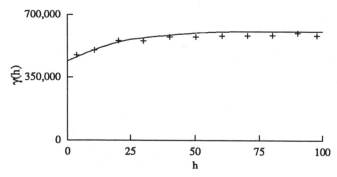

Figure 16.11 The variogram model for U is shown along the minor anisotropy axis, N76°E, in (a), and along the major axis, N14°W, in (b). The average of the intermediate directions, N31°E and N59°W, is shown in (c).

(a) $\gamma_{UV}(h'_x) = 47,000 + 50,000 \; Sph_{25}(h'_x) + 40,000 \; Sph_{50}(h'_x)$

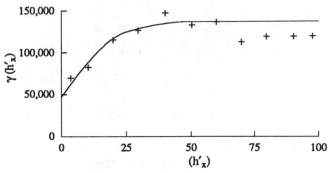

(b) $\gamma_{UV}(h'_y) = 47,000 + 50,000 \; Sph_{30}(h'_y) + 40,000 \; Sph_{150}(h'_y)$

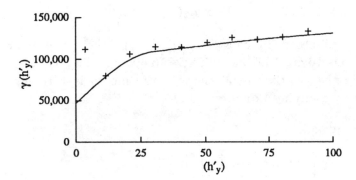

(c)

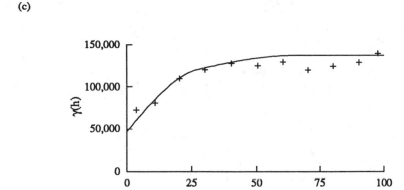

Figure 16.12   The cross-variogram model between U and V is shown along the minor anisotropy axis, N76°E, in (a), and along the major axis, N14°W, in (b). The average of the intermediate directions, N31°E and N59°W, is given in (c).

The length of h'_1 is given by

$$h'_1 = [(h'_{x,1})^2 + (h'_{y,1})^2]^{1/2} \qquad (16.48)$$

and that of h'_2 by

$$h'_2 = [(h'_{x,2})^2 + (h'_{y,2})^2]^{1/2} \qquad (16.49)$$

The anisotropic auto variogram model for U is given by

$$\gamma_U(\mathbf{h}) = 440,000 + 70,000 \; Sph_1(h'_1) + 95,000 \; Sph_1(h'_2) \qquad (16.50)$$

where the reduced distance vectors h'_1, h'_2 and their lengths are given by Equations 16.46, 16.47, 16.48, and 16.49, respectively.

The anisotropic cross-variogram model for U and V is given by

$$\gamma_{VU}(\mathbf{h}) = 47,000 + 50,000 \; Sph_1(h'_1) + 40,000 \; Sph_1(h'_2) \qquad (16.51)$$

where the reduced distance vectors h'_1, h'_2 and their lengths are given by Equations 16.46, 16.47, 16.48, and 16.49, respectively.

Finally the positive definiteness of the model is verified for each structure by applying Equations 16.44.

- The nugget.

$$\det \begin{bmatrix} 22,000 & 47,000 \\ 47,000 & 440,000 \end{bmatrix} = 7,471,000,000 > 0 \qquad (16.52)$$

- The second structure.

$$\det \begin{bmatrix} 40,000 & 50,000 \\ 50,000 & 70,000 \end{bmatrix} = 300,000,000 > 0 \qquad (16.53)$$

- The third structure.

$$\det \begin{bmatrix} 45,000 & 40,000 \\ 40,000 & 95,000 \end{bmatrix} = 2,675,000,000 > 0 \qquad (16.54)$$

Since the determinants of the matrices of the coefficients are all positive and all the diagonal elements are positive, the linear model of coregionalization is positive definite.

Notes

[1] The correspondence between the variogram and the covariance that we discussed in Chapter 9 allows us to go from one to the other. Throughout this chapter, we refer to the modeling of variogram yet show all of our matrices in terms of covariances. Once the variogram model is specified, the covariance for a particular distance and direction, $\mathbf{h}$, can be calculated by subtracting $\gamma(\mathbf{h})$ from the sill of the variogram model.

[2] Even if the solution of the ordinary kriging equations is unique, it may be very sensitive to small changes in the kriging matrices. Some of the kriging weights may be considerably larger than 1 and others considerably less than 0, if the matrices are not positive definite. By respecting the positive definiteness condition, one not only guarantees that the solution exists and is unique, but also that the solution is stable.

[3] Although Equation 16.2 may be true for any choice of weights, it does not provide a very useful way to check for positive definiteness of a matrix in practice. The Notes to Chapter 12 provided three alternative methods that are much more practical.

[4] Elsewhere in the geostatistical literature, the exponential and Gaussian models are given without the factor of 3. While this may make the expression of the basic model more aesthetically pleasing, it has the disadvantage that the parameter a defined this way is not the practical range. We have chosen to present all of the transition models in such a way that a is the range or the practical range. If one is using charts of auxiliary functions (see Chapter 19) for exponential or Gaussian variograms, one should check to see how the chart is calibrated. It is likely that the parameter a, which appears on the chart, is not the range, but rather is one third of the practical range.

[5] The linear variogram model is not strictly positive definite. There are combinations of weights that will make the expression in Equation (16.2) negative. Fortunately, it is positive definite if we insist that the weights in Equation 16.2 must sum to 0. Under these conditions the linear model is a valid model for ordinary kriging and for cokriging which we present in Chapter 17.

[6] The condition that the coefficients must be greater than 0 is sufficient to guarantee positive definiteness, but it is not necessary. A
linear combination of positive definite models with negative coefficients may or may not be positive definite. For example, negative
coefficients are required for modeling cross-variograms when the
two variables are negatively cross-correlated.

[7] In data sets where the structural control is strong, the data values
might be equally continuous along two different directions that are
not mutually perpendicular (for example, the two directions corresponding to conjugate shears) or perhaps the direction of maximum
continuity follows a folded stratigraphic unit and thus changes as
the folded unit changes. The procedures described in this chapter
are appropriate only for those situations where the directions of
minimum and maximum continuity are perpendicular to one another. In data sets where the structural or stratigraphic maximum
and minimum influences are very strong and not necessarily mutually perpendicular it may be best to model the spatial continuity
in a coordinate system that is geologically relevant. Although the
unfolding of structure or the calculation of stratigraphic distances
is certainly more tedious than the straightforward use of a perpendicular data coordinate system, there are several examples in the
literature that demonstrate the advantages of using a coordinate
system relevant to the reality of the data set.

Further Reading

Dagbert, M. et al., "Computing variograms in folded strata-controlled
deposits," in *Geostatistics for Natural Resources Characteriza-
tion*, (Verly, G. , David, M. , Journel, A. G. , and Marechal,
A. , eds.), pp. 71–90, NATO Advanced Study Institute, South
Lake Tahoe, California, September 6-17, D. Reidel, Dordrecht,
Holland, 1983.

Johnson, R. A. and Wichern, D. W. , *Applied Multivariate Statistical
Analysis*. Englewood Cliffs, N.J.: Prentice-Hall, 1982.

Journel, A. G. and Huijbregts, C. J. , *Mining Geostatistics*. London:
Academic Press, 1978.

McArther, G. J. , "Using geology to control geostatistics in the Hellyer deposit," *Mathematical Geology*, vol. 20, no. 4, pp. 343–366, 1968.

Strang, G. , *Linear Algebra and Its Applications*. New York: Academic Press, 1980.

17

COKRIGING

In all of the estimation methods we have previously studied, all estimates were derived using only the sample values of one variable. For example, estimates of V were derived using only the available V data; however, a data set will often contain not only the primary variable of interest, but also one or more secondary variables. These secondary variables are usually spatially cross-correlated with the primary variable and thus contain useful information about the primary variable. We have already seen an example of such cross-correlation in Figure 4.14. The cross h-scatterplots in this figure clearly showed that U values were correlated with nearby V values. In Chapter 12 we saw how we could exploit the spatial correlation of a variable to produce good estimates. Intuitively, it seems we should also be able to exploit the cross-correlation between variables to improve these estimates. It seems reasonable that the addition of the cross-correlated information contained in the secondary variable should help to reduce the variance of the estimation error even further. In this chapter we present *cokriging*, a method for estimation that minimizes the variance of the estimation error by exploiting the cross-correlation between several variables; the estimates are derived using secondary variables as well as the primary variable.

The usefulness of the secondary variable is often enhanced by the fact that the primary variable of interest is undersampled. For example, in the mining industry all available core samples may be assayed for one particular mineral while only those core samples providing a high assay value for that mineral are assayed for a second mineral. Typically, the sampling pattern of the more frequently sampled variable is more regular than that of the undersampled variable. A posting of these variables will most likely reveal large areas where only the more frequently sampled variable exists. In such areas the only information we have about the undersampled variable is the cross-correlated information contained by the other variable.

We begin this chapter with the development of the cokriging system and then follow with an example detailing the construction of the cokriging matrices. The remainder of the chapter consists of a case study that compares estimates of U calculated by ordinary kriging to those calculated by cokriging.

The Cokriging System

In order to simplify the notation, we have chosen to develop the cokriging system in terms of two variables rather than in its full generality with any number of variables.

The cokriging estimate is a linear combination of both primary and secondary data values and is given by:

$$\hat{u}_0 = \sum_{i=1}^{n} a_i \cdot u_i + \sum_{j=1}^{m} b_j \cdot v_j \qquad (17.1)$$

$\hat{u}_0$ is the estimate of U at location 0; $u_1, \ldots, u_n$ are the primary data at n nearby locations; $v_1, \ldots, v_n$ are the secondary data at m nearby locations; $a_1, \ldots, a_n$ and $b_1, \ldots, b_m$ are the cokriging weights that we must determine.

The development of the cokriging system is identical to the development of the ordinary kriging system. For example, we begin by defining the estimation error as

$$\begin{aligned} R &= \hat{U}_0 - U_0 \\ &= \sum_i^n a_i U_i + \sum_j^m b_j V_j - U_0 \end{aligned} \qquad (17.2)$$

where $U_1, \ldots, U_n$ are the random variables representing the U phenomenon at the n nearby locations where U has been sampled and $V_1, \ldots, V_m$ are the random variables representing the V phenomenon at the m nearby locations where V has been sampled. Equation 17.2 can be expressed in matrix notation as

$$R = \mathbf{w}^t \mathbf{Z} \qquad (17.3)$$

where $\mathbf{w}^t = (a_1,\ldots,a_n,b_1,\ldots,b_m,-1)$ and $\mathbf{Z}^t = (U_1,\ldots,U_i,V_1,\ldots,$ $V_m,U_0)$. Equation 17.2 is a linear combination of $n+m+1$ random variables, $U_1,\ldots,U_n$, $V_1,\ldots,V_m$ and U_0. In Chapter 9, we gave an expression for the variance of a linear combination of random variables Equation 9.14 that allows us to express the variance of R as

$$Var\{R\} = \mathbf{w}^t\mathbf{C_Z}\mathbf{w} \qquad (17.4)$$

where $\mathbf{C_Z}$ is the covariance matrix of $\mathbf{Z}$. Expanding and simplifying Equation 17.4 we obtain an expression for the variance of the estimation error in terms of the cokriging weights and the covariances between the random variables:

$$
\begin{aligned}
Var\{R\} &= \mathbf{w}^t\mathbf{C_Z}\mathbf{w} \\
&= \sum_i^n\sum_j^n a_ia_jCov\{U_iU_j\} \;+\; \sum_i^m\sum_j^m b_ib_jCov\{V_iV_j\} \\
&+ 2\sum_i^n\sum_j^m a_ib_jCov\{U_iV_j\} \;-\; 2\sum_i^n a_iCov\{U_iU_0\} \\
&- 2\sum_j^m b_jCov\{V_jU_0\} \;+\; Cov\{U_0U_0\}
\end{aligned}
$$
$$(17.5)$$

where $Cov\{U_iU_j\}$ is the auto covariance between U_i and U_j, $Cov\{V_iV_j\}$ is the auto covariance between V_i and V_j and $Cov\{U_iV_j\}$ is the cross-covariance between U_i and V_j.

The set of cokriging weights we are looking for must satisfy two conditions. First, the weights must be such that the estimate given in Equation 17.1 is unbiased. Second, the weights must be such that the error variances given in Equation 17.5 are the smallest possible.

First, we will tackle the unbiasedness condition. The expected value of the estimate given in Equation 17.1 is

$$
\begin{aligned}
E\{\hat{U}_0\} &= E\{\textstyle\sum_{i=1}^n a_iU_i + \sum_{j=1}^m b_jV_j\} \\
&= \textstyle\sum_{i=1}^n a_iE\{U_i\} + \sum_{j=1}^m b_jE\{V_j\} \qquad (17.6)\\
&= \tilde{m}_U\cdot\textstyle\sum_{i=1}^n a_i + \tilde{m}_V\cdot\sum_{j=1}^m b_j
\end{aligned}
$$

where $E\{U_i\} = \tilde{m}_U$ and $E\{V_j\} = \tilde{m}_V$.

From this equation, it appears that one way of guaranteeing unbiasedness is to ensure that the weights in the first term sum to 1 while those in the second sum to 0:

$$\sum_{i=1}^n a_i = 1 \quad \text{and} \quad \sum_{j=1}^m b_j = 0 \qquad (17.7)$$

Though the conditions given in Equation 17.7 are certainly the most commonly used nonbias conditions, it should be noted that other nonbias conditions are possible; in the cokriging case study we present an alternative nonbias condition and compare the results to those obtained using the conditions given here.

We are now faced with a classical minimization problem subject to two constraints. We are looking for the set of weights that minimizes the error variance given in Equation 17.5 and also fulfills the two nonbias conditions, $\sum_i^n a_i = 1$ and $\sum_j^m b_j = 0$. As in Chapter 12, the Lagrange multiplier method may be used to minimize a function with two constraints. To implement the method we simply equate each nonbias condition to 0, multiply by a Lagrange multiplier, and add the result to Equation 17.5. This gives us the following expression:

$$Var\{R\} = \mathbf{w}^t \mathbf{C_Z} \mathbf{w} + 2\mu_1 \left(\sum_{i=1}^n a_i - 1\right) + 2\mu_2 \left(\sum_{j=1}^m b_j\right) \qquad (17.8)$$

where μ_1 and μ_2 are the Lagrange multipliers. Note that the two additional terms are both equal to 0 and do not contribute to the error variance $Var\{R\}$.

To minimize Equation 17.8 we compute the partial derivatives of $Var\{R\}$ with respect to the $n + m$ weights and the two Lagrange multiplicrs:

$$\frac{\partial(Var\{R\})}{\partial a_j} = 2\sum_{i=1}^n a_i Cov\{U_i U_j\} + 2\sum_{i=1}^m b_i Cov\{V_i U_j\}$$
$$-2Cov\{U_0 U_j\} + 2\mu_1 \qquad \text{for } j = 1, n$$

$$\frac{\partial(Var\{R\})}{\partial b_j} = 2\sum_{i=1}^n a_i Cov\{U_i V_j\} + 2\sum_{i=1}^m b_i Cov\{V_i V_j\}$$
$$-2Cov\{U_0 V_j\} + 2\mu_2 \qquad \text{for } j = 1, m$$

$$\frac{\partial(Var\{R\})}{\partial \mu_1} = 2\sum_{i=1}^n a_i - 1$$

$$\frac{\partial(Var\{R\})}{\partial \mu_2} = 2\sum_{i=1}^m b_i$$

The cokriging system is finally obtained by equating each of these

$n + m + 2$ equations to 0 and rearranging the individual terms:

$$\sum_{i=1}^{n} a_i Cov\{U_i U_j\} + \sum_{i=1}^{m} b_i Cov\{V_i U_j\} + \mu_1 = Cov\{U_0 U_j\} \qquad \text{for } j = 1, n$$

$$\sum_{i=1}^{n} a_i Cov\{U_i V_j\} + \sum_{i=1}^{m} b_i Cov\{V_i V_j\} + \mu_2 = Cov\{U_0 V_j\} \qquad \text{for } j = 1, m$$

$$\sum_{i=1}^{n} a_i = 1$$

$$\sum_{i=1}^{m} b_i = 0 \qquad\qquad (17.9)$$

The corresponding minimized error variance can be calculated using Equation 17.5 or, for this particular set of nonbias conditions, Equation 17.5 can be simplified by making substitutions using the Lagrange multipliers. The simplified version is:

$$Var\{R\} = Cov\{U_0 U_0\} + \mu_1 - \sum_{i=1}^{n} a_i Cov\{U_i U_0\} - \sum_{j=1}^{m} b_j Cov\{V_j U_0\}$$

$$(17.10)$$

The cokriging system given in Equation 17.9 is valid only for point estimation. If an estimate of the mean is required over a local area A, two options are available:

1. Estimate a number of point values on a regular grid within A and average them together to obtain an estimate of the mean within the area.

2. Replace all the covariance terms $Cov\{U_0 U_i\}$ and $Cov\{U_0 V_j\}$ in the cokriging system Equation 17.9, with average covariance values $\overline{Cov}\{U_A U_i\}$ and $\overline{Cov}\{U_A V_j\}$, where $\overline{Cov}\{U_A U_i\}$ is the average covariance between U_i and the point U values within A and $\overline{Cov}\{(U_A V_j\}$ is the average cross-covariance between the V_j and the point U values in A.

The cokriging system can be written in terms of the semivariogram provided the cross-covariance is symmetric, $Cov\{U_i V_j\} = Cov\{V_j U_i\}$. Though the cross-covariance may be nonsymmetric, it is most often modeled in practice as a symmetric function. The spatial continuity is modeled using semivariograms that are then converted to

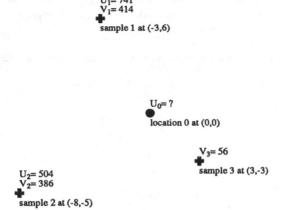

Figure 17.1 A small cokriging data configuration consisting of two primary and three secondary data values.

covariance values for the cokriging matrices using the relationship
$C_{UV}(h) = \gamma_{UV}(\infty) - \gamma_{UV}(h)$

In order for the solution of the cokriging equations to exist and be unique, the set of auto- and cross-variograms must be positive definite. The use of the linear model of coregionalization (Chapter 16) with positive definite matrices of coefficients satisfies this condition.

There are certain situations where cokriging will not improve an ordinary kriging estimate. If the primary and secondary variables both exist at all data locations and the auto- and cross-variograms are proportional to the same basic model then the cokriging estimates will be identical to those of ordinary kriging. Thus if all the variogram models are "quite similar" in shape and the primary variable is not noticeably undersampled, cokriging will not improve things very much.

A Cokriging Example

Our goal in this example is to illustrate the actual construction of a cokriging system. A cokriging data configuration is given in Figure 17.1 and consists of two primary U and three secondary V data surrounding a point to be estimated. The spatial continuity is provided by the linear model of coregionalization obtained from the 275

Table 17.1 Tabulation of the covariance and cross-covariance values for the data configuration given in Figure 17.1 using the linear model of coregionalization given in Equation 17.11.

Variable pair	Grid distance	Structural distance	$C_U(h)$	$C_V(h)$	$C_{UV}(h)$
$U_1 U_1$	0.	0.	605,000		
$U_1 U_2$	12.1	9.1	99,155		
$U_2 U_2$	0.	0.	605,000		
$V_1 V_1$	0.	0.		107,000	
$V_1 V_2$	12.1	9.1		49,623	
$V_1 V_3$	10.8	5.0		57,158	
$V_2 V_2$	0.	0.		107,000	
$V_2 V_3$	11.2	11.2		45,164	
$V_3 V_3$	0.	0.		107,000	
$U_1 V_1$	0.	0.			137,000
$U_1 V_2$	12.1	9.1			49,715
$U_1 V_3$	10.8	5.0			57,615
$U_2 V_1$	12.1	9.1			49,715
$U_2 V_2$	0.	0.			137,000
$U_2 V_3$	11.2	11.2			45,554
$U_0 U_1$	6.7	2.6	134,229		
$U_0 U_2$	9.4	9.0	102,334		
$U_0 V_1$	6.7	2.6			70,210
$U_0 V_2$	9.4	9.0			52,697
$U_0 V_3$	4.2	2.5			75,887

U and 470 V sample data. This model is developed in the last part of Chapter 16 and is given again in Equation 17.11. All covariance and cross-covariance values for the data configuration in Figure 17.1 have been computed using this model and are listed in Table 17.1. Note that the covariances are all symmetric; $C(h) = C(-h)$, although $C_{UV}(h)$ need not necessarily equal $C_{UV}(-h)$.

Using the covariance values shown in Table 17.1 the matrix form of the cokriging system given in Equation 17.9 is constructed as follows [1]:

Table 17.2 Cokriging weights and the solution for the example shown in Figure 17.1.

Variable	Cokriging Weight	Estimate	Cokriging Variance
U_1	0.512		
V_1	-0.216		
U_2	0.488		
V_2	-0.397		
V_3	0.666		
μ_1	205,963		
μ_2	13,823		
U_0		398	681,549

$$
\begin{array}{c}
\quad\;\; U_1 \quad\;\; U_2 \quad\;\; V_1 \quad\;\; V_2 \quad\;\; V_3 \\
\begin{array}{c} U_1 \\ U_2 \\ V_1 \\ V_2 \\ V_3 \\ \\ \end{array}
\begin{pmatrix}
605000 & 99155 & 137000 & 49715 & 57615 & 1 & 0 \\
99155 & 605000 & 49715 & 137000 & 45554 & 1 & 0 \\
137000 & 49715 & 107000 & 49623 & 57158 & 0 & 1 \\
49715 & 137000 & 49623 & 107000 & 45164 & 0 & 1 \\
57615 & 45554 & 57158 & 45164 & 107000 & 0 & 1 \\
1 & 1 & 0 & 0 & 0 & 0 & 0 \\
0 & 0 & 1 & 1 & 1 & 0 & 0
\end{pmatrix}
\cdot
\begin{pmatrix}
a_1 \\ a_2 \\ b_1 \\ b_2 \\ b_3 \\ -\mu_1 \\ -\mu_2
\end{pmatrix}
=
\begin{pmatrix}
134229 \\ 102334 \\ 70210 \\ 52697 \\ 75887 \\ 1 \\ 0
\end{pmatrix}
\begin{array}{c} \\ \\ \\ \\ \\ \\ U_0 \end{array}
\end{array}
$$

The weights obtained from the solution of the cokriging system are given in Table 17.2 along with the final estimate, 398 ppm. The ordinary kriging estimate of U_0 for this small example is 630 ppm.

A Case Study

This case study provides a comparison between cokriging and ordinary kriging. For the cokriging we estimated the undersampled variable U using both the 275 U and 470 V sample data; for ordinary kriging, we used only the 275 U data. The linear model of coregionalization obtained from the sample variograms of these data is given by:

$$\gamma_U(\mathbf{h}) = 440,000 + 70,000 \ Sph_1(\mathbf{h}'_1) + 95,000 \ Sph_1(\mathbf{h}'_2)$$

$$\gamma_V(\mathbf{h}) = 22,000 + 40,000 \ Sph_1(\mathbf{h}'_1) + 45,000 \ Sph_1(\mathbf{h}'_2)$$

$$\gamma_{VU}(\mathbf{h}) = 47,000 + 50,000 \ Sph_1(\mathbf{h}'_1) + 40,000 \ Sph_1(\mathbf{h}'_2)$$

$$(17.11)$$

where the vectors $\mathbf{h}'_n$ are calculated as follows:

$$\mathbf{h}'_1 = \begin{bmatrix} h'_{x,1} \\ h'_{y,1} \end{bmatrix} = \begin{bmatrix} \frac{1}{25} & 0 \\ 0 & \frac{1}{30} \end{bmatrix} \cdot \begin{bmatrix} cos(14) & sin(14) \\ -sin(14) & cos(14) \end{bmatrix} \cdot \begin{bmatrix} h_x \\ h_y \end{bmatrix} \quad (17.12)$$

and

$$\mathbf{h}'_2 = \begin{bmatrix} h'_{x,2} \\ h'_{y,2} \end{bmatrix} = \begin{bmatrix} \frac{1}{50} & 0 \\ 0 & \frac{1}{150} \end{bmatrix} \cdot \begin{bmatrix} cos(14) & sin(14) \\ -sin(14) & cos(14) \end{bmatrix} \cdot \begin{bmatrix} h_x \\ h_y \end{bmatrix} \quad (17.13)$$

The lengths of $\mathbf{h}'_1$ and $\mathbf{h}'_2$ are given by:

$$\mathbf{h}'_1 = [(h'_{x,1})^2 + (h'_{y,1})^2]^{1/2}$$

$$\mathbf{h}'_2 = [(h'_{x,2})^2 + (h'_{y,2})^2]^{1/2}$$

$$(17.14)$$

For the ordinary kriging exercise we used the variogram model $\gamma_U(\mathbf{h})$ as it is given here.

The cokriging plan calls for point estimation on a regular 10 x 10 m^2 grid. The search radius was restricted to 40 m, and a quadrant search was used to limit the total number of data in each quadrant to no more than 6. Within each quadrant, no more than the three closest U samples were used. The closest V samples were also retained, up to a total of 6 minus the number of U samples already retained.

A final restriction limited the extrapolation of the primary variable. Figure 17.2 is a posting of the 275 sample U data that was shown earlier in Chapter 6 with the actual sample values posted. No point estimations were made further than 11 m from the closest U sample value; only 285 of the 780 possible grid points were kriged.

Two cokrigings were actually done, with each using a different set of nonbias conditions. The first cokriging was done using the familiar conditions $\sum_{i=1}^{n} a_i = 1$ and $\sum_{j=1}^{m} b_j = 0$ for the primary and secondary weights, respectively. The second cokriging used only one nonbias

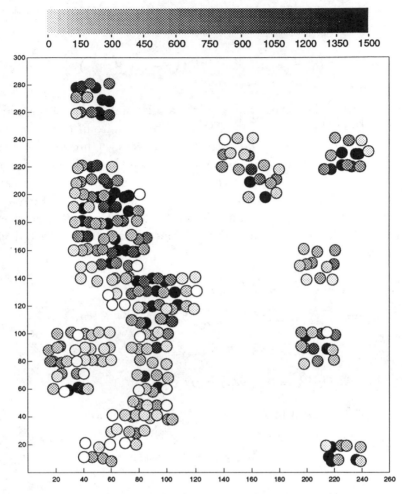

Figure 17.2 A posting of the 275 U sample data. The actual sample value corresponds to degree of shading as indicated by the gray scale at the top of the figure.

condition, which required that the sum of all the weights must equal 1:

$$\sum_{i=1}^{n} a_i + \sum_{j=1}^{m} b_j = 1 \qquad (17.15)$$

With this alternative unbiasedness condition, the estimator must be slightly modified. The unknown U value is now estimated as a weighted linear combination of the nearby U values plus a weighted linear combination of the nearby V values adjusted by a constant so that their mean is equal to the mean of the U values:

$$\hat{U}_0 = \sum_{i=1}^{n} a_i U_i + \sum_{j=1}^{m} b_j (V_j - \hat{m}_V + \hat{m}_U) \qquad (17.16)$$

As one can see, this estimator requires additional information, namely an estimate of the mean value of U and an estimate of the mean value of V over the area that the estimation covers. One simple way to estimate the means m_V and m_U is to compute the arithmetic averages of the 275 sample U values and the corresponding 275 V sample values. These are probably reasonable estimates since the sampling pattern for both variables within this area is more or less free from clustering. Assuming both these estimates are unbiased, then the expected value of the point estimate is:

$$\begin{aligned} E\{\hat{U}_0\} &= E\{\sum_{i=1}^{n} a_i U_i + \sum_{j=1}^{m} b_j (V_j - \hat{m}_V + \hat{m}_U)\} \\ &= \sum_{i=1}^{n} a_i E\{U_i\} + \sum_{j=1}^{m} b_j (E\{V_j\} - E\{\hat{m}_V\} + E\{\hat{m}_U\}) \\ &= m_U \cdot \sum_{i=1}^{n} a_i + m_U \cdot \sum_{j=1}^{m} b_j \\ &= m_U (\sum_{i=1}^{n} a_i + \sum_{j=1}^{m} b_j) \end{aligned}$$

$$(17.17)$$

where $E\{U_i\} = m_U$ and $E\{V_j\} = m_V$. The condition

$$\sum_{i=1}^{n} a_i + \sum_{j=1}^{m} b_j = 1 \qquad (17.18)$$

therefore ensures that the cokriging estimates are unbiased.

The search strategy for the ordinary kriging was similar to the cokriging plan. The same point locations were kriged, the search radius was 40 m and the quadrant search was employed with a maximum of six data per quadrant.

The summary statistics in Table 17.3 reveal a global bias in all the estimates. The reason for this bias is due largely to the extrapolation of high sample values from the Wassuk range anomaly into the bordering area, that contains small U values. Recall that U samples were obtained only in areas of high V values, and if we refer to the exhaustive indicator maps in Figures 5.9a-i and 5.10a-i we see that the anomalous high areas of U correspond with those of V; thus, the 275

Table **17.3** Summary statistics for the ordinary and cokriging estimates of U as well as for the true values of U.

	True values	Ordinary kriging	Cokriging with 1 nonbias condition	Cokriging with 2 nonbias conditions
n	285	285	285	285
m	434	566	493	489
σ	564	357	279	392
CV	1.30	0.63	0.57	0.80
min	0	81	56	-156
Q_1	30	275	289	192
M	205	502	447	468
Q_3	659	796	680	737
max	3,176	1,613	1,496	1,702
$\rho_{\hat{U}U}$		0.48	0.57	0.52

sample values of U are from areas of anomalously high values. This is confirmed by the large difference between the sample mean of U, 434 ppm, and the exhaustive mean value of U, 266 ppm. Note that the ordinary kriged estimates are the most severely biased and that the cokriged estimates with one nonbias condition are the most smoothed.

Perhaps the most important thing to notice in this table is the large negative cokriging estimate, -156 ppm; of the 285 estimates, 17 were negative. The reason for the negative estimates originates with the nonbias condition $\sum_{j=1}^{m} b_j = 0$. In order for these weights to sum to 0, some of them must necessarily be negative. When these negative weights are multiplied by large V sample values, negative estimates can occur. If the sum of the negative products is larger in absolute value than the sum of the positive weights times their sample values, then the estimate is negative. Note that the cokriging with one nonbias condition does not produce any negative estimates.

Summary statistics for the three distributions of estimation errors are tabulated in Table 17.4. Of the three, the error distribution resulting from the cokriging with one nonbias condition is the most accept-

Table 17.4 Summary statistics for the distribution of ordinary and cokriging estimation errors of U.

	Ordinary kriging	Cokriging with 1 nonbias condition	Cokriging with 2 nonbias conditions
n	285	285	285
m	133	59	55
σ	502	466	493
min	-1,979	-2,469	-2,287
Q_1	-89	-60	-121
M	164	152	73
Q_3	427	328	362
max	1553	941	1412
MAE	394	346	356
MSE	268,264	219,701	245,681

able. It has the smallest spread of errors as measured by the standard deviation, interquartile range, mean absolute error, and mean squared error. It also has the smallest maximum error.

Perhaps the most informative comparison of the three estimations can be made using three maps showing the locations of the estimation errors. Figure 17.3 is a posting of the ordinary kriging estimation errors. At first glance the residuals seem to be more or less evenly distributed; however, a closer examination shows a border of positive residuals around the Wassuk range anomaly. This is indicative of overestimation caused by the extrapolation of the high sample values from the Wassuk range anomaly into the bordering areas that contain relatively lower values of both U and V. Such overestimation is prevalent along the northeast border of the Wassuk anomaly, as indicated by the dark plus signs.

Figure 17.4 is a posting of the cokriging estimation errors using two nonbias conditions. This map is quite similar to the map of the ordinary kriged residuals, although some of the overestimations bordering the Wassuk anomaly are smaller, indicated by the slightly lighter shad-

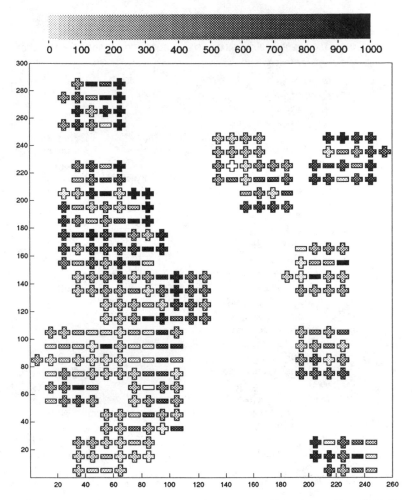

Figure 17.3 A posting of the 285 ordinary kriging estimation errors. Positive and negative errors are indicated by the + and − signs, respectively. The value of the error corresponds to the degree of shading as shown by the gray scale.

ing of the plus signs. The reduction in the extrapolated overestimations is due to the influence of the smaller secondary V values bordering the Wassuk anomaly.

The cokriging estimation errors using one nonbias condition are posted in Figure 17.5. The map contains noticeably fewer severe overestimations along the border of the Wassuk anomaly than the previous

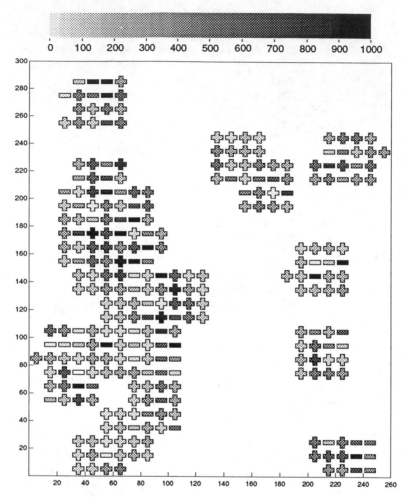

Figure 17.4 A posting of the 285 cokriging estimation errors using 2 nonbias conditions. Positive and negative errors are indicated by the + and − signs, respectively.

two maps do. Most of the symbols are lightly shaded indicating relatively small estimation errors. This is especially true along the northeast border of the Wassuk anomaly. Again, this improvement in the estimations is due to the influence of the smaller V sample values bordering the Wassuk anomaly. The stronger influence of these values in

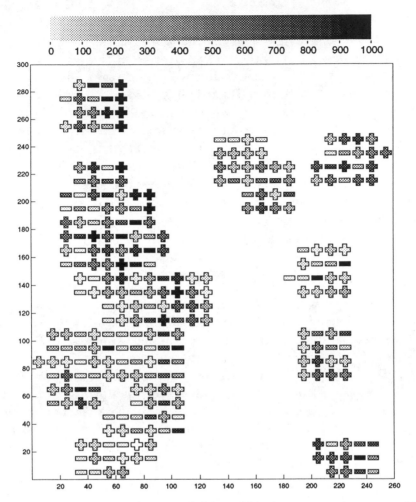

Figure 17.5 A posting of the 285 cokriging estimation errors using one nonbias condition. Positive and negative errors are indicated by the + and − signs, respectively. Note the large overestimations along the northeast border of the Wassuk Range in Figure 17.3 are considerably smaller in this figure.

this estimation is due to the alternative nonbias condition that results in more weight being attributed to the secondary variable.

To summarize, it appears that cokriging with two nonbias conditions is less than satisfactory. Consider the case where only two secondary data values are found equidistant from the point of estimation and from all primary data. Since they are equidistant, they must be weighted equally, and since the nonbias condition requires the weights to sum to 0, one weight must be negative, the other positive. Although this solution is mathematically correct it is difficult to imagine a physical process for which such a weighting scheme is appropriate. Cokriging with two nonbias conditions is also prone to negative estimates; 17 of the 285 estimates were negative. Though this method did reduce the bias, it did not reduce the spread of the errors by much.

Using one nonbias condition, however, gave us considerable improvements, not only in the bias and the spread of the errors, but also in the lower incidence of negative estimates. Though this approach required a prior estimation of the global means of U and V, it is clear from the case study that even with a rather simple estimate of these means we can substantially improve the estimation.

Notes

[1] A word of caution: some algorithms designed for solving systems of equations may develop numerical problems with the covariance matrices as they are given in this example. The large differences of up to five orders of magnitude between matrix elements may lead to problems in precision and provide bad results. One way around this is to rescale all the covariance values. For example, we can divide all the covariance entries (except the 1s) by 10,000 without altering the correct solution. Then the elements in the covariance matrix are all closer to the same order of magnitude and the solution less prone to numerical instability.

Further Reading

Edwards, C. and Penney, D. , *Calculus and Analytical Geometry.* Englewood Cliffs N.J.: Prentice-Hall, 1982.

18

ESTIMATING A DISTRIBUTION

The estimation methods we have examined in earlier chapters are appropriate for the estimation of a mean value. In Chapter 10 we looked at techniques for estimating a global mean, while in Chapter 13 we looked at techniques for estimating a local mean. There are many important problems, however, that call for estimates of other characteristics of a distribution of unknown values; in Chapter 8 we gave several examples of such problems. In this chapter, we will address the issue of estimating the complete distribution of unknown values, both globally and locally.

As with the methods we discussed for estimating a mean, the global and local estimation of a complete distribution calls for different approaches. If we have many sample data within an area of interest, we typically approach it as a global estimation problem. If there are few available samples within an area of interest, we view it as a local estimation problem. As we will see shortly, the tools we use to estimate global and local distributions are the same as those we used to estimate global and local means: the global problem can be addressed by finding appropriate declustering weights for the available samples, while the local problem can be addressed by finding weights that account not only for clustering, but also for the distance from the area being estimated to each nearby sample. The only difference between the methods discussed here and those that we proposed for estimating a mean value is that instead of applying them to the actual sample value, we apply them to a transformed value known as an *indicator*.

Cumulative Distributions

The estimation of a complete distribution can be accomplished either by estimating the proportion of values that fall within particular classes or by estimating the proportion of values that fall above or below certain thresholds. In the first approach, we are estimating the frequency distribution; in the second we are estimating the cumulative frequency distribution. From an estimate of either of these, the other one can easily be calculated.

In this chapter, we will be estimating the cumulative frequency distribution directly. We will define the cumulative distribution function, denoted $F(v_c)$, to be the proportion of values below the value v_c. The cumulative frequency below the minimum value is 0 and the cumulative frequency below the maximum value is 1:

$$F(v_{min}) = 0 \qquad F(v_{max}) = 1$$

There are two general approaches to estimating cumulative distributions. The first, usually referred to as the *nonparametric* approach, is to calculate estimates of $F(v)$ at several values of v: $\hat{F}(v_1), \ldots, \hat{F}(v_n)$. The second, usually referred to as the *parametric* approach, is to determine a function that completely describes the cumulative distribution for any value of v. While the parametric approach gives us the entire function $\hat{F}(v)$, it depends very heavily on the use of random function model in which the multivariate distribution is assumed to be known. The nonparametric approach does not lean as heavily on the random function model, but does not produce a complete estimate of the cumulative distribution. If the cumulative distribution is needed at thresholds other than those at which it was actually estimated, some interpolation or extrapolation between the available estimates is required. Such interpolation or extrapolation always involves some assumptions about the nature of the cumulative distribution; particularly for extrapolation beyond the last threshold at which cumulative distribution was actually estimated, these assumptions can have a large impact.

In this chapter, we will adopt a nonparametric approach to the problem of estimating a complete distribution. With such an approach, the estimation of a complete distribution involves the estimation of the proportion above or below several cutoff values. The basic problem in estimating a complete distribution, therefore, is the estimation of the

Table 18.1   The number of sample V values below 500 ppm in each of the three sampling campaigns.

Campaign	Total Number	Number Below 500 ppm	Percentage
1	195	160	82
2	150	69	46
3	125	39	31

proportion of an unknown exhaustive distribution that lies above or below a particular threshold.

The Inadequacy of a Naïve Distribution

Let us begin our discussion of the problem by looking at the estimation of the proportion of the the exhaustive Walker Lake area that has a V value below 500 ppm. Of the 78,000 V values in the exhaustive data set, 63,335 (roughly 80%) are below 500 ppm. In practice, however, we do not have an exhaustive data set to which we can refer; our estimate of the proportion above 500 ppm must be based only on the available samples.

A straightforward but naïve approach is to use the histogram of our samples. Of the 470 available samples, 268 have V values below 500 ppm; it would clearly be a mistake to conclude from this simple calculation that only 57% of the Walker Lake area has V values below 500 ppm. Even without the privilege of knowing the correct answer, we should be suspicious of this simple counting of the samples below 500 ppm since we already know that our samples have been preferentially located in areas with high V values.

Table 18.1 shows the number of samples below 500 ppm for each of the three sampling campaigns. In the first campaign, the only one in which the samples were located on a pseudoregular grid, about 80% of the samples are below 500 ppm; in the second campaign, in which additional samples were located near the highest from the first campaign, less than 50% of the samples are below 500 ppm; in the third campaign, this proportion is barely 30%. The only campaign in which a simple counting of the samples produces a reasonable answer is the first one. As subsequent samples are more likely to be located in areas with high values, the simple counting of samples increasingly underestimates the actual proportion of values below 500 ppm.

Though the first campaign is more reliable since its samples are not preferentially clustered, it seems wasteful to completely ignore the contribution of the other two campaigns; despite their preferential clustering, they still contain useful information that we should be able to use. Earlier, when we looked at the estimation of the global mean in Chapter 10, we ran into a similar problem. By itself, the first campaign gave us a more reasonable estimate for the actual exhaustive mean V value than either of the last two campaigns. The same tools that we used then to incorporate the clustered information in an estimate of the global mean can also be used to incorporate the clustered information in an estimate of the global proportion below any particular threshold. Before we look at how to adapt those earlier methods, let us first take a look at why point estimates are inadequate for our purpose.

The Inadequacy of Point Estimates

Having noted that the clustering of the available samples makes a simple counting of the available samples a poor estimate of the true proportion, it might seem that a global distribution could be estimated by first calculating point estimates on a regular grid then combining these point estimates into a global distribution. Unfortunately, this method is also inadequate since point estimates typically have less variability than true values. When we compared various point estimation procedures in Chapter 11, we noticed that the standard deviation of our estimates was less than that of the true values. This reduction in variability is often referred to as the *smoothing effect* of estimation, and is a result of the fact that our estimates are weighted linear combinations of several sample values. In general, the use of more sample values in a weighted linear combination increases the smoothness of the estimates. For example, Table 11.5 showed that the triangulation estimates, each of that incorporated three sample values, were more variable (less smoothed) than the inverse distance squared estimates which incorporated all samples within 25 m of the point being estimated.

Table **18.2** The number of point estimates for which the estimated *V* value was below 500 ppm in each of three point estimation methods studied earlier.

Method	Total Number	Standard Deviation	Number Below 500 ppm	Percentage
True	780	251	627	80
Polygonal	780	246	628	81
Triangulation	672	211	580	86
Ordinary Kriging	780	202	671	86

Table 18.2 shows the hazards of using point estimation to estimate the proportion of true values below 500 ppm. For three of our earlier point estimation case studies, this table shows the number of estimates for which the estimated *V* value was below 500 ppm. By virtually all of the criteria we discussed in Chapter 11 for evaluating sets of point estimates, both the triangulation estimates and the ordinary kriging estimates were better point estimates than the polygonal ones. As Table 18.2 shows, however, the reduced variability of these point estimates makes them unreliable as estimates of a global distribution. With the actual values being more variable than their corresponding estimates, the proportion of estimates below a particular threshold will not accurately reflect the proportion of actual values below that same threshold. There will be a greater proportion of estimates than true values below high cutoffs such as the 500 ppm cutoff we were looking at earlier. For low cutoffs, the reverse is true; the proportion calculated from the distribution of point estimates will be too small.

Cumulative Distributions, Counting, and Indicators

The solution to the problem of estimating the proportion below a certain cutoff from a sample data set lies in understanding what it is we do when we calculate the actual proportion from an exhaustive data set. With access to the exhaustive Walker Lake data set, we calculated the proportion of values below 500 ppm by counting the number of values below this cutoff and dividing by the total number of values

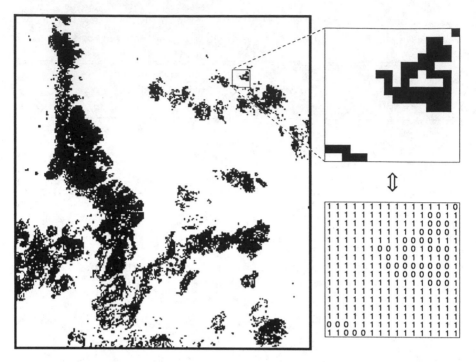

Figure 18.1 The indicator map of the exhaustive Walker Lake data set for the 500 ppm cutoff. All values below 500 ppm have an indicator of 1 and are shown in white; all values above 500 ppm have an indicator of 0 and are shown in black.

in the data set:

$$F(500) = \frac{\text{Number of values below 500 ppm}}{\text{Total number of values}} = \frac{63,335}{78,000} = 0.81 \quad (18.1)$$

The simple notion of counting can be described mathematically by an indicator variable. For the 500 ppm threshold, we could transform each one of our 78,000 exhaustive values into an indicator as follows:

$$i_j = \begin{cases} 1 & \text{if } v_j \leq 500 \\ 0 & \text{if } v_j > 500 \end{cases}$$

The displays of the exhaustive data set that we presented in Figure 5.9 were maps of this indicator variable for different thresholds. Figure 18.1 shows the map of the indicator variable for the 500 ppm cutoff. All of the locations at which the indicator is 0 are shown in black and all of the locations at which the indicator is 1 are shown in white. The number of values below 500 ppm is the sum of all of the indicators:

$$\text{Number of values below 500 ppm} = \sum_{j=1}^{n} i_j$$

Equation 18.1, which gave the cumulative proportion of samples below 500 ppm, can be written as

$$F(500) = \frac{\sum_{j=1}^{n} i_j}{n} = \frac{63,335}{78,000} = 0.81 \tag{18.2}$$

This procedure can be repeated for any cutoff. To calculate the proportion of values below the cutoff v_c, we can transform the values $v_1, \ldots, v_n$ to a set of corresponding indicator variables $i_1(v_c), \ldots, i_n(v_c)$, with the following:

$$i_j(v_c) = \begin{cases} 1 & \text{if } v_j \leq v_c \\ 0 & \text{if } v_j > v_c \end{cases} \tag{18.3}$$

The cumulative proportion of values below any cutoff can then be expressed as

$$F(v_c) = \frac{1}{n} \sum_{j=1}^{n} i_j(v_c) \tag{18.4}$$

This equation shows that like the exhaustive mean, m, which is an equally weighted average of the 78,000 V values, the exhaustive proportion below any threshold, $F(v_c)$, can also be expressed as an equally weighted average. By translating the notion of counting into the concept of an indicator, we have managed to translate the notion of a proportion below a certain cutoff into the concept of an average indicator.

In fact, the recognition that the proportion below a certain cutoff is an average indicator allows us to adapt our previous estimation tools to handle the problem of estimating a complete distribution. In the global and local procedures we described earlier, we were trying to estimate a mean value. Though the true mean could be expressed as a simple average of the true values, our estimate was expressed as a weighted average of the available sample values. The weights we chose for each

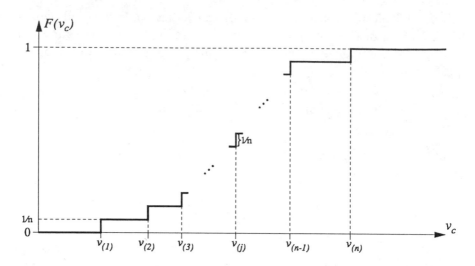

Figure 18.2   The cumulative distribution function as defined by Equation 18.4. At each of the values in the exhaustive data set, the cumulative proportion below that cutoff value increases by $\frac{1}{n}$.

sample value accounted for clustering, in the case of global estimation, and for both clustering and distance, in the case of local estimation. For the estimation of the proportion below a particular cutoff v_c, the only adaptation we have to make to these earlier procedures is that instead of dealing with the sample values $v_1, \ldots, v_n$, we will deal with the corresponding indicators $i_1(v_c), \ldots, i_n(v_c)$.

Figure 18.2 shows how $F(v_c)$ behaves. For values of v_c less than the minimum $v_{(1)}$, the cumulative distribution is 0; at $v_{(1)}$ it jumps to $\frac{1}{n}$. It continues in this staircase pattern, jumping by $\frac{1}{n}$ at every cutoff value that coincides with one of the values in the data set. Its last jump is at the maximum value in the data set, $v_{(n)}$, at which point the cumulative frequency is 1.

Estimating a Global Cumulative Distribution

For any cutoff v_c, we can transform the continuous values of the variable V into an indicator $I(v_c)$ using Equation 18.3. The actual proportion of true values below v_c will be the average of all the true indicators. In practice, we do not have access to all of the true values, but only

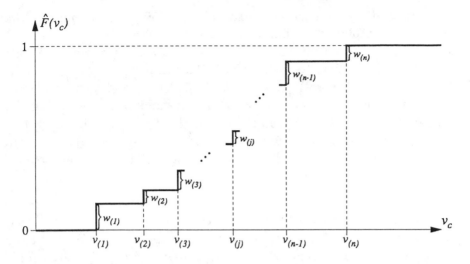

Figure 18.3 The estimated cumulative distribution function as defined by Equation 18.5. At each of the sample values, the estimated proportion below that cutoff value increases by the weight assigned to that sample.

to a set of samples. We can apply the indicator transformation to our available samples and estimate the actual proportion by taking a weighted average of these sample indicators:

$$\hat{F}(v_c) = \sum_{j=1}^{n} w_j \cdot i_j(v_c) \qquad (18.5)$$

As with our earlier estimates, the n weights are standardized so that they sum to 1.

Figure 18.3 shows how the estimated cumulative distribution function behaves. Like its exhaustive counterpart, $\hat{F}(v_c)$ starts at 0 and rises to 1 in a series of steps. The height of the steps, however, is not the same at each of the sample values $v_1, \ldots, v_n$. At each sample value, v_j, the estimated cumulative distribution function increases by w_j, the weight assigned to that particular sample.

If the available samples cover the entire area of interest, with no clustering in anomalous areas, then the sample indicators can be equally weighted. The equal weighting of sample indicators is identical to the procedure of counting the number of sample below the chosen cutoff and dividing by the total number of samples. For example, earlier we estimated the proportion of values below 500 ppm by counting the number of samples in the first campaign that were below 500 ppm and dividing by 195, the total number of samples in that campaign. This procedure is identical to taking those 195 samples, assigning an indicator of 1 those whose V value is below 500 ppm and an indicator of 0 to those whose V value is above 500 ppm, and then calculating the simple average of the 195 indicators. For this first campaign, in which the sampling was on a pseudoregular grid, this equal weighting of the sample indicators produced a good estimate of the true proportion below the 500 ppm cutoff.

If the samples are preferentially located in areas with anomalous values, then the weights assigned to the indicators in Equation 18.5 should account for this clustering. For example, in the Walker Lake sample data set the second and third sampling campaigns located additional samples in areas with high V values. This entails that our samples are clustered in areas where the indicators tend to be 0. Naïvely averaging such clustered sample indicators will produce an estimate that is too low. Earlier, with the results of Table 18.1, we saw that the simple averaging of sample indicators produced very poor estimates for the second and third campaigns.

When estimating a global proportion below a certain cutoff from a clustered sample data set, it is often difficult in practice to extract a subset that is regularly gridded. Even if such a subset can be determined, it is unsatisfying to completely disregard the information contained in the additional clustered samples. Earlier, when we tackled the problem of estimating the global mean, we handled this dilemma by using a weighted linear combination that gave less weight to sample values in densely sampled areas. This same approach can be used with clustered indicators.

To estimate the proportion of V values below 500 ppm using the entire sample data set, we begin by assigning indicators to each of the 470 available samples. Figure 18.4 shows the indicator map for the 500 ppm threshold. At every sample location (represented by the small dot in Figure 18.4) where the value is below 500 ppm, we have an indicator of 1; at all the remaining locations we have an indicator of 0.

The naïve estimate of $F(500)$ that we calculated by simply counting the number of samples that were less than 500 ppm can be written in

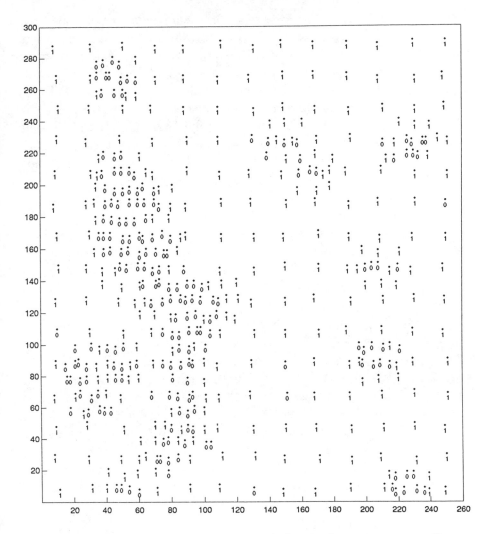

Figure 18.4 A posting of the sample indicators for a 500 ppm cutoff.

terms of indicators as

$$\hat{F}(500) = \frac{1}{470} \sum_{j=1}^{470} i_j(500) = \frac{268}{470} = 0.57$$

As we noted earlier, the sample indicators shown in Figure 18.4 are preferentially clustered in the Wassuk Range area where the indicator tends to be 0. The naïve estimate given earlier fails to take this into account and therefore considerably underestimates the true proportion below 500 ppm. A good estimate of the true proportion below 500 ppm must take this into account by giving less weight to those indicators that are in the densely sampled areas.

We are already familiar with two procedures for allocating declustering weights to a sample data set, the polygonal method and the cell declustering method. Though we could calculate declustering weights for the sample indicators using either of these methods, this is not necessary. Both of these declustering procedures produce weights that depend only on the locations of the sample data. Since the 470 sample indicators are at exactly the same locations as the 470 V samples, their declustering weights will be the same as those we calculated earlier for the case studies on the estimation of the global mean in Chapter 10. Using the polygonal declustering weights in Equation 18.5, the estimated proportion of values below 500 ppm is 0.817; using the cell declustering weights calculated using 20 x 20 m^2 cells, the estimate is 0.785. Both of these are much closer to the true value of 0.812 than the naïve estimate of 0.570 that we obtained earlier.

To estimate the complete global distribution of V values, we need only to repeat the estimation of $F(v_c)$ for several cutoffs. Table 18.3 shows estimates of $F(v_c)$ for several cutoffs spanning the complete range of V values. All three of the estimates shown at each cutoff are calculated using Equation 18.5; the only difference between the estimates is the choice of weights for the 470 sample indicators. The results from Table 18.3 are also shown graphically in Figure 18.5, in which the true cumulative distribution is plotted along with each of the estimated distributions.

These results show that with appropriate declustering weights, the global distribution can be estimated very well. The cell declustering method and the polygonal method are both good procedures for calculating declustering weights. In this particular example, the polygonal method produces slightly better results.

Estimating Other Parameters of the Global Distribution

In the previous section we have seen how the global cumulative distribution can be estimated. In many practical problems, an estimate

Table 18.3 Estimates of the proportion of V values below various cutoffs calculated using three different weighting procedures in Equation 18.5.

Cutoff	True	Polygons[*]	Cells[†]	Naïve[‡]
0	0.076	0.091	0.086	0.047
50	0.215	0.224	0.216	0.119
100	0.311	0.302	0.288	0.164
150	0.392	0.374	0.366	0.209
200	0.469	0.475	0.461	0.274
250	0.541	0.543	0.527	0.332
300	0.607	0.597	0.572	0.370
350	0.667	0.669	0.635	0.423
400	0.721	0.725	0.691	0.479
450	0.769	0.771	0.740	0.532
500	0.812	0.817	0.785	0.570
550	0.849	0.862	0.836	0.630
600	0.881	0.899	0.875	0.685
650	0.908	0.925	0.909	0.762
700	0.930	0.938	0.927	0.806
750	0.947	0.950	0.943	0.843
800	0.961	0.962	0.958	0.877
850	0.971	0.973	0.969	0.911
900	0.979	0.983	0.981	0.943
950	0.985	0.987	0.985	0.955
1000	0.989	0.991	0.990	0.970
1050	0.992	0.994	0.994	0.979
1100	0.995	0.995	0.995	0.983
1150	0.996	0.996	0.996	0.987
1200	0.997	0.996	0.996	0.987
1250	0.998	0.997	0.998	0.991
1300	0.999	0.998	0.998	0.994
1350	0.999	0.998	0.998	0.994
1400	1.000	0.999	0.999	0.996
1450	1.000	0.999	0.999	0.996
1500	1.000	0.999	0.999	0.996

[*] Weights proportional to the area of the polygon of influence.

[†] Weights inversely proportional to the number of samples falling within the same 20x20 m^2 cell.

[‡] All samples given equal weight.

(a) Polygonal estimates

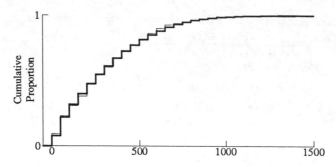

(b) Cell declustering estimates

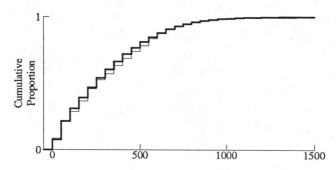

(c) Naive estimates from equal weighting

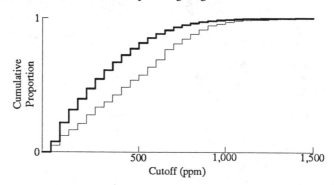

Figure 18.5 A comparison of each of the estimated distributions given in Table 18.3 with the true cumulative distribution. In each figure the true cumulative distribution appears as the thick line; the corresponding estimate appears as the thin line.

of the complete global distribution is not necessary; rather, all that is needed are estimates of a few of its summary statistics.

Since the global variance is a simple average of the squared differences from the mean, the expression for its estimate is similar to the other global estimates we have studied:

$$\hat{\sigma}^2 = \sum_{j=1}^{n} w_j \cdot (v_j - \hat{m})^2$$

It is estimated by taking a weighted average of the squared differences between the available samples, $v_1, \ldots, v_n$, and an estimate of the global mean, $\hat{m}$. As with our other global estimates, the weights used in this formula allow us to account for the possibility that the squared differences to which we have access in the sample data set are not representative of the exhaustive data set due to clustering. The same weights that were used earlier for declustering the sample values for an estimate of the mean and declustering the sample indicators for an estimate of the cumulative proportion can be used again in the estimation of the global variance.

An estimate of the standard deviation of the global distribution can be obtained from the estimate of the variance. An estimated coefficient of variation can be calculated from the estimated mean and the estimated standard deviation.

The global coefficient of skewness is also expressed as a simple average; with clustered sampling, it is therefore estimated by the corresponding weighted average:

$$\textit{Estimated coefficient of skewness} = \frac{\sum_{j=1}^{n} w_j \cdot (v_j - \hat{m})^3}{\hat{\sigma}^3}$$

The weights applied to each of the cubed differences from the mean are the same as those used to estimate the mean $\hat{m}$ and to estimate the standard deviation $\hat{\sigma}$.

An estimate of the cumulative distribution allows the estimation of any quantile. For example, the median, M, is the same as $q_{0.5}$ and its estimate is the value at which the estimated cumulative distribution reaches 0.5:

$$\hat{F}(\hat{M}) = 0.5$$

This value can be calculated either by graphing the estimated cumulative distribution, as was done in Figure 18.5, or by sorting the sample values in ascending order and accumulating the declustering weights until they reach 0.5.

In a similar manner, the lower and upper quartiles can be estimated by determining where the estimated cumulative distribution reaches 0.25 and 0.75, respectively:

$$\hat{F}(\hat{Q}_1) = 0.25 \qquad \hat{F}(\hat{Q}_3) = 0.75$$

Using only the information contained in the sample data set, the global minimum and maximum have to be estimated by the corresponding sample statistics. Though this may be an adequate solution for the minimum, since the variables in many earth science data sets are strongly positively skewed, it is often unsatisfying for the maximum. For variables whose exhaustive distribution has a long tail of high values, it is very likely that the true maximum value is not one of the sample values. Unfortunately, there is little we can do unless we make some further assumptions or supplement our sample data set with physical or chemical information.

Finally, the proportion of the distribution that falls between any two values, v_a and v_b, can be calculated from the estimated cumulative distribution by subtraction:

Estimated proportion between v_a and v_b = $\hat{F}(v_b) - \hat{F}(v_a)$

Using the cumulative distribution estimated by using polygonal weights (Table 18.3), the proportion of values falling within the interval 100 to 300 ppm is

$$\hat{F}(300) - \hat{F}(100) = 0.597 - 0.302 = 0.295$$

The histograms corresponding to each of the cumulative distributions tabulated in Table 18.3 are shown in Figure 18.6; Table 18.4 provides their univariate statistics. With the exception of the minimum and maximum, the estimates calculated using the naïve weights bear little resemblance to the actual exhaustive statistics. With the use of appropriate declustering weights, the estimates improve. The estimates calculated using the polygonal weights are very close to the corresponding exhaustive values. In practice, however, one should not expect the agreement to be this good; in this particular example, we are quite lucky to get such good global estimates. The estimates based

Table 18.4   A comparison of estimated exhaustive statistics of the global distribution for three different weighting procedures.

	True	Polygons	Cells	Naïve
n	78,000	470	470	470
m	277.9	276.8	292.0	436.5
σ	249.9	245.3	253.6	299.9
$\frac{\sigma}{m}$	0.90	0.89	0.84	0.69
min	0	0	0	0
Q_1	67.8	67.5	77.1	184.3
M	221.3	224.2	234.3	425.3
Q_3	429.4	430.8	455.8	645.4
max	1,631.2	1,528.1	1,528.1	1,528.1
$skewness$	1.02	1.03	0.90	0.45

on the cell declustering weights are more typical of the kinds of discrepancies one might see in other situations.

In general, the use of declustering weights is a tremendous improvement over the naïve weighting. The exhaustive statistics that are usually the best estimated are the measures of the location of the center of the distribution. Extreme quantiles are often difficult to estimate accurately, as are those statistics that involve squared terms or cubed terms.

Estimating Local Distributions

Many practical problems require not only an estimate of the global distribution but also estimates of the distribution of the unknown values over small areas. For example, in the exploration phase of an ore deposit an estimate of the global distribution provides some rough idea of the total tonnage of ore and quantity of metal above various cutoffs. In the feasibility and development phases, these global estimates are no longer sufficient. For long- and short-range planning one typically needs estimates of tonnage of ore and quantity of metal for smaller blocks corresponding to several weeks or months of production. Local distributions are also important in environmental applications. In the

(a) True exhaustive histogram

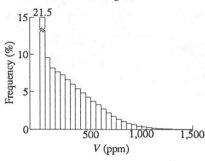

(b) Polygonal

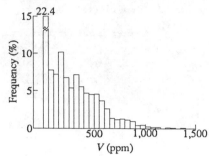

(c) Cell declustering

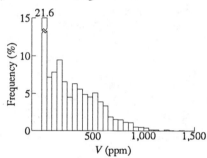

(d) Naive histogram from equal weighting

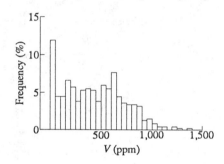

Figure 18.6   A comparison of the histograms calculated from each of the distributions given in Table 18.3.

estimation of the concentration of a pollutant over some area of interest, an estimate of the global distribution can tell us the proportion of the area in which the pollutant exceeds some specified threshold. If clean-up or removal of the pollutant is planned, estimates of the distributions of the pollutant concentration over small areas are also required.

Our recognition that the cumulative proportion below a given cutoff can be expressed as an average indicator leads us to consider the same estimation tools that we used earlier for estimating the local mean. Rather than use weighted linear combinations of the nearby sample values to estimate the local mean, however, we will use weighted linthe nearby sample indicators to estimate the local proportion below a specified threshold.

Figure 18.7 shows an example of estimation of a local distribution using indicators. The goal is to estimate the distribution of values within the rectangular area using nearby samples shown in the figure. By assigning indicators to each of the samples shown in Figure 18.7a, we can treat this problem as a series of estimations of the average indicator value at several cutoffs. For the 65 ppm cutoff shown in Figure 18.7b, only three of the nearby sample indicators are 1s, so a local estimate of the average indicator at this cutoff will be quite low. For example, the naïve estimation procedure, which simply averages the nearby values, would produce an estimate of $\hat{F}(65) = 0.273$. For a threshold of 225 ppm, slightly less than half of the indicators shown in Figure 18.7c are equal to 1; naïve local averaging would produce an estimate of $\hat{F}(225) = 0.455$. At the 430 ppm cutoff, most of the sample indicators shown in Figure 18.7d are 1s, and a local estimate of the average indicator will be quite high. At this cutoff, naïve local averaging would produce an estimate of $\hat{F}(430) = 0.818$.

As we discussed earlier in this chapter when we introduced indicator variables, an estimate of an average indicator also serves as an estimate of a cumulative proportion. Using the three naïve local averages given in the previous paragraph, Figure 18.8a shows the three estimated points of the cumulative distribution for the example shown in Figure 18.7. From these three estimates of the cumulative proportion below the 65 ppm, 225 ppm, and 430 ppm cutoffs, we can calculate the proportion falling within each of the following four classes: < 65 ppm, 65-225 ppm, 225-430 ppm and > 430 ppm. The proportion falling within the third class, for example, is

$$\text{Proportion between 225 ppm and 430 ppm} = \hat{F}(430) - \hat{F}(225)$$
$$= 0.818 - 0.455 = 0.363$$

Figure 18.8b presents the estimated distribution shown in Figure 18.8a in the form of a histogram.

Choosing Indicator Thresholds

The estimated cumulative distribution and histogram shown in Figure 18.8 are quite crude since we have performed the estimation at only three thresholds. By increasing the number of thresholds at which

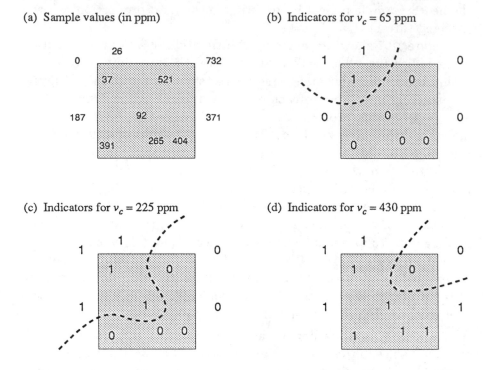

Figure 18.7 An illustration of how nearby samples can be transformed to indicators to estimate the distribution of unknown values over a small area.

we estimate the cumulative proportion, we can refine the appearance of our estimated cumulative distribution and the corresponding histogram. Our ability to refine the estimated cumulative distribution is limited, however, by the number of nearby samples. If there are only a few available samples, then the estimated distribution will appear quite crude regardless of the number of cutoffs we choose.

In actual practice, we should carefully consider the number of cutoffs at which we need estimates. Though the use of several cutoffs may allow us to draw visually satisfying histograms, this is rarely the real goal of a study. For most practical problems that require indicator techniques, a careful consideration of the final goal allows us to use a few well-chosen thresholds. For example, in mining applications there are typically a few cutoff values that have practical and economic significance. The mine plan may call for the separation of material into

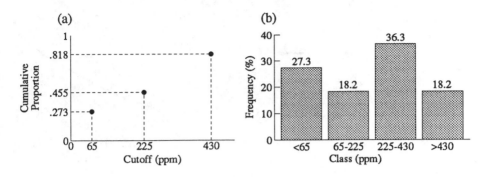

Figure 18.8 Local distribution estimated by simple averaging of the nearby indicators for the the example shown in Figure 18.7. The cumulative distribution is shown in (a) and the corresponding histogram is shown in (b).

ore and waste based on a particular ore grade; the ore material may be separated into a few stockpiles based on other cutoff grades. In such cases, the cutoffs at which indicator estimation is performed should be the same as those that have practical relevance to the proposed mining operation. Many environmental applications also have thresholds that are significant for health or safety reasons, and indicator estimates at these cutoffs may be sufficient to address the goals of an environmental study.

If there are no thresholds that have special significance to the problems being addressed, the most common practice is to perform indicator estimation at the nine cutoffs corresponding to the nine deciles of the global distribution. Despite being conventional, this choice is still arbitrary; if there is a particular part of the distribution for which accurate estimation is more important, then one should choose more cutoffs in that important range. For example, in many precious metal deposits most of the metal is contained in a small proportion of very high grade ore. In such situations, it makes sense to perform indicator estimation at several high cutoffs since the accurate estimation of the upper tail is more important than the estimation of the lower portion of the distribution.

No matter how many cutoffs one chooses with the indicator approach, the cumulative distribution curve will be estimated at only a finite number of points. For an estimate of the complete curve,

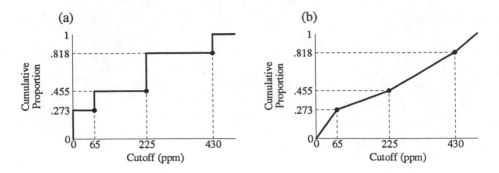

Figure 18.9 Two functions which satisfy the conditions of a cumulative distribution function and also pass through the estimated points shown in Figure 18.8a.

there will be a need to interpolate between the estimated points and to extrapolate beyond the first and last of the estimated points. This interpolation and extrapolation necessarily involves some assumptions about how the distribution behaves at points where it has not been directly estimated. We know that it is a nondecreasing function and that it cannot be less than 0 or greater than 1; however, even with these constraints there are many different functions that can pass through the estimated points. Figure 18.9 shows two different functions, both of which pass through the three estimated points shown in Figure 18.8a and also satisfy the conditions of a cumulative distribution function.

For most applications that require an estimate of the complete distribution, the extrapolation beyond the first and last available estimated points is a more important issue than the interpolation between estimated points. A knowledge of the minimum or maximum value provides some constraint for this extrapolation. For example, in Figure 18.9 we made use of the fact that the minimum value is 0 ppm.

Case Studies

For the case studies, we will estimate the local distributions of the V values within 10 x 10 m^2 blocks using the indicator approach with three of the local estimation methods discussed earlier; In all of these case studies, indicator estimation is performed at the 65 ppm, 225 ppm, and 430 ppm cutoffs. These three values correspond roughly to the declustered median and the quartiles of the global distribution that

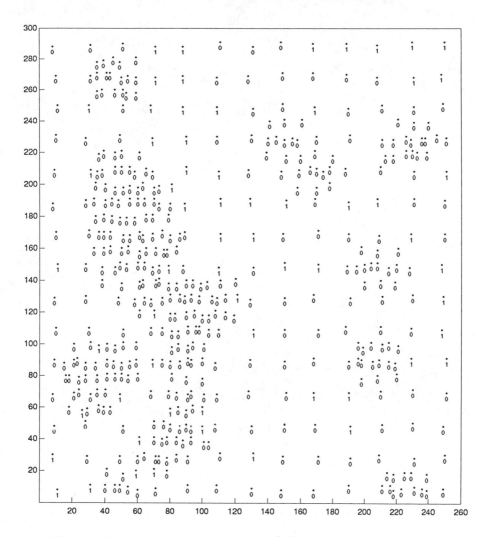

Figure 18.10 A posting of the sample indicators for a 65 ppm cutoff.

was given earlier in Table 18.4. The 470 sample indicators for these three cutoffs are shown in Figure 18.10 through Figure 18.12.

At each of the three cutoffs, the local mean of the indicators is estimated for each of the 780 10 x 10 m^2 blocks covering the Walker Lake area. The estimates of the local mean indicator were calculated using polygons, inverse distance squared, and kriging.

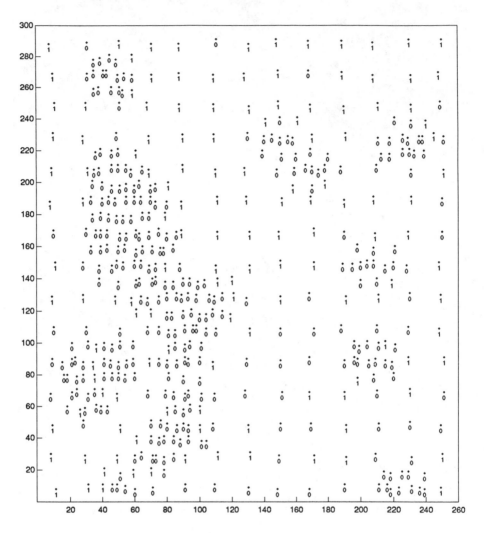

Figure 18.11 A posting of the sample indicators for a 225 ppm cutoff.

Polygons. In the polygonal estimation procedure, first discussed in Chapter 11, the estimated value at any point is equal to the nearest sample value. This can also be seen as a weighted linear combination of sample values in which all of the weight is given to the nearest sample. The average indicator within each 10 x 10 m^2 block was estimated by calculating the polygonal estimate at each of the 100 points within the block and averaging the resulting 100 estimates.

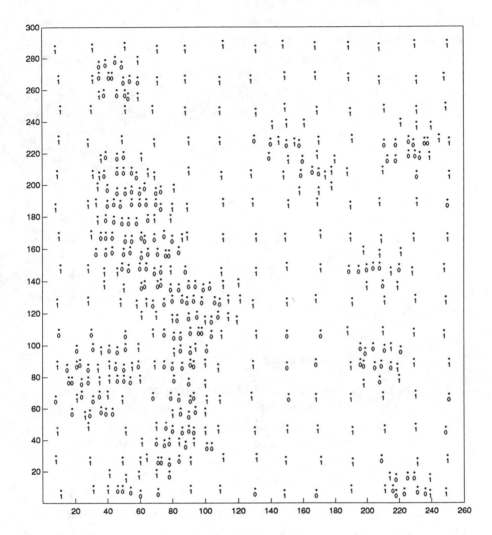

Figure 18.12 A posting of the sample indicators for a 430 ppm cutoff.

Inverse Distance Squared. In the inverse distance squared procedure, first discussed in Chapter 11, the value at any point is estimated
by a weighted linear combination of the nearby sample value. The
weight assigned to each nearby sample is inversely proportional to the
square of its distance from the point being estimated. For the following case study, the same search strategy was used for all three cutoffs:
all available samples within 25 m of the point being estimated were

grouped into quadrants, and the nearest 10 samples in each quadrant were used in the weighted linear combination. The average indicator within each 10 x 10 m^2 block was estimated by calculating the inverse distance squared estimate at each of the 100 points within the block and averaging the resulting 100 estimates.

Ordinary Kriging. In the ordinary kriging procedure, first discussed in Chapter 12, the value at any point is estimated by a weighted linear combination of the nearby sample indicators. The weights are chosen so that the resulting estimate is unbiased and has a minimum estimation variance. The average indicator within each 10 x 10 m^2 block was estimated using ordinary kriging of a block mean, as discussed in Chapter 13. All points within 25 m of the center of the block were grouped into quadrants, and the nearest 10 samples in each quadrant were used in the weighted linear combination. The average covariances between each sample point and the 10 x 10 m^2 block were calculated by discretizing the block into a 6 x 6 grid of points and averaging the resulting 36 point-to-point covariances.

Indicator Variograms

Ordinary kriging requires a model of the variogram or the covariance function of the variable being estimated. The estimation of V values called for a model of the spatial continuity of V values. Now that we are estimating indicators rather than the original V values, we should use a variogram model of the spatial continuity of the indicators. When applying the ordinary kriging procedure to the estimation of an indicator at a particular cutoff, we should ideally use a variogram model that reflects the pattern of spatial continuity for that particular cutoff. For example, the ordinary kriging of the local mean indicator for the 65 ppm cutoff should be done with a variogram that captures the spatial continuity of the indicators for the 65 ppm cutoff; the estimation of the local mean indicator for the 225 ppm cutoff, on the other hand, should use a variogram model that describes the spatial continuity of the indicators for the 225 ppm cutoff.

This ability to use different patterns of spatial continuity for different thresholds distinguishes ordinary kriging from other procedures for estimating the local average of an indicator. With the polygonal and inverse distance squared procedures, the weights assigned to nearby samples are the same at all cutoffs. With ordinary kriging, however, the weights assigned to nearby indicators at a particular cutoff will depend on the variogram model chosen for that cutoff.

The ability to customize the estimation procedure to the pattern of spatial continuity appropriate for each threshold makes ordinary kriging a more powerful technique for the estimation of indicators than the other local estimation procedures we have studied. There are many situations in which the pattern of spatial continuity of the high values is not the same as that of the low values. From the maps of the indicators for each of the nine deciles of the exhaustive Walker Lake data set shown in Chapter 5, for example, it is clear that the very highest values (the black portions of Figure 5.9i) are not as continuous as the very lowest (the white portions of Figure 5.9a). While the very highest values do tend to be grouped together, their clusters are neither as large nor as solid as the clusters of the very lowest values.

Similar observations can be made in many real data sets. For example, in petroleum reservoirs the very highest permeabilities may correspond to fractures, while the very lowest may be due to lenses of shale. An indicator map at a very high threshold would separate the fracture system from the remainder of the reservoir, showing the high permeability values in the fractures as 0s and the remainder of the reservoir as 1s; on such a map, the 0s would likely appear in long connected strings. An indicator map at a very low threshold, on the other hand, would separate the shales from the remainder of the reservoir, showing the low permeability values in the shales as 1s and the remainder of the reservoir as 0s; on such a map, the 1s would likely be connected in elongated lenses. The appearance of the two maps would be quite different. The pattern of spatial continuity for the indicator map at the high cutoff would reflect the structural character of the reservoir's fracture system; the pattern of spatial continuity for the indicator map at the low cutoff would reflect the depositional character of the shale units. Estimation of the average indicators at the two cutoffs should take into account the differing patterns of spatial continuity.

While making the ordinary kriging of indicators more powerful, weights that change from one cutoff to the next also make the technique more demanding. To exploit the power of this technique, we have to develop variogram models for each cutoff at which we intend to do estimation. Sample variograms must be calculated and modeled for each cutoff. Fortunately, sample variograms calculated from indicator data are usually fairly well behaved. Since an indicator variable is either 0 or 1, indicator variograms do not suffer from the adverse effects of erratic outlier values. In fact, even in studies where indicator kriging is not going to be used, indicator variograms are often used to reveal the pattern of spatial continuity of the original variable. Despite being more easily interpreted and modeled than variograms of the original variable, indicator variograms are easily affected by preferential clustering of the sample data. The structure revealed by indicator variograms may not be due to the pattern of spatial continuity but rather to the clustering of the sample data set [1].

The ordinary kriging of indicators at several cutoffs, using a separate variogram model for each cutoff, is usually referred to simply as *indicator kriging.* There is an approximation to indicator kriging that, in many situations, produces very good results. This approximation consists of using the same variogram model for the estimation at all cutoffs. The variogram model chosen for all cutoffs is most commonly developed from the indicator data at a cutoff close to the median. Practice has shown that the variogram based on indicators defined at a median cutoff is often better behaved than the variogram based on indicators defined at other cutoffs. Since the variogram used in this approximation to indicator kriging is often based on the median indicator, this procedure is usually referred to as *median indicator kriging.* One should not feel compelled, however, to use the median indicator variogram model. One can use whatever variogram model is felt to more representative, even if it is based on indicators at some cutoff other than the median.

With one variogram model for all cutoffs, the weights assigned to each sample no longer depend on the cutoff. Once the weights have been calculated for the estimation of the indicator at the first cutoff, they can be used again for the estimation any other cutoff. This makes median indicator kriging computationally faster than indicator kriging, which requires that the weights be recalculated for every cutoff since the variogram model changes from one cutoff to the next.

Before adopting the median indicator kriging approach, it is wise to compute indicator variograms for several thresholds to determine whether they all can be adequately modeled by a common shape. If there are noticeable differences between the sample indicator variograms for different thresholds, one should be cautious about using median indicator kriging. In particular, differences in the nugget ef-

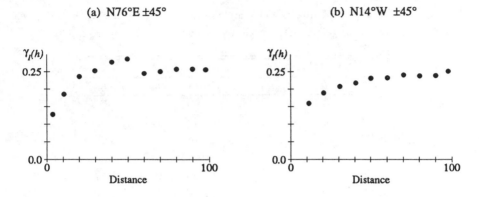

Figure 18.13 Directional sample indicator variograms at the 430 ppm cutoff. The variogram in the N76°E direction, perpendicular to the axis of the Wassuk Range anomaly, is shown in (a); the variogram in the direction parallel to the anomaly, N14°W, is shown in (b).

fects at the different thresholds will cause indicator kriging with different variogram models at different cutoffs to produce results than are quite different from those produced by median indicator kriging.

For the case study presented here, the same indicator variogram model was used for all three cutoffs. This model was based on the variogram of the sample indicators for the 430 ppm cutoff. This cutoff is close to the naïve median for the sample data set. Using the 470 indicator data shown in Figure 18.12, sample variograms were calculated in the N14°W and N76°E directions. Figure 18.13 shows the sample variograms in each direction, and Table 18.5 provides the details of the calculations for each lag. The model fit to the sample variogram is shown in Figure 18.14.

Like the original V values, the indicators for the 430 ppm cutoff are more continuous in a direction that runs parallel to the axis of the Wassuk Range anomaly. In the N76°E direction the variogram reaches its sill value of 0.25 at 30 m. The variogram climbs more slowly in the N14°W direction, reaching 0.25 at 100 m. The variogram models fit to the directional sample variograms are shown in Figure 18.14. In both directions, the variogram model consists of a nugget effect plus two spherical structures [2]. In the N76°E direction, the model

Table 18.5 Details of the directional sample indicator variograms at the 430 ppm cutoff, with an angular tolerance of $\pm$ 45 degrees as shown in Figure 18.13.

N76°E			N14°W		
Pairs	h	$\gamma_I(h)$	Pairs	h	$\gamma_I(h)$
82	3.6	0.128	0	—	—
716	10.5	0.185	806	11.4	0.159
1,242	20.1	0.236	1,328	20.7	0.189
1,471	29.7	0.253	1,648	30.7	0.207
1,701	40.2	0.278	1,993	40.8	0.217
1,699	49.8	0.287	2,278	50.4	0.230
2,009	60.1	0.244	2,882	60.4	0.231
1,983	70.2	0.250	3,047	70.4	0.239
2,080	80.2	0.258	3,234	80.4	0.236
2,101	89.9	0.257	3,126	90.2	0.237
1,093	97.7	0.255	1,335	97.8	0.250

is:

$$\gamma_I(h) = 0.09 + 0.08 Sph_{20}(h) + 0.08 Sph_{30}(h)$$

and in the N14°W direction, the model is:

$$\gamma_I(h) = 0.09 + 0.08 Sph_{25}(h) + 0.08 Sph_{100}(h)$$

The complete two dimensional anisotropic variogram model can be expressed as:

$$\gamma_I(\mathbf{h}) = 0.09 + 0.08 Sph_1(\mathbf{h}_1) + 0.08 Sph_1(\mathbf{h}_2)$$

where the vectors $\mathbf{h}_1$ and $\mathbf{h}_2$ are calculated as follows:

$$\mathbf{h}_1 = \begin{bmatrix} h_{x,1} \\ h_{y,1} \end{bmatrix} = \begin{bmatrix} \frac{1}{20} & 0 \\ 0 & \frac{1}{25} \end{bmatrix} \cdot \begin{bmatrix} h'_x \\ h'_y \end{bmatrix}$$

and

$$\mathbf{h}_2 = \begin{bmatrix} h_{x,2} \\ h_{y,2} \end{bmatrix} = \begin{bmatrix} \frac{1}{30} & 0 \\ 0 & \frac{1}{100} \end{bmatrix} \cdot \begin{bmatrix} h'_x \\ h'_y \end{bmatrix}$$

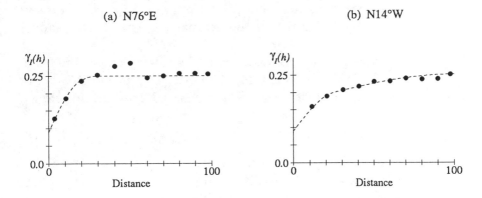

Figure 18.14 Models for sample indicator variograms at a 430 ppm cutoff. The variogram model in the N76°E direction, perpendicular to the axis of the Wassuk Range anomaly, is shown in (a) along with the sample points; the variogram model and the sample variogram points in the N14°W direction is shown in (b).

The distances h'_x and h'_y are calculated in the coordinate system aligned with the directions of minimum and maximum continuity:

$$\left[\begin{array}{c} h'_x \\ h'_y \end{array} \right] = \left[\begin{array}{cc} cos(14°) & sin(14°) \\ -sin(14°) & cos(14°) \end{array} \right] \cdot \left[\begin{array}{c} h_x \\ h_y \end{array} \right]$$

where h_x is the distance calculated in the east-west direction and h_y is the distance calculated in the north-south direction.

Order Relation Corrections

There are several constraints on indicator estimates that we do not explicitly observe in the calculation of our estimates. First, the proportion below any cutoff cannot be less than 0 or greater than 1. One way of meeting this constraint is to use weighted linear combinations in which the weights are positive and sum to 1. The polygonal method and the inverse distance squared method always satisfy this condition; however, kriging does not constrain the weights to be positive. Whether we use complete indicator kriging with different variogram models for different cutoffs or median indicator kriging with only one variogram model, there is a possibility that we will produce a negative estimate or an estimate above 1. In these situations, it is proper to adjust these estimates to the appropriate lower or upper bound; negative estimates should be set to 0 and estimates greater than 1 should be set to 1.

There is a second constraint that must be met by our estimates of the proportion below various cutoffs. The estimated proportion below one cutoff cannot be greater than the estimated proportion below a higher cutoff:

$$\hat{F}(v_i) \leq \hat{F}(v_j) \qquad \text{if } v_i \leq v_j$$

Since our estimated proportions below various cutoffs are calculated independently, there is a possibility that an estimate of the proportion below one cutoff might not be consistent with an estimate of the proportion below another cutoff.

One way of satisfying this second constraint is to use only positive weights that sum to 1, and to use the same weights for the estimation at all cutoffs. Once again, the polygonal and inverse distance procedures do observe this condition while the procedures based on kriging do not. Since indicator kriging uses a different weighting scheme at each cutoff, it is common to observe small inconsistencies in the estimated proportions below closely spaced thresholds. Even with median indicator kriging, in which the same weights are used at all cutoffs, the possibility of negative weights gives rise to the possibility of inconsistent estimates. In both procedures, however, these inconsistencies are typically very small and are easily corrected by small adjustments to the estimated proportions.

For indicator kriging and median indicator kriging, there are several ways to adjust the estimates so that they satisfy the order relations [3]. These order relation problems are usually very small, however, and there is little point in using a sophisticated correction procedure if the largest correction is in the third decimal place. A simple approach is to check each pair of estimates at successive cutoffs and to adjust both estimates to their average if they do not satisfy the correct order relation.

Case Study Results

For each of the 780 10 x 10 m^2 blocks, estimates of the proportion of V values below the three cutoffs have been calculated using three different methods: polygonal weighting, inverse distance squared weighting, and median indicator kriging. For the median indicator kriging, less than

Table 18.6 Summary statistics for the estimates of the local distributions by different methods: the mean residual, m_R; the standard deviation of the residuals, σ_R; and the correlation coefficient between true and estimated proportions, $\rho_{F,\hat{F}}$.

		Polygons	Inverse Distance Squared	Median Indicator Kriging
65 ppm	m_R	0.005	-0.016	-0.007
	σ_R	0.274	0.213	0.201
	$\rho_{F,\hat{F}}$	0.724	0.774	0.793
225 ppm	m_R	0.002	-0.040	-0.020
	σ_R	0.318	0.261	0.248
	$\rho_{F,\hat{F}}$	0.715	0.751	0.769
430 ppm	m_R	-0.002	-0.054	-0.029
	σ_R	0.259	0.212	0.199
	$\rho_{F,\hat{F}}$	0.731	0.779	0.801

3% of the estimates required order relation corrections, with the largest correction being 0.006. The 780 estimated proportions below cutoff are compared to the corresponding true values for each of the three cutoffs in Figures 18.15 through 18.17. In these figures, the shading within each of the 10 x 10 m² blocks records the proportion of the block that is below the specified cutoff; solid black records that none of the block is below the cutoff, whereas solid white records that all of the block is below the cutoff.

Table 18.6 presents some summary statistics for the estimates for each cutoff. At each cutoff, the estimates calculated by median indicator kriging are slightly better than the estimates calculated by the other two methods. Though these statistics present a good summary of the overall performance of the various methods at each cutoff, they do not give us an appreciation of how well the various estimation methods do in terms of the original goal of the study; namely, the estimation of local distributions.

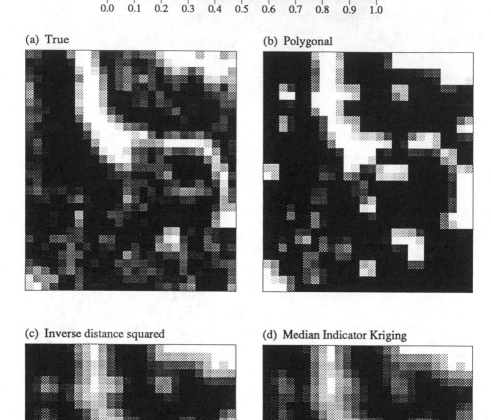

Figure 18.15 A comparison of the true proportion below 65 ppm in 10 x 10 m^2 blocks to the proportion estimated by different methods.

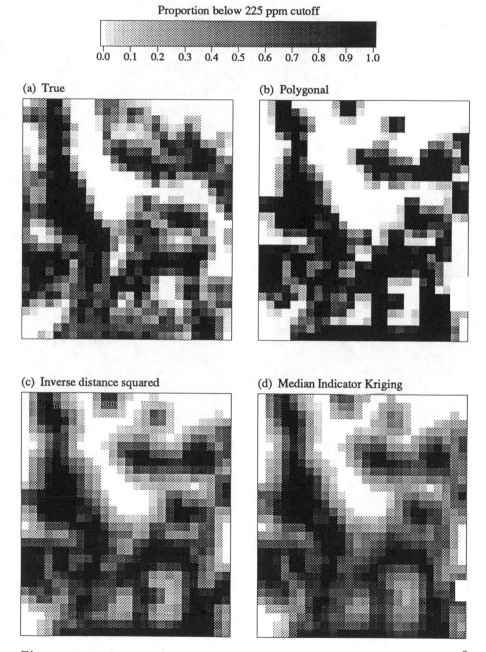

Figure 18.16 A comparison of the true proportion below 225 ppm in 10 x 10 m^2 blocks to the proportion estimated by different methods.

Proportion below 430 ppm cutoff

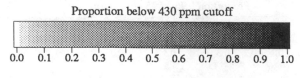

(a) True

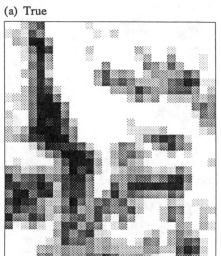

(b) Polygonal

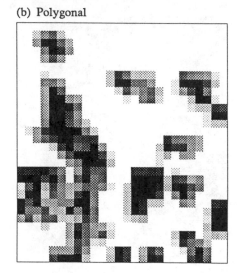

(c) Inverse distance squared

(d) Median Indicator Kriging

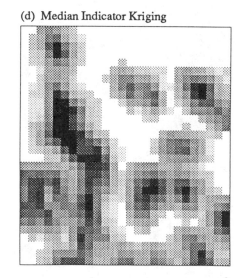

Figure 18.17 A comparison of the true proportion below 430 ppm in 10 x 10 m^2 blocks to the proportion estimated by different methods.

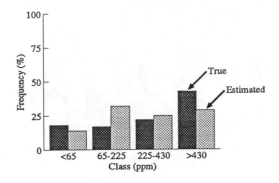

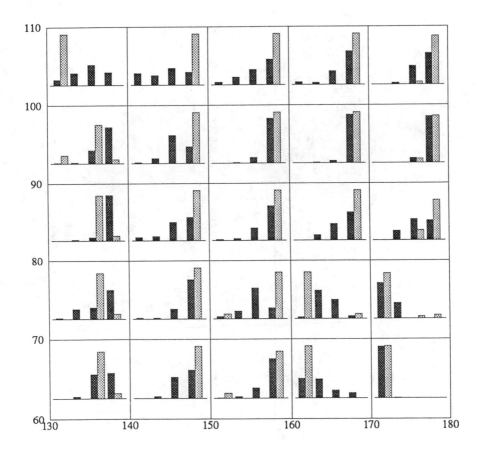

Figure 18.18 A comparison of the true local distributions to those estimated by applying the polygonal method of local estimation to indicators. As the legend at the top shows, within each 10 x 10 m² block, the frequency of true values falling within each of the four classes is shown by the solid dark bars, while the frequency estimated by the polygonal weighting of nearby indicators is shown by the light stippled bars.

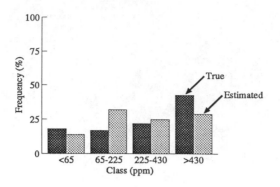

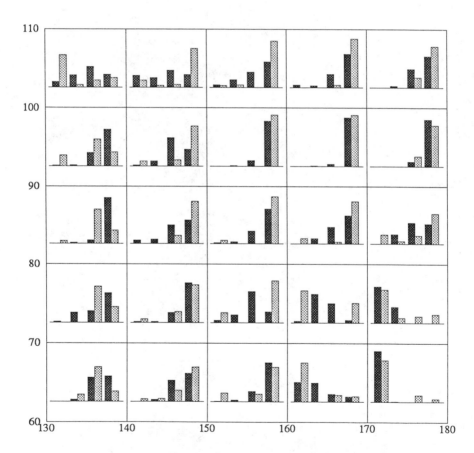

Figure 18.19 A comparison of the true local distributions to those estimated by applying the inverse distance squared method of local estimation to indicators. As the legend at the top shows, within each 10 x 10 m^2 block, the frequency of true values falling within each of the four classes is shown by the solid dark bars, while the frequency estimated by the inverse distance squared weighting of nearby indicators is shown by the light stippled bars.

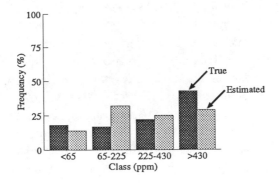

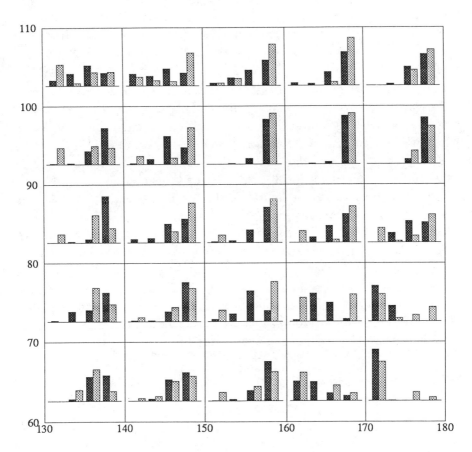

Figure 18.20 A comparison of the true local distributions to those estimated by median indicator kriging. As the legend at the top shows, within each 10 x 10 m^2 block, the frequency of true values falling within each of the four classes is shown by the solid dark bars, while the frequency estimated by median indicator kriging is shown by the light stippled bars.

Figures 18.18 through 18.20 show the estimated local distributions versus the true local distributions for a few of the blocks near the center of the Walker Lake area. These figures show that for the purpose of estimating the local distribution, the polygonal estimation procedure produces erratic results. Since the polygonal estimate of the indicator at each point depends solely on the closest available sample, the estimates in sparsely sampled areas may all be depending on the same sample value. The result is that the estimated distribution appears as a single spike, with all values falling within the same class as the nearest available sample. In blocks where the true values are fairly constant, such as the flat area of Walker Lake itself, the polygonal estimation procedure will do a reasonable job. For blocks in which there is a broad range of values, however, the local distribution estimated by the polygonal approach can be quite poor. While it is a good procedure for estimating a global distribution of values, the polygonal approach is not suited for the estimation of local distributions.

By applying the inverse distance squared method to the sample indicators, we produce estimated local distributions that are almost as good as those estimated by median indicator kriging. As we noted when we first compared inverse distance methods to ordinary kriging, the major drawback of the inverse distance weighting schemes is that they do not account for local clustering of the available data. If there is no significant clustering of the data, or if the search strategy accounts for the redundancy of closely spaced samples, inverse distance estimates can be very nearly as good as those calculated by ordinary kriging. It should be emphasized, however, that the median indicator approximation has been used in the case study results shown here. If there is a noticeable difference between the patterns of spatial continuity at different cutoffs, complete indicator kriging with different variogram models for different cutoffs may produce significantly better estimates than the median indicator and the inverse distance squared approaches.

Notes

[1] For more details on how preferential clustering affects indicator variograms, see Appendix E of:
Srivastava, R. , *A non-ergodic framework for variograms and covariance functions*. Master's thesis, Stanford University, 1987.

[2] The modeling of indicator variograms is often made easier by the fact that the sill is a function of the quantile at which the indicator is defined. For an indicator variable whose mean is m, the variance is $m(1-m)$. Since the sill of the variogram model is roughly equal to the variance in most practical studies, the sill can be determined simply by knowing the global proportion of ones and zeros. For a median cutoff, half the sample data have an indicator of 1 and half have an indicator of 0. The mean indicator, therefore, is 0.5 and the sill is roughly $0.5(1-0.5) = 0.25$. The fact that indicator variograms have sills that depend on the proportion of values below and above the chosen cutoff makes it difficult to compare directly the indicator variograms for different cutoffs. When comparing indicator variograms for different thresholds, it is common to standardize them to a sill of one by dividing each by $m(1-m)$.

[3] Several methods for correcting order relation problems are discussed in:
Sullivan, J. , "Conditional recovery estimation through probability kriging— theory and practice," in *Geostatistics for Natural Resources Characterization*, (Verly et al., eds.), pp. 365–384, Proceedings of the NATO Advanced Study Institute, South Lake Tahoe, California, September 6-17, D. Reidel, Dordrecht, Holland, 1983.

Further Reading

Journel, A. , "Non-parametric estimation of spatial distributions," *Mathematical Geology*, vol. 15, no. 3, pp. 445–468, 1983.

19

CHANGE OF SUPPORT

In Chapter 8 we began the discussion of change of support with an example that showed the practical importance of the problem. In this chapter we will return to that particular example and explore two methods for change of support.

It should be noted at the outset that the solutions offered here apply only to quantities that average arithmetically, such as porosities, ore grades, or pollutant concentrations. For other quantities, such as permeability or soil strength, whose averaging process is not arithmetic, the techniques described here are not appropriate. A discussion of change of support models appropriate for such variables is beyond the scope of this book since these methods require an understanding of the differential equations that govern their behavior. The Further Reading section at the end of the chapter provides references to some of the pertinent literature.

The Practical Importance of the Support Effect

We begin our discussion of the support effect by continuing the example from Chapter 8. In this example, we imagine that the V variable in the Walker Lake data set is the concentration of some metal that we intend to mine. Though we have cast this discussion in terms of a mining problem, similar considerations apply in many other applications. In the assessment of the concentration of some pollutant, for example,

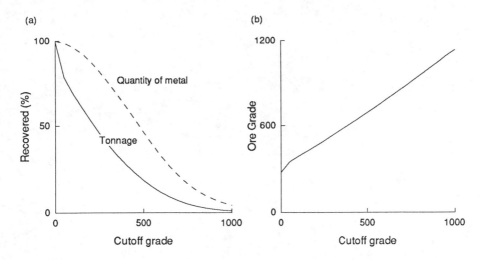

Figure 19.1 The tonnage of ore, quantity of metal and ore grade as functions of cutoff grade for the 78,000 point V values in the exhaustive Walker Lake data set.

the proportion of the area under study that exceeds a certain limit will vary with the volume of the sample considered.

In our Walker Lake Mine, we will process all material above a certain grade as ore and reject the remaining material as waste. The profitability of our mine will depend on several factors, among them the tonnage of material we process as ore, the average grade of this ore, and the resulting quantity of metal we recover. Each of these three depends on the cutoff grade we choose for discriminating between ore and waste.

If we use a cutoff grade of 0 ppm, then everything will be treated as ore. The ore tonnage at this 0 ppm cutoff will be T_0, the tonnage of material in the entire area. The average ore grade, $\bar{v}_{ore}$, will be the same as the overall global mean of 278 ppm, and the quantity of metal recovered will be

$$Q_0 = T_0 \frac{\bar{v}_{ore}}{10^6}$$

As we increase the ore/waste cutoff, some of the material will be rejected as waste and the tonnage of ore will decrease. The waste will contain a small amount of unrecovered metal so the quantity of re-

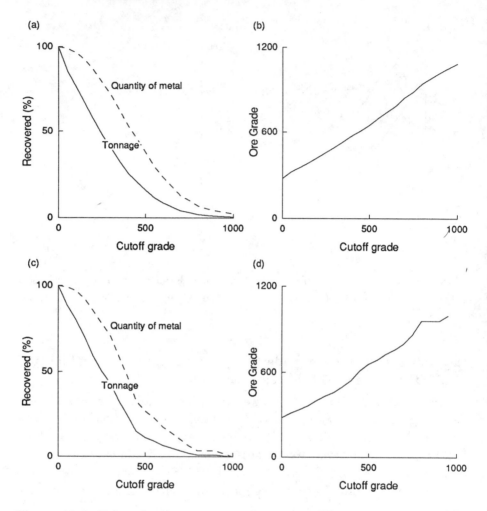

Figure 19.2 Exhaustive recovery curves for tonnage (%), quantity of metal (%), and recovered grade based on block averages of V. The curves in (a) and (b) are for 10 x 10 m^2 blocks while the curves for 20 x 20 m^2 blocks are shown in (c) and (d).

covered metal will also decrease slightly. Since we have rejected the material with the lowest grade, the average ore grade will increase.

Figure 19.1a and b show how these three quantities behave for the 78,000 point values in the Walker Lake data set. In these figures, the tonnage of ore is given as a percentage of the total quantity of material, T_0, and the quantity of recovered metal is given as a percentage of the total quantity of metal. Figure 19.1a, which shows the tonnage of ore as a function of cutoff, is in fact a cumulative distribution curve in which we show the cumulative distribution above various thresholds; the cumulative distribution below various thresholds that we introduced in Chapter 2 would correspond to the tonnage of waste.

The curves shown in Figure 19.1a and b are based on point values, with the implicit assumption that we can discriminate between ore and waste on a very fine scale. In practice, the ore/waste classification will likely be made on much larger volumes. In an open pit operation, our selective mining unit (the minimum volume we can classify as ore or waste) might be the volume of a single truck load; in an underground operation, our selective mining unit might be the volume of an entire stope.

Figure 19.2 shows how the tonnage of ore, quantity of metal, and ore grade for V are affected if we combine the point values into larger units before making the classification. As the size of the selective mining unit increases, the average ore grade decreases. The quantity of metal also generally decreases, although for very low cutoffs we may actually pick up a small amount of the metal that was rejected using a more selective operation. For cutoff grades below the mean of 278 ppm, an increase in the size of the selective mining unit usually results in an increase in the tonnage of ore; above the mean, an increase in the size of the selective mining unit usually results in a decrease in the ore tonnage.

For the planning of a mining operation, this support effect plays an important role. Using the results in Figures 19.1 and 19.2, we can see that projections based on the point support curves will be quite misleading if the actual mining operation uses a 20 x 20 m^2 selective mining unit. For example, at a cutoff of 500 ppm the curves based on point support (Figure 19.1) indicate that we will recover nearly half of the total metal and process about 20% of the total material as ore. In the actual operation with 20 x 20 m^2 selective mining units, however, we would recover only 25% of the total quantity of metal while processing only 10% of the total material (Figure 19.2).

When we estimate distributions of unknown variables, we have to be careful how we use our estimates. Since our estimated distributions are typically based on point sample values [1], they may be representative of a very different support than the one in which we are actually interested. For example, the global distribution of V values that we estimated in Chapter 10 may produce misleading predictions about global truncated statistics (such as tonnage of ore, quantity of metal, and ore grade). Similarly, our estimates of the local distribution within 10 x 10 m² panels from Chapter 17 may produce misleading predictions about local truncated statistics. If later decisions will be made on a support different from that on which the estimated distribution is based, it is important to account for the effect of support in our estimation procedures.

The best way to handle the support effect is to use sample data that have the same support as the volume we intend to estimate; unfortunately, this is rarely possible. Without such data, we must make some correction based on assumptions about how the distribution of values changes as their support increases. As we will see, the correction is usually rather imprecise and heavily dependent on our assumptions. Nevertheless, it is certainly better to make a coarse correction (and to carefully document the assumptions) than to ignore the problem. In the next section we will look at how the change of support affects various summary statistics, notably the mean and the variance. We will then consider mathematical procedures that allow us to make a correction consistent with our observations.

The Effect of Support on Summary Statistics

Figure 19.3 shows the effect of support on the distribution of U. The distribution of the 78, 000 point values of U is shown in Figure 19.3a, the distribution of the 780 10 x 10 m² block averages of U is shown in Figure 19.3b, and the distribution of the 195 20 x 20 m² block averages is shown in Figure 19.3c.

As the support of the data increases we notice that the maximum value decreases, from more than 9, 000 ppm for the point values to less than 2, 000 ppm for the 20 x 20 m² block averages. This makes sense since the most extreme point values will certainly be diluted by smaller values when they are combined in large blocks. Similarly, one can expect the minimum value to increase, though in this example the skewness of the original point distribution and its large spike of 0 ppm values makes this tendency less noticeable.

Increasing the support has the effect of reducing the spread; the interquartile range decreases, as does the standard deviation. As the

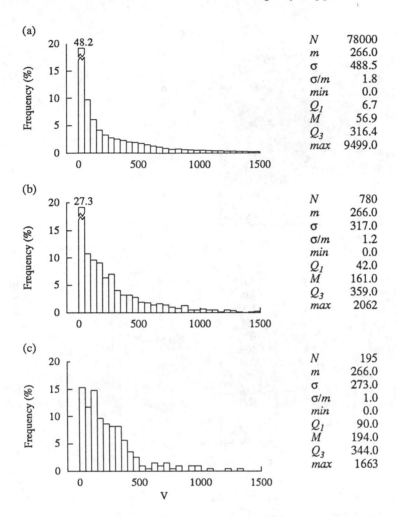

Figure 19.3 The effect of support on the distribution of U. The exhaustive distribution of the 78,000 point U values is shown in (a); the exhaustive distribution of 10 x 10 m^2 block averages (b), and of 20 x20 m^2 block averages in (c).

support increases, the distribution also gradually becomes more symmetric. For example, the difference between the median and the mean becomes smaller. The only summary statistic that is unaffected by the support of the data is the mean. For all three distributions shown in Figure 19.3, the mean is 266 ppm.

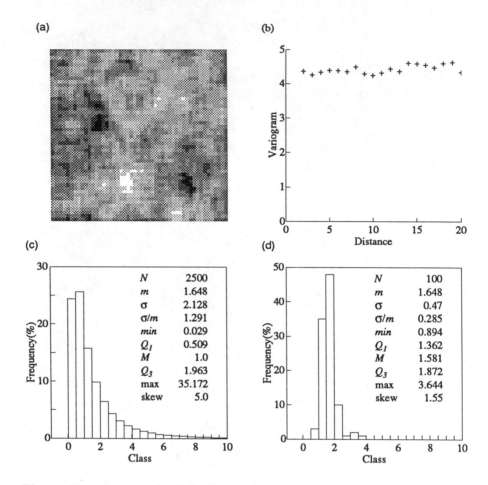

Figure 19.4 An example of the change of support effect on spatially uncorrelated data. A grey scale map of $2,500$ point values is shown in (a). The variogram of these $2,500$ values is shown in (b), the point histogram is shown in (c), and the histogram of 5 x 5 block averages is shown in (d).

The tendencies seen in the three histograms in Figure 19.3 will persist as we group the points into even larger blocks. For the largest possible support, the entire Walker Lake area, the histogram will be a single spike at 266 ppm, with no spread and no asymmetry.

The rate at which the spread decreases and the distribution becomes more symmetric depends on the spatial arrangement of the data values. The two data sets shown in Figures 19.4 and 19.5 have the same univariate distribution but they have very different patterns of spatial continuity. In Figure 19.4a, there is no apparent spatial continuity; extremely high values may be very close to extremely low ones. In Figure 19.5a, on the other hand, there is some spatial continuity; there is an evident zonation, with high values being located in certain areas and low values being located in others. These patterns of spatial continuity are also evident in the variograms of each data set shown in Figures 19.4b and 19.5b.

The histograms and summary statistics of the point values and 5 x 5 block averages for these two data sets are given in Figures 19.4c,d and Figures 19.5c,d. As noted earlier, the distributions of the point values for the two data sets are identical; however, the distributions of the block averages are quite different. For homogeneous data sets in which the extreme values are pervasive, the reduction of spread and asymmetry will occur more rapidly than for heterogeneous data sets in which there is a strong zonation, with extreme values tending to be located in a few select areas.

The effect of the support of the volume over which we are averaging will be greatest for data sets in which the data are spatially uncorrelated. For such data sets, classical statistics tells us that the standard deviation of block averages is inversely proportional to their area; for such data sets the distribution of block averages will rapidly become symmetric. As the data values become more continuous, the support effect decreases; the reduction in the spread and the symmetrization both occur less rapidly [2].

This link between spatial continuity and the support effect makes the variogram a useful tool in assessing the effect of change of support. The variograms shown in Figures 19.4b and 19.5b document the lack of spatial continuity in the first data set and the presence of it in the second. As the variogram approaches a pure nugget effect, the support effect will become stronger in the sense that the decrease in the spread of the distribution is more noticeable. Later in this chapter we will see that we can go beyond these qualitative observations, using the variogram model to estimate the reduction in variance.

It is important to realize that the variogram alone does not capture all of the relevant spatial characteristics of a data set. In Figure 19.6 we show another data set that has the same univariate distribution and the same variogram as the data set in Figure 19.5; however, a comparison

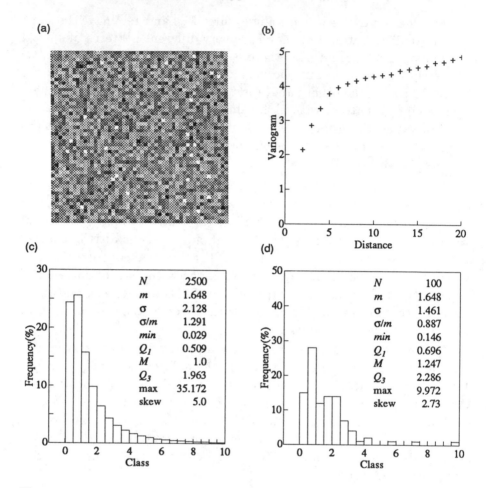

Figure 19.5 An example of the change of support effect on spatially correlated data. A grey scale map of 2,500 point values is shown in (a). The variogram of these 2,500 values is shown in (b), the point histogram is shown in (c), and the histogram of 5 x 5 block averages is shown in (d). The point histogram of this figure is identical to those in Figures 19.4c and 19.6c; however, the 5 x 5 block histograms are different in each figure.

of Figures 19.5a and 19.6a clearly reveals a distinct pattern of spatial continuity for each data set. While the reduction in the variance due to the support effect is practically the same for both data sets, the increase in the symmetry is not.

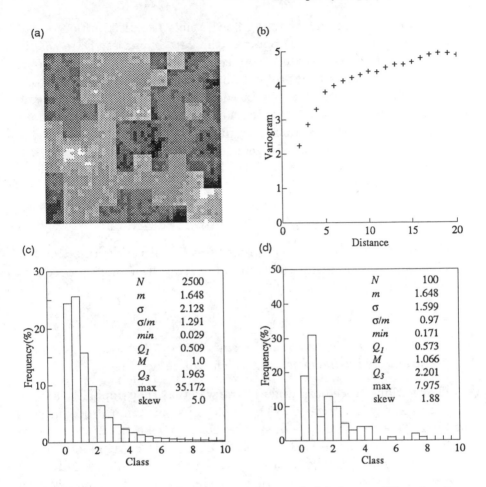

Figure 19.6 In this example the grey scale map in (a) shows a different pattern of spatial correlation than that shown in Figure 19.5a, even though their respective variograms and point histograms (c) are practically the same. Note, however, that the histograms of the 5 x 5 block averages shown in (d) of these two figures are quite different, illustrating that the variogram alone does not capture all of the relevant information concerning the effect of a change of support. In particular, it tells us very little about the amount of symmetrization that will take place.

The common summaries of spatial continuity—the variogram, the covariance function, and the correlogram—do not capture the details of how the extreme values are connected. This connectivity of extreme values is important in determining how rapidly the distribution of values will symmetrize as the support of the data increases. Data sets in which the extreme values are poorly connected, such as the one shown previously in Figure 19.5a, are often described as having high entropy or maximum disorder; data sets in which the extreme values are well connected are described as having low entropy. Earth science data tend to have low entropy; there is typically some structure to extreme values. Indeed, the instinct of a geologist is to delineate structure. Regrettably, the most common statistical models work toward producing maximum entropy.

Correcting for the Support Effect

There are several mathematical procedures for adjusting an estimated distribution to account for the support effect. All of these procedures have two features in common:

1. They leave the mean of the distribution unchanged [3].

2. They adjust the variance by some factor that we will call the *variance adjustment factor*, which will be denoted by f.

The various procedures differ in the way that they implicitly handle the degree of symmetrization. The choice of a particular procedure depends largely on the degree of symmetrization we expect. As we have seen in the previous section, the effect of support on symmetry is related to the entropy—the connectedness of extreme values—and is not adequately described by the variogram. The degree of symmetrization we expect is therefore necessarily a question of informed judgment. Qualitative information about the spatial arrangement of values must be brought to bear on the problem. If past experience in similar environments suggests that the extreme values are poorly connected, then we should choose a procedure that implicitly increases the symmetry of the distribution as the support increases. If the extreme values tend to be well connected, then we might prefer a procedure that does not implicitly symmetrize the distribution.

In the next sections we will present two of the simpler techniques of support effect correction, one of which does not change the symmetry and one that does. Both of these techniques require that we already have some variance adjustment factor in mind. For the moment we will not worry about how we choose this factor; at the end of this chapter

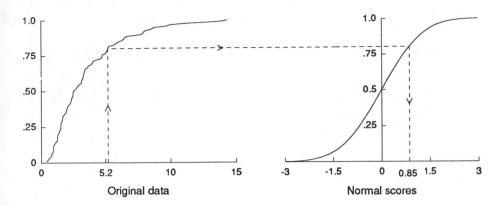

Figure 19.7 A graphical procedure for transforming the values of one distribution into those of another. In this particular example the original distribution is transformed into a standard normal distribution.

we will show one way of estimating this adjustment factor from the variogram model.

Transforming One Distribution to Another

Before looking at the specific details of support effect correction procedures, it will be useful to discuss their common thread, the transformation of one distribution to another.

Figure 19.7 shows a graphical procedure for transforming the values from one distribution to those of another. On the left-hand side of this figure, we begin with a cumulative distribution curve for the distribution of the original untransformed values. With this curve, we can calculate the cumulative proportion, $p(v_0)$, that corresponds to a particular value v_0. Starting on the x axis with the value v_0, we move up to the cumulative distribution curve and then across to the corresponding cumulative proportion. If we then want to transform the value v_0 to a corresponding value from another distribution, we can reverse the procedure, calculating a value v_0' that corresponds to $p(v_0)$. Instead of using the same cumulative distribution curve (which would simply take us back to the value from which we started) we use the cumulative distribution curve of the distribution to which we want to transform the original values. The dotted lines in Figure 19.7 show an

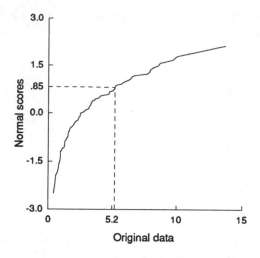

Figure 19.8 The graphical procedure of Figure 19.7 shown as a transformation of the quantiles of one distribution to those of another.

example of this procedure. The 5.2 ppm value from the distribution on the left corresponds to a cumulative proportion of 0.803 that, in turn, corresponds to the value 0.85 from the distribution on the right.

By relating values that share the same cumulative proportion, this graphical transformation procedure is in fact mapping the quantiles of one distribution to those of another. In the example from Figure 19.7, $q_{0.803}$ of the distribution on the right was transformed to $q'_{0.803}$ of the distribution on the left. For any cumulative proportion, p, q_p of the distribution on the left will be transformed to q'_p of the distribution on the right.

Rather than show the graphical transformation as a two-step procedure:

$$original\ value \Rightarrow cumulative\ proportion \Rightarrow transformed\ value$$

we can show it as a one-step procedure that maps quantiles of one distribution to those of another using their q-q plot. In Figure 19.8 we have shown the q-q plot of the two distributions shown earlier in Figure 19.7. Any value from the original distribution can be mapped onto a corresponding value from the transformed distribution by moving up from the original value, v_0, on the x axis to the q-q curve then across to the corresponding transformed value v_0' on the y axis.

Starting with an untransformed distribution and an intended distribution for the transformed values, one can calculate the q-q curve that accomplishes the transformation. Unfortunately, for the support effect problem we do not know the intended distribution of the block values, and therefore can not calculate the appropriate q-q curve. So rather than accomplish the transformation by calculating a q-q curve from two known distributions, we will accomplish it by assuming a particular shape for the q-q curve. The fact that we want the mean to remain unchanged and the variance to be changed by a prescribed amount will allow us to calculate the necessary parameters of the curve.

Affine Correction

The affine correction is probably the most simple of the various support effect correction procedures. The basic idea behind it is that the variance of a distribution can be reduced without changing its mean simply by squashing all of the values closer to the mean. How much we squash will depend on how much we want to reduce the variance.

The affine correction transforms q, a quantile (or value) of one distribution, to q', a quantile (or value) of another distribution using the following linear formula:

$$q' = \sqrt{f} * (q - m) + m \tag{19.1}$$

The mean of both distributions is m; if the variance of the original distribution is σ^2, the variance of the transformed distribution will be $f \cdot \sigma^2$.

Figure 19.9 shows Equation 19.1 on a q-q plot. When we first presented q-q plots in Chapter 3, we noted that the q-q curve of two distributions that have the same shape will be a straight line. By using a linear equation to relate the values of the point support distribution to those of the block support distribution, the affine correction preserves the shape of the original distribution. It implicitly assumes that there is no increase in symmetry with increasing support.

Figure 19.10 shows how the affine correction alters a distribution. In Figure 19.10a we begin with the global distribution that we estimated in Chapter 10 using only the 470 available sample values. To produce this declustered estimate of the global histogram, we assigned

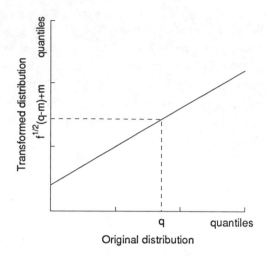

Figure 19.9 The affine correction plotted on a q-q plot.

declustering weights to each sample value and then calculated the proportion in a particular class by accumulating the weights of the sample values falling with that class. Each sample value was then transformed using Equation 19.1 with a variance adjustment factor of 0.8; the histogram of the resulting transformed values was then calculated using the same declustering weights that were calculated earlier. The result is shown in Figure 19.10b. The results of repeating this procedure with $f = 0.6$ and $f = 0.4$ are shown in Figure 19.10c and d. Note that while the various measures of spread decrease, the mean and the skewness remain unchanged.

The main advantage of the affine correction is its simplicity. Its main disadvantage is that it produces a minimum value that may not be realistic. If the variance adjustment factor is not too small [4] and if the cutoffs of interest are close to the mean, then the affine correction procedure is often adequate.

Indirect Lognormal Correction

The indirect lognormal correction is a method that borrows the transformation that would have been used if both the original point support distribution and the transformed block support distribution were lognormal [5]. The idea behind it is that while skewed distributions may

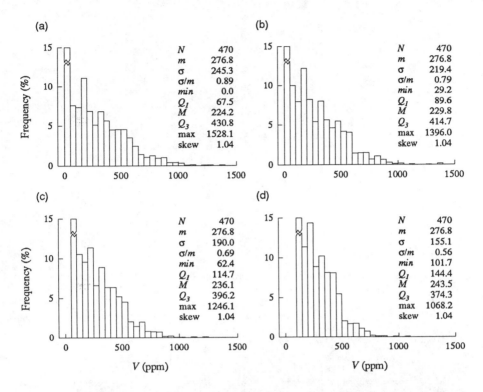

Figure 19.10 Histograms and summary statistics of three distributions obtained by applying the affine correction. The point histogram is given in (a) while the distributions resulting from the variance reduction factors of 0.8, 0.6, and 0.4 are shown in (b), (c), and (d), respectively.

differ in important respects from the lognormal distribution, change of support may affect them in a manner similar to that described by two lognormal distributions with the same mean but different variances [6].

The q-q curve that transforms the values of one lognormal distribution to those of another with the same mean but a different variance has an exponential form:

$$q' = aq^b \tag{19.2}$$

The coefficient, a, and the exponent, b, are given by the following

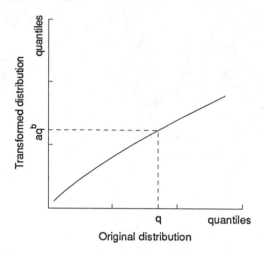

Figure 19.11   The indirect lognormal correction as a *q-q* plot.

formulas:

$$a = \frac{m}{\sqrt{f \cdot CV^2 + 1}} \left[\frac{\sqrt{CV^2 + 1}}{m} \right]^b \qquad b = \sqrt{\frac{\ln(f \cdot CV^2 + 1)}{\ln(CV^2 + 1)}} \qquad (19.3)$$

In these formulas, m is the mean and f is the variance adjustment factor as before; CV is the coefficient of variation.

The problem with the direct application of Equation 19.2 is that it does not necessarily preserve the mean if it is applied to values that are not exactly lognormally distributed. The unchanging mean is one of the few things of which we are reasonably certain in the support effect correction, and it is better to adjust our transformation so that it works as we intended it to. The indirect lognormal correction, therefore, rescales all of the values from the transformation given in Equation 19.2 so that their mean is m:

$$q'' = \frac{m}{m'} \cdot q' \qquad (19.4)$$

where m' is the mean of the distribution after it has been transformed by Equation 19.2. This procedure requires two steps. First, the values are transformed according to Equation 19.2. Second, they are rescaled to the correct mean. The magnitude of the rescaling is a reflection

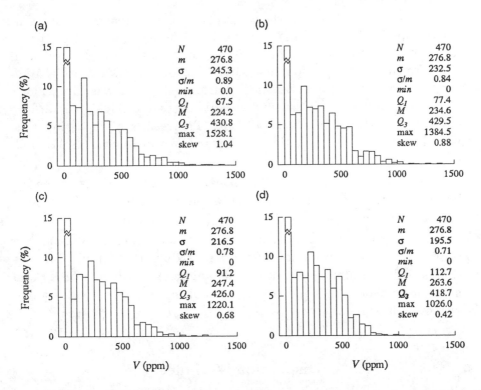

Figure 19.12 Histograms and summary statistics of three distributions obtained by applying the indirect lognormal correction. The point histogram is given in (a) while the distributions resulting from the variance reduction factors of 0.8, 0.6, and 0.4 are shown in (b), (c), and (d), respectively.

of the similarity between the original distribution and a lognormal distribution. If the original distribution is close to lognormal, then the factor $\frac{m}{m'}$ in Equation 19.4 will be close to one.

Figure 19.11 shows Equation 19.1 on a q-q plot. For the low quantiles of the original distribution, where the curve is steep, the transformation does not squash values together as much as it does for high quantiles, where the curve is flatter. The extreme values in the tail therefore get pushed toward the mean more than the median values in the hump of the distribution, with the result that the indirect lognormal correction implicitly lowers the skewness and increases the symmetry.

Figure 19.12 shows how the indirect lognormal correction alters a distribution. In Figure 19.12a we begin with the same global distribution that we showed in Figure 19.10a. Each sample value was then transformed using Equation 19.4 with a variance adjustment factor of 0.8; the histogram of the resulting transformed values was then calculated using the same declustering weights that were calculated earlier. The result is shown in Figure 19.12b. The results of repeating this procedure with $f = 0.6$ and $f = 0.4$ are shown in Figures 19.12c and d.

There are two important differences between the results of the indirect lognormal correction and the affine correction: the skewness decreases as the variance is reduced and the minimum stays at 0.

Though somewhat more complex than the affine correction, the indirect lognormal correction offers a procedure that often produces more sensible results if extreme cutoffs are being considered or if there is reason to believe that the preservation of shape implicit in the affine correction is unrealistic.

Dispersion Variance

Both of the support correction procedures we have explored require a variance adjustment factor. In this section we introduce the concept of dispersion variance that will lead, in the next section, to a method that uses the variogram to estimate the variance adjustment factor.

To introduce dispersion variance, let us return to the original definition of variance:

$$\sigma^2 = \frac{1}{n}\sum_{i=1}^{n}(v_i - m)^2 \tag{19.5}$$

It is the average squared difference between a set of values and a mean. Throughout the previous chapters, we have tacitly understood that the v_is were point values and that m was the mean calculated over the entire set of all the v_is. Dispersion variance allows us to generalize this definition.

Before we try to give a definition of dispersion variance, let us look at the small example shown in Figure 19.13. In Figure 19.13a we show 12 samples on a regular grid. The mean of these 12 samples is 35 ppm,

(a)

26	27	30	33
•	•	•	•

30	34	36	32
•	•	•	•

43	41	42	46
•	•	•	•

(b)

26	27	30	33
•	•	•	•
(33)	(34)	(36)	(37)

30	34	36	32
•	•	•	•
(33)	(34)	(36)	(37)

43	41	42	46
•	•	•	•
(33)	(34)	(36)	(37)

Figure 19.13   An example for dispersion variance calculations. The 12 point values shown in (a) are grouped into 1 x 3 blocks in (b) with the average block values shown in parentheses beneath each sample location.

and the variance can be calculated as follows:

$$
\begin{aligned}
\text{Variance of point values} = \frac{1}{12}[\ &(26-35)^2 + (30-35)^2 + (43-35)^2 + \\
&(27-35)^2 + (34-35)^2 + (41-35)^2 + \\
&(30-35)^2 + (36-35)^2 + (42-35)^2 + \\
&(33-35)^2 + (32-35)^2 + (46-35)^2\ ] \\
= &\ 40.0 \text{ ppm}^2
\end{aligned}
$$

In Figure 19.13b we have grouped the samples into four 1 x 3 blocks. Beneath each sample value we have shown in parentheses the corresponding block average. To describe the variability of the four block values, we could calculate the variance of the four block averages:

$$
\begin{aligned}
\text{Variance of block values} = \frac{1}{4}[\ &(33-35)^2 + (34-35)^2 + \\
&(36-35)^2 + (37-35)^2\ ] \\
= &\ 2.5 \text{ ppm}^2
\end{aligned}
$$

Another source of variability in which we may be interested is the variability of point values within their corresponding blocks:

$$
\begin{aligned}
\text{Variance of point values within blocks} = \frac{1}{12}[\ &(26-33)^2 + (30-33)^2 + (43-33)^2 + \\
&(27-34)^2 + (34-34)^2 + (41-34)^2 + \\
&(30-36)^2 + (36-36)^2 + (42-36)^2 + \\
&(33-37)^2 + (32-37)^2 + (46-37)^2\ ] \\
= &\ 37.5 \text{ ppm}^2
\end{aligned}
$$

These three calculations are all similar in the sense that they all are calculations of some average squared deviation. They differ only in the support of the individual values (the v_is in Equation 19.5) and in the support of the mean that is subtracted from each individual value.

Dispersion variance is an average squared difference that has the support of the individual values and the support of the mean explicitly stated:

$$\sigma^2(a,b) = \frac{1}{n}\sum_{i=1}^{n}(\underbrace{v_i}_{\text{Support } a} - \underbrace{m_i}_{\text{Support } b})^2 \qquad (19.6)$$

In the calculation of the variance between the 12 values in Figure 19.13, the support of the data was individual points and the mean that was subtracted from each data value was calculated over the entire area. The dispersion variance of point values within the entire area, $\sigma^2(\cdot, A)$, was 40 ppm^2. In the calculation of the variance between the four block average values in Figure 19.13, the support of the data was 1 x 3 block averages and the mean value that was subtracted from each data values was calculated over the entire area. The dispersion variance of 1 x 3 block averages within the entire area, $\sigma^2(1\text{x}3, A)$, was 2.5 ppm^2. In the calculation of the variance of point values within 1 x 3 blocks, the data had point support and the means had 1 x 3 block support. The dispersion variance of point values within 1 x 3 blocks, $\sigma^2(\cdot, 1\text{x}3)$, was 37.5 ppm^2.

So far, we have not actually made much progress in our change of support problem. In fact, all we have done is come up with a name for some of the important parameters in the problem. When we have a distribution of point values, its variance will be the dispersion variance of point values within a particular area. If we want a distribution of block values instead, we recognize that this point variance is too large; we would like some way of adjusting this distribution of point values so that its variance becomes the dispersion variance of block values within the same area. The reason that we have introduced the concept of dispersion variance is that it provides a means for estimating the reduction in variance due to the support effect [8].

An important relationship involving dispersion variances is the following:

$$\underbrace{\sigma^2(a,c)}_{\substack{\text{Total}\\\text{variance}}} = \underbrace{\sigma^2(a,b)}_{\substack{\text{Variance}\\\text{within blocks}}} + \underbrace{\sigma^2(b,c)}_{\substack{\text{Variance}\\\text{between blocks}}} \tag{19.7}$$

This relationship expresses the fact that the variance of point values within a certain area can be seen as the variance of point values within blocks plus the variance of block values within the area. Our 12 point example provides a good example of this relationship:

$$
\begin{array}{ccccc}
\sigma^2(\cdot, A) & = & \sigma^2(\cdot, 1 \; x \; 3) & + & \sigma^2(1 \; x \; 3, A) \\
40.0 & = & 37.5 & + & 2.5
\end{array}
$$

This numerical example gives us some insight into how we might estimate the variance of the block distribution. We are typically able to estimate a distribution of point values. From our estimated distribution we can calculate $\sigma^2(\cdot, A)$, the variance of points within the entire area, which is the quantity referred to as *total variance* in Equation 19.7. We would like to reduce this variance so that it more accurately reflects the variance of values for blocks of some larger volume B. The variance of these block values is $\sigma^2(B, A)$, which is the quantity referred to as *variance between blocks* in Equation 19.7. Using Equation 19.7, we have the following relationship:

$$
\begin{aligned}
\sigma^2(\cdot, A) &= \sigma^2(\cdot, B) + \sigma^2(B, A) \\
\sigma^2(B, A) &= \sigma^2(\cdot, A) - \sigma^2(\cdot, B)
\end{aligned}
$$

The variance adjustment factor we were discussing earlier is the ratio of the block variance to the point variance:

$$
\begin{aligned}
f &= \frac{\sigma^2(B, A)}{\sigma^2(\cdot, A)} \\
&= \frac{\sigma^2(\cdot, A) - \sigma^2(\cdot, B)}{\sigma^2(\cdot, A)} \\
&= 1 - \frac{\sigma^2(\cdot, B)}{\sigma^2(\cdot, A)} \tag{19.8}
\end{aligned}
$$

We already have an estimate of $\sigma^2(\cdot, A)$, the variance of our point values; we will have a means of estimating the variance adjustment factor if we can find some way to estimate $\sigma^2(\cdot, B)$. The practical value of Equation 19.7 is that it gives us a way of relating the variance of the block distribution (which we don't know) to the variance of the point distribution (which we do know), using the dispersion variance of points within blocks as an intermediary step. Our goal now becomes the estimation of this dispersion variance. As we will see in the next section, the variogram gives us a way of estimating this quantity.

Estimating Dispersion Variances from a Variogram Model

The dispersion variance of point values within any area can be estimated from a variogram model since there is a direct link between the definition we gave for dispersion variance (Equation 19.6) and the definition we gave for the variogram (Equation 7.1). As an example, let us look at the expression for the total variance of point values within a large area, $\sigma^2(\cdot, A)$. Using Equation 19.6 this can be written as

$$\sigma^2(\cdot, A) = \frac{1}{n} \sum_{i=1}^{n} (v_i - m)^2$$

$$= \frac{1}{n} \sum_{i=1}^{n} v_i^2 - m^2$$

where $v_1, \ldots, v_n$ are the n point values in the volume V; m is the arithmetic mean of these values:

$$m = \frac{1}{n} \sum_{i=1}^{n} v_i$$

The dispersion variance of points within the entire area can therefore be written as:

$$\sigma^2(\cdot, A) = \frac{1}{n} \sum_{i=1}^{n} v_i^2 - (\frac{1}{n} \sum_{i=1}^{n} v_i)^2$$

$$= \frac{1}{2n} \sum_{i=1}^{n} v_i^2 + \frac{1}{2n} \sum_{j=1}^{n} v_j^2 - \frac{1}{n^2} \sum_{i=1}^{n} \sum_{j=1}^{n} v_i v_j$$

$$= \frac{1}{2n^2} \sum_{i=1}^{n} \sum_{j=1}^{n} v_i^2 + \frac{1}{2n^2} \sum_{j=1}^{n} \sum_{i=1}^{n} v_j^2 - \frac{2}{2n^2} \sum_{i=1}^{n} \sum_{j=1}^{n} v_i v_j$$

$$= \frac{1}{2n^2} (\sum_{i=1}^{n} \sum_{j=1}^{n} v_i^2 + \sum_{i=1}^{n} \sum_{j=1}^{n} v_j^2 - 2 \cdot \sum_{i=1}^{n} \sum_{j=1}^{n} v_i v_j)$$

$$= \frac{1}{2n^2} \sum_{i=1}^{n} \sum_{j=1}^{n} (v_i - v_j)^2 \qquad (19.9)$$

This final expression is reminiscent of the definition for the variogram:

$$\gamma(\mathbf{h}) = \frac{1}{2N(\mathbf{h})} \sum_{(i,j)|\mathbf{h}_{ij}=\mathbf{h}} (v_i - v_j)^2$$

Both are the average of some squared differences. With the variogram, we average the squared differences of all pairs separated by a particular vector $\mathbf{h}$. With the dispersion variance, we average the squared differences of all pairs within the area A. The dispersion variance, therefore, can be seen as a kind of variogram calculation in which pairs of values are accepted in the averaging procedure as long as the separation vector $\mathbf{h}_{ij}$ is contained within A:

$$\sigma^2(\cdot, A) = \frac{1}{2N(A)} \sum_{(i,j)|\mathbf{h}_{ij} \in A} (v_i - v_j)^2$$

Though this could be estimated from a set of sample data, it is usually derived from a variogram model instead:

$$\tilde{\sigma}^2(\cdot, A) \approx \bar{\bar{\tilde{\gamma}}}(A)$$

where the right-hand side refers to the the variogram model $\tilde{\gamma}(\mathbf{h})$ averaged over all possible vectors contained within A. We have used the $\tilde{\ }$ in this equation to remind ourselves that these are quantities derived from a *model* and to draw attention to the importance of this model. For certain variogram models there are formulas, tables, or graphs for calculating $\bar{\bar{\tilde{\gamma}}}(A)$ for rectangular blocks [7,9]. In practice, the more common procedure for calculating these average variogram values is to discretize the volume A into n points and to approximate the exhaustive average of the variogram within the area by an average of the n^2 variogram values between the n discretized locations:

$$\tilde{\sigma}^2(\cdot, A) \approx \frac{1}{n^2} \sum_{i=1}^{n} \sum_{j=1}^{n} \tilde{\gamma}(\mathbf{h}_{ij}) \qquad (19.10)$$

We now have all of the pieces we need to estimate the variance adjustment factor. Equation 19.8 calls for two quantities: the dispersion variance of point values within the entire area and the dispersion

Table 19.1 An example of the effect of the number of discretizing points on the estimated variance reduction factor.

Discretizing Grid	$\bar{\bar{\gamma}}(10 \times 10)$	$\bar{\bar{\gamma}}(260 \times 300)$	f
1 x 1	0	0	—
2 x 2	29813	80250	0.628
3 x 3	34655	94765	0.634
4 x 4	36265	99456	0.635
5 x 5	36990	101698	0.636
6 x 6	37376	102713	0.636
7 x 7	37607	103267	0.636
8 x 8	37755	103622	0.636
9 x 9	37856	103887	0.636
10 x 10	37928	104070	0.636

variance of point values within a block. Using Equation 19.10, the denominator, $\sigma^2(\cdot, B)$, can be estimated by discretizing the block B into several points and calculating the average variogram value between all possible pairs of points. Though we could take our estimate of $\sigma^2(\cdot, A)$ directly from our estimated distribution, it is preferable in practice to estimate it from the variogram model. With both the numerator and the denominator being estimated from the same variogram model, we rely only on the shape of the variogram and not on its magnitude.

Table 19.1 shows the effect of the number of discretizing points on the estimation of the variance adjustment factor. Using the variogram model for the V variable given in Equation 16.34 this table shows the estimated dispersion variance for points within 10 x 10 m^2, the estimated dispersion variance for points within the 260 x 300 m^2 global area and the resulting variance adjustment factor. While the numerator and the denominator of the ratio involved in Equation 19.8 converge slowly to their corresponding exhaustive values, the ratio itself converges much more rapidly, with its estimate based on a discrete 5 x 5 grid being virtually identical to its exhaustive value.

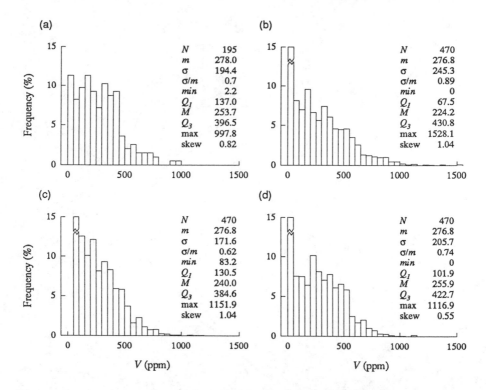

Figure 19.14 Histograms of the distributions obtained by applying a change of support to the declustered sample histogram of V. The exhaustive histogram of 20×20 m^2 block averages is shown in (a) while the declustered sample histogram is shown in (b). The distribution obtained by applying the affine correction to the declustered sample distribution is shown in (c), while the distribution obtained using the indirect lognormal correction is shown in (d).

Case Study: Global Change of Support

In many mining applications the distinction between ore and waste is often made by comparing the average grade of a block of ore to a cutoff grade. If the average grade of the block is estimated to be greater than the cutoff grade, then the block is mined as ore; otherwise it is either mined as waste or left in place. Thus the estimation of recoverable quantities requires knowledge of the distribution of block grades. Typically the sample support is a great deal smaller than that of the block; thus, the distribution of block grades must be derived from

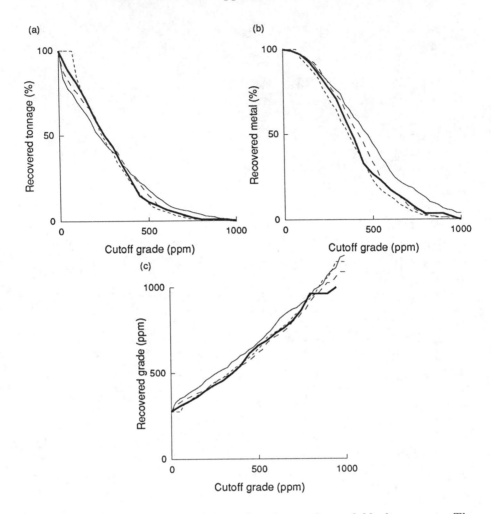

Figure 19.15 Recovery curves for V based on point and block support. The heavy solid line on each graph is the recovery curve for the exhaustive distribution of 20 x 20 m^2 block values of V. The solid light curve shows the recoveries for the declustered point histogram of V, while the short dashed curve shows the recoveries obtained using the affine correction and the long dashed curve shows the recoveries using the indirect lognormal correction.

a sample distribution of quasi-point support using one of the change of support methods.

For this case study discussed in this section we will be interested in a change of support from points to blocks measuring 20 x 20 m^2. The distribution of points is provided by the 470 sample values of V. Recall from Chapter 18 that this sample distribution must be declustered to obtain a reasonable estimate of the global distribution. Using the polygon weights derived in Chapter 10, we obtain the declustered histogram (point support) shown in Figure 19.14a. The 20 x 20 m^2 block histograms shown in Figures 19.14c and d are obtained using the affine and indirect lognormal change of support methods.

To apply either one of these methods we must first determine the variance reduction factor f using Equation 19.8. The dispersion variance of points within the 20 x 20 m^2 block $\sigma^2(\cdot, B)$ is 52,243 ppm^2. This was obtained using Equation 19.10, and the variogram model for V given in Equation 16.34. The dispersion variance of points within the sample domain $\sigma^2(\cdot, A)$ is 104,070 ppm^2 and is given in Table 19.1. Thus the variance reduction factor f is equal to 0.498. Using Equations 19.1, 19.2, 19.3, and 19.4, we then obtained the block distributions given by the affine and indirect lognormal corrections, respectively.

The histogram of true block average values is compared in Figure 19.14 to the uncorrected sample histogram and the sample distribution after correction by the affine and indirect lognormal procedures. Note that while the mean of the declustered sample histogram is quite close, its spread is definitely too large. Both of the correction procedures manage to reduce the spread without changing the mean. For the affine correction, the minimum value 83.2 ppm and the coefficient of skewness is unchanged. For the indirect lognormal correction, the minimum stays at 0 ppm and the coefficient of skewness decreases.

Figure 19.15 provides a further comparison of the methods by showing the tonnage of ore, its grade, and the quantity of metal predicted to be recovered for each of the distributions. The four lines on each plot in this figure correspond to the various distributions as follows:

- Heavy solid line — Exhaustive 20 x 20 m^2 block distribution of V.

- Light solid line — Declustered sample distribution of V.

- Short dashes — Block distribution derived from the affine correction.

- Long dashes — Block distribution derived from the indirect log-normal correction.

Note that for practically all cutoffs, the recoveries predicted using the declustered sample distribution (point support) depart the furthest from the exhaustive recoveries shown by the heavy solid line. Recoveries predicted using the corrected or block distributions are much closer to the true or exhaustive recoveries. For example, at the recoveries for a 500 ppm cutoff on the exhaustive block histogram are 11.28% recovered tonnage, 26.8% recovered quantity of metal, and a recovered grade of 660 ppm. The corresponding figures derived from the sample histogram are 18.3%, 44.8%, and 677 ppm. The predictions based on the uncorrected sample histogram overestimate the recovered quantity of metal by 167%. Such severe overpredictions could prove to be disastrous in the early stages of mine design. The predictions based on the corrected sample distributions, however, are sufficiently accurate for most mine planning or design purposes.

Notes

[1] In practice, even though the support of our samples are averages over some volume—a cylinder of drill core, for example—they are most often treated as "point" samples since the volume they represent is very small compared to the larger volumes whose arithmetic average we are trying to estimate.

[2] The symmetrization of the distribution is a consequence of the Central Limit Theorem, which states that sample means tend toward a Normal distribution regardless of the distribution of the samples. Since this convergence is faster if the samples are independent, a lack of spatial correlation entails a more rapid symmetrization of the point distribution.

[3] The models provided for a change of support in this chapter leave the mean of the distribution unchanged, and are not appropriate for variables whose averaging process is not arithmetic. For example, these procedures are not appropriate for adjusting effective block permeabilities since the effective block permeability is not equal to the arithmetic average of the corresponding point permeabilities.

[4] Experience suggests that some symmetrization is likely to occur for variance reductions greater than 30%. For this reason, the indirect

lognormal correction is likely to be more appropriate than is the affine correction if the variance adjustment factor is less than 0.7.

[5] This is not quite the same thing as the direct lognormal correction, a procedure that fits a lognormal model to the original distribution, and then reduces its variance while preserving its mean.

[6] If the original distribution is not exactly lognormal, then the actual reduction in variance will be different from f. The second step, in which the values from the first step are rescaled by m/m', will change the variance by the square of this factor. Since the mean of the block distribution is better known than its variance, it is felt to be more important to honor the mean exactly than to try to honor the variance exactly. If the reduction in variance is known quite precisely (i.e., from historical information) then one could experiment with several values of f to find the one that produces the desired reduction in variance after the entire correction has been performed.

[7] David, M. , *Geostatistical Ore Reserve Estimation.* Amsterdam: Elsevier, 1977.

[8] Though the dispersion variance approach is more traditional in geostatistics, the estimation of the variance of block values could be accomplished by resorting once again to Equation 9.14, which gave us the variance of a weighted linear combination of random variables. If the block average, V_B, is simply an equal weighting of all the point values within the block:

$$V_B = \frac{1}{n} \sum_{i \mid x_i \in B}^{n} V(x_i)$$

then the variance of the random variable V_B is

$$Var\{V_B\} = \sum_{i \mid x_i \in B}^{n} \sum_{j \mid x_j \in B}^{n} Cov\{V(x_i)V(x_j)\}$$

The variance of block averages is the average covariance between all possible pairs of locations within the block. The covariance of point values, under the assumption of stationarity, can be taken directly from our model of the covariance function:

$$Cov\{V(x_i)V(x_i)\} = \tilde{C}(0)$$

The variance adjustment factor, f, is therefore

$$f = \frac{\frac{1}{n^2} \sum_{i=1}^{n} \sum_{j=1}^{n} \tilde{C}(\mathbf{h}_{ij})}{\tilde{C}(0)}$$

This formula, expressed in terms of the covariance model, gives essentially the same result as Equations 19.8 and 19.10.

[9] Journel, A. G. and Huijbregts, C. J. , *Mining Geostatistics.* London: Academic Press, 1978, pp. 108-148.

Further Reading

Parker, H. , "The volume-variance relationship: a useful tool for mine planning," in *Geostatistics*, (Mousset-Jones, P. , ed.), pp. 61–91, McGraw Hill, New York, 1980.

Shurtz, R. , "The electronic computer and statistics for predicting ore recovery," *Engineering and Mining Journal*, vol. 11, no. 10, pp. 1035–1044, 1959.

20

ASSESSING UNCERTAINTY

In previous chapters we have concentrated on the estimation of various unknown quantities. In this chapter we will look at how we can supplement these estimates with some assessment of their uncertainty. We begin with a qualitative discussion of what we mean by "uncertainty" and the factors that influence it. We follow this with a look at how uncertainty should be reported, a question whose answer depends on the goal of our study. We then propose several methods for deriving these various uncertainty measures and conclude with a case study.

Error and Uncertainty

Qualitatively, we know what we mean by "uncertainty." The many other words used to express this notion: "reliability," "confidence," "accuracy," all carry similar connotations that revolve around the recognition that the single value we report is, in some sense, only a reasonable or useful guess at what the unknown value might be. We hope that this estimate will be close to the true value, but we recognize that whatever estimation method we choose there will always be some error. Though it is not possible to calculate this error exactly, we do hope that we can assign it an "uncertainty," some indication of its possible magnitude.

A useful first step in assessing the uncertainty is to consider the factors that influence the error. One obvious factor is the number of the nearby samples; additional nearby samples should help to make the estimate more reliable. Another important consideration is the proximity of the available samples; the closer the samples to the point we are trying to estimate, the more confident we are likely to be in our estimate.

There are two other factors that are less obvious but may be equally as important as the number and proximity of the samples. The first is the spatial arrangement of the samples. The additional confidence that additional samples might bring is influenced by their proximity to existing samples. Consider the example shown in Figure 20.1. We already have one nearby sample, as shown in Figure 20.1a. Our confidence in our estimate will likely increase if we have other nearby samples. If those additional samples are extremely close to the existing sample (as in Figure 20.1b), however, our confidence will not increase as much as it would if the additional samples were more evenly distributed about the point we are trying to estimate (as in Figure 20.1c).

The second factor which complicates the question of uncertainty is the nature of the phenomenon under study. If we are dealing with an extremely smooth and well-behaved variable, our estimates are going to be more reliable than if we are dealing with a very erratic variable. For example, the configuration of four samples shown in Figure 20.1c may produce extremely accurate estimates of the thickness of some sedimentary horizon and yet not be enough to produce good estimates of the gold grade within that horizon.

It is important to recognize that the nature of the phenomenon under study may vary from one locality to the next. It is common in practice to use one variogram model to describe the pattern of spatial continuity for an entire region. Since the estimates are unchanged by a rescaling of the variogram model, the shape of a single variogram model may be adequate for the purposes of estimation. If there are fluctuations in the local variability, however, a single variogram model is not adequate for the purposes of assessing uncertainty. Since the ordinary kriging variance is affected by the magnitude of the variogram model, the possibility of fluctuations in the magnitude of the variogram must be taken into account when assessing uncertainty. Many earth science data sets exhibit the proportional effect we described earlier, with the local variability being proportional to the local mean. In such cases, the uncertainty of a particular estimate is linked to the magnitude of the data values that are used in the weighted linear combination.

These various factors interact. For example, for very smooth and

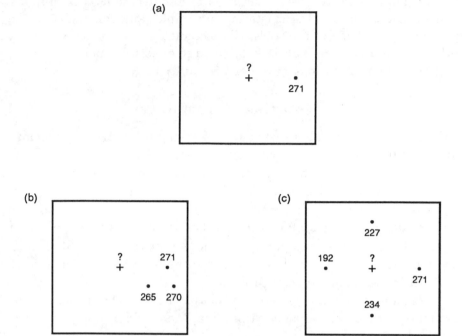

Figure 20.1 The effect of additional sampling on uncertainty of the estimate. The estimate of the unknown value at the plus sign in the center of (a) will become more reliable with additional sampling. The clustered samples samples in (b), however, will not improve the reliability as much as the more evenly distributed samples in (c).

well-behaved phenomena proximity may be more important than number of samples; we might prefer to have one very close sample than several samples slightly further away. For very erratic phenomena, the reverse might be true; even at a greater distance, several samples might produce a more reliable estimate than a single nearby sample.

The continuity of the phenomenon also interacts with the effect of clustering. With a very smooth and well-behaved phenomenon, two samples very close to each other are not much better than one sample. If the variable being studied fluctuates very wildly, however, two closely spaced samples might not be so redundant.

As we start to explore various methods for characterizing the uncertainty of our estimates, we should keep in mind these factors: number and proximity of samples, clustering of samples and continuity of the phenomenon. Some of the tools will account for some factors but not others. There has been considerable misuse of certain tools due to the fact that their limitations have not been fully understood.

Reporting Uncertainty

Without a clear idea about what we are trying to report, any attempt at quantifying uncertainty will lack a clear objective meaning. One of the frustrating aspects of dealing with uncertainty is that it is usually not clear what the measurement of uncertainty actually means. Though there are certain formats that have become traditional, the *95% confidence interval*, for example, it is worth considering whether or not they are useful or appropriate for the problem at hand. In this section, we propose several approaches to reporting uncertainty, each suited to a particular type of problem.

Common to all the methods discussed here is the notion of the estimation error, which, as in earlier chapters, is defined as follows:

$$error = r = \text{estimated value - true value}$$
$$= \hat{v} - v$$

Though this error has a clear definition, it can not be calculated since it requires that we know the unknown true value.

An Uncertainty Index. Figure 20.2 shows a common problem in earth sciences; we would like to know which sampling pattern gives us more reliable estimates. In Figure 20.2a, we have our samples located on a square grid; in Figures 20.2b and c the samples are arranged on a rectangular grid with different orientations. Later in this chapter we will propose a specific solution for this type of question. For the moment, we will simply note that this type of question calls for an index of uncertainty, some number that permits sets of estimates to be ranked according to their reliability.

The absolute value of this index is unimportant, it is intended to be used only in a relative way to enable comparisons of the possible error of different estimates. This index does not provide a guarantee that one particular estimate is better than another, it merely provides some indication about the possible magnitude of the errors. If we perform a hindsight study, we would hope that there is some obvious correlation

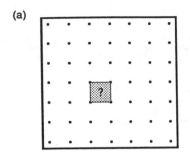

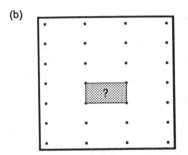

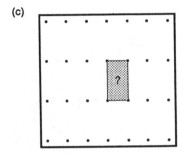

Figure 20.2 Sample configuration and uncertainty. A common problem in earth sciences is the ranking of sampling patterns. Different sampling configurations will produce estimates of differing reliability.

between our uncertainty index and the magnitude of the actual error. Figure 20.3 shows hypothetical plots of some uncertainty indices versus actual errors. The index shown in Figure 20.3a is not very good since it is poorly correlated with the magnitude of the actual error. The uncertainty index shown in Figure 20.3b is very dangerous since it is completely misleading; the estimates with largest errors also have the lowest uncertainty index. The situation shown in Figure 20.3c is the most preferable; the estimates with a high uncertainty index tend to have the larger errors.

Confidence Intervals. Though a good uncertainty index can be tremendously useful in comparing the reliability of estimates, it is not useful for decisions that require an absolute (rather than a relative) indication of the magnitude of the error.

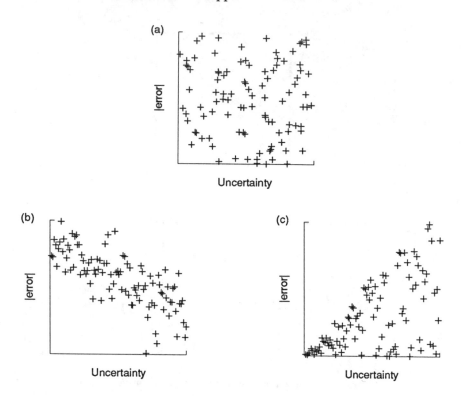

Figure 20.3 Hypothetical scatter plots of uncertainty indices versus the actual magnitude of the error. The uncertainty index in (a) is poor due to its lack of correlation with the actual magnitude of the errors. The index in (b) is dangerously misleading due to its pronounced negative correlation. The positively correlated index in (c) is the most preferable.

Confidence intervals are perhaps the most familiar way of accounting for our inability to pin down the unknown value exactly. Rather than report a single value, we report an interval and a probability that the unknown value falls within this interval. For example, rather than simply report an estimate of 300 ppm, we could convey some idea about the reliability by reporting that there is a 50% probability that the error is somewhere in the range 300 ± 20 ppm. For confidence intervals to be practically useful, it must be made clear what exactly such a probabilistic statement means.

Probabilistic statements are quite common in many aspects of daily life. Weather reports typically provide probabilities of rain; public opinion polls often provide a confidence interval. For most of these familiar probabilistic statements, their objective meaning lies in the fact that there is some repeatability in time. When meteorologists report a 20% chance of rain, they are saying that if we looked at all the times they made such a statement we would see that it rained on about 20% of those occasions. When pollsters report a 95% confidence interval of ±3%, they are saying that if they immediately reconducted their poll many times with different people, 95% of the time their results would be within 3% of what they have reported from their single sampling.

In most earth science applications there is no repeatability in time. For example, in an ore deposit a particular block will be mined only once; in a toxic waste site, a polluted area will be cleaned only once. So what does a probabilistic statement mean when only one true value exists, when there will be no second or third repetition of the same exercise?

In such situations, the meaning of a probabilistic statement, if any, lies in the idea that there is some spatial repeatability. Though we will not mine a particular block twice, we will mine many other blocks from the same mine. If we are willing to believe that the conditions in one area of the mine are similar to those in other areas, then we can choose to group these areas together. A 50% confidence interval of ±20 ppm means that if we look at all estimates in some group then the actual value will be within 20 ppm of its corresponding estimate in about 50% of the cases in that group.

The idea here is that although we cannot calculate the actual magnitude of one individual error, we can group together several estimates from different locations and try to make some statements about the distribution of these errors. As we will discuss later in this chapter, a common misuse of confidence intervals stems from a misunderstanding of the population in question. Though confidence intervals can provide a very useful assessment of uncertainty, we should always keep in mind that they have an objective meaning only in the context of the grouping that we have already decided is appropriate.

Probability Distributions.   Though confidence intervals are the most familiar way of reporting uncertainty, they are rarely used directly in decision making. Typically, they are used only in a comparative way, as an index of uncertainty, wasting the effort that went into giving them some absolute meaning. There are two main reasons for this.

First, they are nearly always expressed as symmetric intervals, as $\pm x$. For many of the very skewed distributions encountered in earth sciences, the magnitude of possible underestimates may be quite different from the magnitude of overestimates. For example, for our Walker Lake data set to report an estimate of 200 ±500 ppm is not particularly useful. It is obvious that if 200 ppm is our estimate, then we have not overestimated by more than 200 ppm, but we might have underestimated by much more than that, perhaps by several thousand parts per million.

The second reason that confidence intervals are rarely used quantitatively is that they provide quite limited information. Even when the variable under study has a symmetric distribution, a single confidence interval provides, at most, an optimistic and a pessimistic alternative to the estimate. Though this is sufficient for some problems, there are many others that call for a more detailed assessment of the likelihood of a broad range of possible values.

The concept of confidence intervals can be extended quite naturally to the concept of a complete probability distribution. If we can report an interval within which there is a certain probability, then we should also be able to report a slightly larger interval within which there is a slightly larger probability. Taken to the limit, we should be able to report the probability that the unknown value falls within any particular interval. In doing so we are describing a probability distribution, a range of possible values, each with an associated probability of occurrence.

Such a probability distribution has considerably more information than a single confidence interval. It can be used to describe asymmetric confidence intervals or to describe the probability of exceeding certain thresholds. Combined with the concept of a "loss function," that describes the impact of errors on the profitability of an operation, a probability distribution can be the vehicle for truly incorporating the effect of uncertainty into a decision making process. Regrettably, this topic is well beyond the scope of this book; the Further Reading section at the end of this chapter provides references to some of the relevant literature.

Like confidence intervals, a probability distribution derives its objective meaning from the concept of spatial repeatability and the belief that certain estimates can be grouped together. The appropriateness of a particular grouping is a choice that we must make beforehand and, having made it, that we must keep in mind as we use our probability distributions to assist in the making of various decisions.

Ranking Uncertainty

Let us first consider the problem of producing an uncertainty index that permits the ranking of estimates in order of reliability. From our discussion at the beginning of this chapter on the factors that influence uncertainty, there are several rather simple indices we could consider.

The first factor we discussed was the number of samples; we expect that estimates that are based on many samples will be more reliable than those based on just a few. One rather simple index of uncertainty, therefore, is n, the number of nearby data. Having already decided to use a particular search neighborhood, we can easily count the number of samples that fall within this neighborhood.

We also recognized that the proximity of the samples was an important consideration, so another straightforward index is $1/\bar{d}$, where $\bar{d}$ is the average distance to the available samples. As this average distance increases, the estimate becomes less reliable.

Though both of these proposals, n and $1/\bar{d}$, do account for two of the important factors that affect uncertainty, they do not capture the interaction between these two factors. Furthermore, they do not account for the sample data configuration nor for the behavior of the variable under study. An index of uncertainty that does account for these two additional factors and also for the interaction between the various factors is the error variance $\tilde{\sigma}_R^2$ that we first encountered in Chapter 12:

$$\tilde{\sigma}_R^2 = \tilde{C}_{00} + \sum_{i=1}^{n} \sum_{j=1}^{n} w_i w_j \tilde{C}_{ij} - 2 \sum_{i=1}^{n} w_i \tilde{C}_{i0} \qquad (20.1)$$

This formula gave us the variance of the random variable R, which represented the error in our random function model, as a function of the following model parameters:

- $\tilde{C}_{00}$, the variance of point values

- $\tilde{C}_{ij}$, the covariance between the ith sample and the jth sample

- $\tilde{C}_{i0}$, the covariance between the ith sample and the unknown value being estimated

Once we have chosen these parameters, Equation 20.1 gives us an expression for the error variance as a function of n variables, namely the weights $w_1, \ldots, w_n$. Let us look at the three terms in this equation and see how they incorporate the various factors that we discussed earlier.

The first term represents the variance of the point values and accounts, in part, for the erraticness of the variable under study. As the variable becomes more erratic, this term increases in magnitude, thus giving us a higher uncertainty index.

The second term is a weighted sum of all the covariances between the various samples pairs. As we noted when we discussed the ordinary kriging system, the covariance function acts like an inverse distance in that it decreases as the distance increases. Rather than express distance in a euclidean sense, however, it expresses it in a statistical sense. Our model of $\tilde{C}(\mathbf{h})$ is customized to the pattern of spatial continuity relevant to the particular data set we are studying. If the samples are far apart, then the second term will be relatively small. As they get closer together, the average distance between them decreases and the average covariance increases. This term therefore accounts for the clustering by increasing the uncertainty if we use samples that are too close together.

The third term is a weighted sum of the covariances between the samples and the value being estimated. It accounts for the proximity (again in a statistical sense) of the available samples. As the average distance to the samples decreases, the average covariance increases and, due to its negative sign, this term decreases the index of uncertainty.

The error variance also takes into account the weighting scheme we have chosen for our estimate. For any estimation method that uses a weighted linear combination whose weights sum to 1, Equation 20.1 provides an index of uncertainty [1]. Though not as simple as counting the nearby samples or calculating their average distance, this index does, at the expense of choosing a covariance model, provide a measure of uncertainty that incorporates the relevant factors that we discussed earlier.

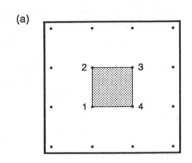

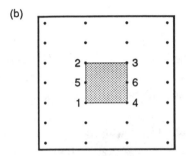 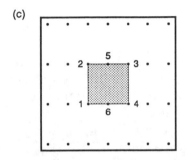

Figure 20.4 The alternative sample data configurations for the case study on ranking uncertainty. Additional samples are going to be added to the 10 x 10 m^2 grid shown in (a). The question we want to address is whether these additional samples should be added on existing north-south lines, as in (b), or on existing east-west lines, as in (c).

Case Study: Ranking Sample Data Configurations

In this section we use the estimation variance to help us decide which of two sample data configurations will produce more reliable estimates. We will imagine that we already have samples on a 10 x 10 m^2 grid throughout the Walker Lake area (as shown in Figure 20.4a) and we are interested in the effect of adding some more samples. In this example, we will look at the choices shown in Figures 20.4b and c. In both of these alternatives, additional samples are located half way between existing holes. In one, the additional samples are located on north-south lines, while in the other, they are located on east-west lines.

We intend to do block estimation with the samples and would like to know which configuration will give us better estimates. In comparing the two alternatives, simple indices such as n or $1/\bar{d}$ are not helpful since both configurations would receive identical rankings.

The error variance offers us a more discriminating uncertainty index if we are willing to take the time to choose a covariance model. In this case study we will use the covariance model we developed in Chapter 16 and repeat in Equation 20.4.

The formula we gave for the error variance in Equation 20.1 was for point estimates. The corresponding formula for block estimates is

$$\tilde{\sigma}_R^2 = \tilde{\bar{C}}_{AA} + \sum_{i=1}^{n}\sum_{j=1}^{n} w_i w_j \tilde{C}_{ij} - 2\sum_{i=1}^{n} w_i \tilde{\bar{C}}_{iA} \qquad (20.2)$$

The first term on the right hand side is no longer $\tilde{C}_{00}$, the variance of point values, but rather $\tilde{\bar{C}}_{AA}$, the average covariance within the block A that we are estimating. The third term consists of a weighted sum of the average covariances between the sample data and the block being estimated.

If we use the ordinary kriging weights [2], Equation 20.2 can be expressed in a form that is computationally more convenient:

$$\tilde{\sigma}_{OK}^2 = \tilde{\bar{C}}_{AA} + \mu - \sum_{i=1}^{n} w_i \tilde{\bar{C}}_{iA} \qquad (20.3)$$

where $\tilde{\bar{C}}_{AA}$ and $\tilde{\bar{C}}_{iA}$ have the same meanings as before and μ is the Lagrange parameter whose value we obtain by solving the ordinary kriging equations.

In Chapter 16, we developed the following variogram model

$$\gamma(\mathbf{h}) = 22,000 + 40,000 \ Sph_1(\mathbf{h}_1') + 45,000 \ Sph_2(\mathbf{h}_2') \qquad (20.4)$$

where the vectors $\mathbf{h}_1'$ and $\mathbf{h}_2'$ are calculated as follows:

$$\mathbf{h}_1' = \begin{bmatrix} h_{x,1}' \\ h_{y,1}' \end{bmatrix} = \begin{bmatrix} \frac{1}{25} & 0 \\ 0 & \frac{1}{30} \end{bmatrix} \cdot \begin{bmatrix} \cos(14) & \sin(14) \\ -\sin(14) & \cos(14) \end{bmatrix} \cdot \begin{bmatrix} h_x \\ h_y \end{bmatrix}$$

$$\mathbf{h}_2' = \begin{bmatrix} h_{x,2}' \\ h_{y,2}' \end{bmatrix} = \begin{bmatrix} \frac{1}{50} & 0 \\ 0 & \frac{1}{150} \end{bmatrix} \cdot \begin{bmatrix} \cos(14) & \sin(14) \\ -\sin(14) & \cos(14) \end{bmatrix} \cdot \begin{bmatrix} h_x \\ h_y \end{bmatrix}$$

We can turn this model upside down by subtracting it from its sill value to obtain the corresponding covariance model:

$$C(\mathbf{h}) = 107,000 - \gamma(\mathbf{h}) \qquad (20.5)$$

Table 20.1 Point-to-point and average point-to-block covariances for the sample data configuration shown in Figure 20.4b.

			C				D
	1	2	3	4	5	6	A
1	107,000	60,072	39,302	49,483	72,243	45,575	59,843
2	60,072	107,000	49,483	46,442	72,243	49,614	62,805
3	39,302	49,483	107,000	60,072	45,575	72,243	59,843
4	49,483	46,442	60,072	107,000	49,614	72,243	62,805
5	72,243	72,243	45,575	49,614	107,000	49,483	64,736
6	45,575	49,614	72,243	72,243	49,483	107,000	64,736

Using this covariance model, Table 20.1 gives all of the point-to-point covariances between the various samples and the point-to-block average covariances between the samples and the block being estimated for the sample data configuration shown in Figure 20.4b (with the additional samples on north-south lines). The first six columns of this table show the covariances for the left-hand side $\mathbf{C}$ matrix; the final column shows the covariances for the right-hand side $\mathbf{D}$ vector. The average covariances were calculated by discretizing the block into 100 points on a 10 x 10 grid. Using the same discretization [3] the average covariance within the block, $\tilde{C}_{AA}$, is $67,400$ ppm^2.

Solving the following system of equations:

$$\mathbf{C} \cdot \mathbf{w} = \mathbf{D}$$

we obtain the ordinary kriging weights and the Lagrange parameter:

$$\mathbf{w} = \begin{bmatrix} w_1 \\ w_2 \\ w_3 \\ w_4 \\ w_5 \\ w_6 \\ \mu \end{bmatrix} = \begin{bmatrix} 0.147 \\ 0.174 \\ 0.147 \\ 0.174 \\ 0.179 \\ 0.179 \\ -1,810 \end{bmatrix}$$

which gives us an error variance of $6,600$ ppm^2 using Equation 20.3.

Repeating this procedure for the sample configuration shown in Figure 20.4c (with the additional samples on east-west lines), we obtain

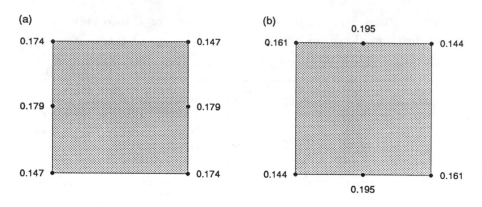

Figure 20.5   The two alternative data configurations along with their kriging weights.

a slightly lower error variance of $5,500$ ppm^2. This index of uncertainty therefore suggests that additional sampling on east-west lines will yield more reliable estimates than the same number of samples on north-south lines.

Before we check with our exhaustive data set to see if this conclusion is in fact correct, let us look at the kriging weights for the two configurations to see if we can make sense of this conclusion. Figure 20.5 shows the kriging weights for the two data configurations. The additional samples on the north-south lines get less weight than those on the east-west lines. The reason for this is that our variogram model is anisotropic, with N14°W being the direction of greatest continuity. For the purposes of estimation, samples that fall in the direction of maximum continuity are more useful than those that fall in the perpendicular direction. The two additional samples shown in Figure 20.5b are more useful than those shown in Figure 20.5a since they fall closer to the direction of maximum continuity. It makes sense, therefore, that the additional samples on the east-west lines get more weight and that this configuration produces more reliable estimates.

To check this theoretical result, we have estimated the average V value of the 780 10×10 m^2 blocks using the two alternative sample data configurations shown in Figure 20.5. Figure 20.6 shows the histograms of the estimation errors for the two alternatives. Both are quite symmetrical, with means and medians close to 0. While one has a mean

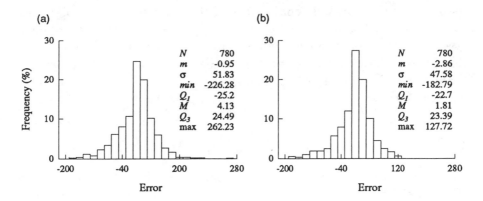

Figure 20.6 The histograms of the errors of block kriging with the two sample data configurations. The distribution of the errors from the north-south configuration is shown in (a), and the corresponding distribution for the east-west configuration is shown in (b).

slightly closer to 0, the other has the better median. Both sample data configurations are therefore reasonably unbiased. In terms of spread, however, the errors from performing estimation with the additional east-west samples are clearly better than those with the north-south samples. The greatest underestimate and the greatest overestimate both have smaller magnitudes with the east-west sampling. The interquartile range and the standard deviation are also smaller for the east-west configuration than for the north-south configuration.

With its error distribution having a lower spread, the east-west sampling can be said to produce more reliable estimates in average. The conclusion that we reached earlier is valid; in this case, the ordinary kriging variance is a good index of uncertainty. It is important to note that the usefulness of the ordinary kriging variance depends entirely on our prior choice of a covariance model; a good model of the pattern of spatial continuity is crucial. Time spent exploring the spatial continuity through the various tools presented in the first section of this book will be rewarded by an improved ability to assess uncertainty.

It is also important to emphasize that the statement that one sample configuration produces more reliable estimates than another makes sense only with respect to the entire population. Despite the overall superiority of the east-west configuration, there are some particular cases where the north-south sampling may do a better job. Of the 780 block averages, the north-south configuration produced an estimate closer to the true block average for 358 of them. An index of uncertainty does not make definitive statements about the relative magnitudes of specific local errors.

Assigning Confidence Intervals

A ranking of the reliability of estimates may not be enough for some of the decisions that need to be made in earth sciences. There are many problems that require a measure of the uncertainty whose absolute magnitude is meaningful. For example, it is often important to be able to establish a range within which the unknown true value is likely to fall. For such problems, we typically supplement the estimate with a confidence interval. Such a confidence interval consists of a minimum and a maximum value and a probability that the unknown value falls within this range.

The most traditional way to establish such confidence intervals involves two important assumptions. The first is that the actual errors follow a Normal or Gaussian distribution. The second is that the error variance $\tilde{\sigma}_R^2$ from the random function model is an accurate estimate of the variance of the actual errors. If we are willing to make these assumptions, then the estimate $\hat{v}$ can be combined with the error variance, $\tilde{\sigma}_R^2$, to produce the 95% confidence interval: $\hat{v} \pm 2\tilde{\sigma}_R$.

The first of these assumptions is largely a matter of convenience. The Normal distribution is the most well-known and thoroughly studied probability distribution. As a model for error distributions, its most relevant features are that 68% of the values fall within one standard deviation of the mean and that 95% fall within two standard deviations. The 95% confidence interval is a fairly common standard for reporting uncertainty, though poor estimates are occasionally quoted with the 68% confidence interval to make the possible error seem less alarming.

Global distributions of errors, even for very skewed data, do tend to be symmetric. This does not mean, however, that they are necessarily well modeled by a Normal distribution. Unfortunately, there has been very little work on alternative models, and the $\pm 2\sigma$ 95% confidence interval will likely remain a standard for reporting uncertainty.

Even if we do decide to go with the Normality assumption, it is important to consider our second assumption— that we are able to predict the error variance. Whether $\tilde{\sigma}_R^2$ is an accurate estimate of the actual error variance depends heavily on the variogram model. As we saw when we first looked at ordinary kriging in Chapter 12, a rescaling of the variogram model does not affect the kriging weights but does affect the error variance. If we intend to use the error variance to derive confidence intervals, we had better be sure that we have it correctly scaled.

From the results of the case study we performed in the previous section, it is clear that we have some problems interpreting the error variance of the random function model as the variance of the actual errors. For both of the sample data configurations that we studied, the actual variance of the errors was much smaller than the error variance predicted by our random function model. For example, for the north-south sampling the value of $\tilde{\sigma}_R^2$ was 6,600 ppm^2, while the variance of the actual errors was 2,686 ppm^2.

The use of $\tilde{\sigma}_R^2$ for the definition of Normal confidence intervals requires, in part, that the sill of the variogram is an accurate estimate of the global variance. One of the reasons for discrepancies between the actual error variance and the error variance predicted by the random function model is that the sill of the sample variogram may not be a good estimate of the global variance. We noted earlier that preferential sampling in high-valued areas makes the naive arithmetic mean a poor estimate of the global mean. If there is a proportional effect, then preferential sampling also affects measures of variability such as the sample variance and the sample variogram.

We can see examples of such discrepancies from our analysis of the exhaustive and the sample data sets in the first section of this book. We saw that the sample variance of 89,940 ppm^2 was considerably larger than the exhaustive variance of 62,450 ppm^2. The same is true of the sills of the sample and exhaustive variograms. The sample V variogram had a sill of 107,000 ppm^2, while the exhaustive V variogram had a sill of about 62,000 ppm^2.

Though the magnitude of our variogram model does not affect our estimates, it does affect our estimation variance. We should therefore try to adjust our variogram model so that its sill more accurately reflects the true global variance. One way of doing this is to rescale the variogram so that its sill agrees with a good declustered estimate of the global variance. In Chapter 18 we looked at the problem of declustering estimates of global parameters. Using the areas of the polygons of influence as declustering weights, our estimate of the global variance was $60,172$ ppm^2. For the purposes of establishing good global confidence intervals, we could rescale the variogram model we gave earlier in Equation 20.4 so that its sill is about $60,000$. With an original sill of $107,000$, the coefficients of the original model should all be multiplied by $60,000/107,000 = 0.56$, giving us the following model:

$$\gamma(\mathbf{h}) = 12,400 + 22,400 \; Sph_1(\mathbf{h}_1') + 25,200 \; Sph_2(\mathbf{h}_2') \qquad (20.6)$$

Repeating our earlier study of the reliability of the ordinary kriging block estimates from the sample data configuration shown in Figure 20.5a, we find that the kriging weights are unchanged but that the ordinary kriging variance is now $3,700$ ppm^2. Though certainly an improvement over our initial value of $6,600$ ppm^2, this is still considerably larger than the variance of the actual errors, $2,686$ ppm^2. The reason that our ordinary kriging variance still overestimates the actual error variance is that our variogram model poorly describes the short scale variability.

The preferential clustering of samples in areas with high values affects not only the magnitude of the variogram, but also its shape. In the Walker Lake example, the sample variograms have higher nugget effects than do the corresponding exhaustive variograms. With all of our closely spaced samples in areas with high variability, our sample variogram indicates greater short scale variability than actually exists over the entire area. Though the effect of preferential sampling on the magnitude of our sample variogram can be accounted for simply by rescaling the sill of the variogram model to a good declustered estimate of the global variance, it is much more difficult to account for the effect of preferential sampling on the shape of the variogram. Unfortunately, this problem has received very little attention; the Further Reading section at the end of the chapter provides references to some of the very recent work on this important problem.

Case Study: Confidence Intervals for An Estimate of The Global Mean

In Chapter 10, we looked at the problem of estimating the global mean. In this section, we will show how this estimate can be supplemented by a confidence interval.

Before we get into the actual calculations, it is important to consider the objective meaning of such a global confidence interval. Earlier in this chapter, we discussed probabilistic statements and how their meaning lies in the concept of repeatability. We replaced repeatability in time with repeatability in space, arguing that probabilistic statements in earth sciences might have some meaning if the same (or almost the same) estimation will be performed at several other locations throughout the area of interest. Since there is only one true global mean and only one set of sample data from which to estimate it, we will not be repeating a similar estimation anywhere else. How, then, should our statement of uncertainty be interpreted?

Of the several possible interpretations commonly offered, most are quite tenuous; however, there are two that have some practical relevance. In one, the repeatability comes from the idea that there are other areas that are statistically similar; in the other, the repeatability comes from the idea that other sample data sets could have been obtained.

If the area under study is part of a larger region, then the sense of repeatability may come from repeating the same exercise on other areas in the same region. For example, a petroleum reservoir typically falls within some larger basin that contains several other reservoirs. A probabilistic statement about the volume of oil in place may derive its meaning from the fact that the estimation will be repeated in several other reservoirs in the same basin, all producing from the same formation. The population over which we are averaging (and therefore assuming some statistical homogeneity) includes all other reservoirs whose geological characteristics are similar to the one currently under study.

There are other applications in which the notion of other statistically similar areas may apply. For example, in environmental applications a single source may be responsible for several polluted sites. In mining applications, genetically similar deposits with similar geological settings may have similar statistical characteristics. However, the phenomenon under study is often unique and there are no other phenomena that lend sense to the idea of averaging over statistically similar areas. In such cases, the meaning of a probabilistic statement comes from the notion that the available sampling could be entirely discarded and another set of samples collected. The probabilistic statement can be viewed as describing the possible fluctuations in the estimation error of the global mean from one sampling to the next.

The error variance given by the random function model is often used to establish global confidence intervals. The formula given earlier for the error variance of block estimates:

$$\tilde{\sigma}_R^2 = \bar{\tilde{C}}_{AA} + \sum_{i=1}^{n}\sum_{j=1}^{n} w_i w_j \tilde{C}_{ij} - 2\sum_{i=1}^{n} w_i \bar{\tilde{C}}_{iA} \qquad (20.7)$$

can be used once the weights, $w_1, \ldots, w_n$, and a model for the covariance, $\tilde{C}(\mathbf{h})$, have been chosen. When this formula is used to establish a global error variance, A is the entire area under study and not a particular block within it. $\bar{\tilde{C}}_{AA}$ is the average covariance within the entire area and can be calculated by discretizing A into several points and averaging the covariances between all possible pairs of points; $\tilde{C}_{ij}$ is, as before, the covariance between the sample value at the ith location and the jth location; $\bar{\tilde{C}}_{iA}$ is the average covariance between the sample at the ith location and the entire region A.

Though this formula can always be applied, it is important to consider the many assumptions behind it. First, the set of weights should sum to 1; for the two methods we discussed in Chapter 10, polygonal weighting and cell declustering, this is indeed the case. Second, the statistical characteristics of the variable under study should not vary from one location to another. Specifically, we assume that the random variables that model the real data values all have the same expected value and that the covariance between any two of these random variables does not depend on their specific locations, but rather only on the separation vector between them. In practice, these theoretical considerations translate into an assumption that there is no noticeable trend in the local means and variances. Finally, the covariance model $\tilde{C}(\mathbf{h})$ is assumed to be correctly chosen.

If all of these assumptions are legitimate, then the error variance predicted by the random function model may be a useful tool in establishing global confidence intervals. Before it is used, however, we should be clear on what this modeled error variance represents and whether or not this has any relevance to reality.

Recall that we invoked the random function model to help us solve the problem of choosing weights for local estimation. We had a real and important criterion—the minimization of the actual error variance—but we were unable to solve the resulting equations because they called for knowledge of the unknown true values. We therefore adopted the random function model, arguing that the real data set that we are studying can be viewed as one possible outcome of the random function model we have built. We then forgot the particular details of our one actual outcome and used the entire ensemble of all possible outcomes to help us choose appropriate weights for local estimation.

Though $\tilde{\sigma}_R^2$ is *not* the actual error variance we had in mind when we began the problem, it is a convenient intermediary crutch. Experience has shown that by minimizing the error variance over the entire ensemble of all possible outcomes of our conceptual random function model, we usually do a pretty good job of minimizing the actual error variance for our one real outcome. If we now want to use $\tilde{\sigma}_R^2$ to make meaningful statements about uncertainty, we should make sure that there is some valid interpretation in reality for the entire ensemble of all possible outcomes of the random function model.

Earlier, we offered two interpretations of a global confidence interval. In one, we imagined other possible areas that were statistically similar to the area under study; in the other, we imagined other possible sample data sets within the area under study. Under the first interpretation, the other possible outcomes of the random function model may correspond to the other areas we have in mind. Under the second interpretation, however, the other possible outcomes of the random function model have no real significance.

If by a global confidence interval we intend to describe how the estimation error of the global mean might fluctuate if other sample data sets had been collected, then the error variance given by our random function model will likely be too large. Rather than calculate the error variance over all possible outcomes, we should be calculating it only over the one outcome that has real significance. The range of possible fluctuations over all outcomes is generally larger than the range of possible fluctuations over a single outcome.

In the case study that follows, we have calculated the error variance of the global mean using Equation 20.7. We then compare this modeled variance to the actual variance of the various estimation errors when the sample data set is changed. The point of this case study is to demonstrate that the variance predicted by the random function model is indeed too large if we intend that our global variance reflect fluctuations due to resampling. Unfortunately, we do not have access to alternative data sets with which we could check the validity of $\tilde{\sigma}_R^2$

under the interpretation that it reflects the error variance over all similar areas within the same geological environment.

We will estimate the global mean of the V variable using a weighted linear combination in which the weights are determined by ordinary kriging. With 470 samples, the solution to this problem would require the solution of 471 simultaneous linear equations with 471 unknowns. Though there are computers that can perform the necessary calculations, the complete and exact solution of the ordinary kriging equations is usually prohibitively time consuming and expensive.

A shortcut that is often used in practice is to do local block kriging throughout the area and, for each sample, to accumulate the various kriging weights it receives in the estimation of nearby blocks. In Table 20.2, we show such calculations for a few of the samples. The results shown in this table were obtained by estimating the mean grade of 10 x 10 m^2 blocks with each such block discretized into 100 points on a regular square grid. The search strategy considered all points within 25 m of the center of the block and kept no more than six samples within each quadrant. The variogram model used in these calculations was the one given in Equation 20.6. Table 20.2 shows all the block estimates that made use of the first three samples in the sample data set. The accumulated weights can be standardized so that they sum to one by dividing each accumulated weight by the total number of block estimates; in the example given here, a total of 780 blocks were kriged.

In Table 20.3, we compare the weights calculated by accumulating local kriging weights to those calculated by the polygonal method. A glance at this table confirms that the two procedures produce very similar weights. There is an excellent correlation, $\rho = 0.996$, between the two sets of weights. For the Walker Lake data set, it is clear that for the problem of global estimation, ordinary kriging is not necessary; the polygonal alternative produces essentially the same weighting scheme with much less computational effort.

We will assume that the variogram model given in Equation 20.6 correctly describes the pattern of spatial continuity and is applicable throughout the area. Once our variogram model is chosen, Equation 20.7 will produce an error variance for any weighted linear combination whose weights sum to 1. Due to the strong similarity between the two sets of weights shown in Table 20.3, the actual estimates and their corresponding error variances will both be very similar. For this

Table 20.2 Examples of the calculation of global weights from the accumulation of local ordinary kriging weights.

Block Center	Sample Locations		
	(11,8)	(8,30)	(9,48)
(5,5)	1.00	—	—
(5,15)	0.59	0.41	—
(5,25)	0.26	0.57	0.09
(5,35)	—	0.59	0.31
(5,45)	—	0.28	0.54
(5,55)	—	—	0.51
(5,65)	—	—	0.23
(15,5)	0.71	—	—
(15,15)	0.43	0.27	—
(15,25)	0.17	0.40	0.11
(15,35)	—	0.35	0.30
(15,45)	—	0.12	0.40
(15,55)	—	—	0.27
(15,65)	—	—	0.05
(25,5)	0.31	—	—
(25,15)	0.16	0.11	—
(25,25)	0.07	0.13	—
(25,35)	—	0.09	0.11
(25,45)	—	0.03	0.10
(25,55)	—	—	0.04
(25,65)	—	—	-0.01
(35,5)	0.08	—	—
Accumulated Weight	3.78	3.35	3.05
Standardized Weight	0.00485	0.00429	0.00391

case study, we have chosen to use the weighting scheme produced by accumulating local ordinary kriging weights. For this weighting scheme, the estimated global mean is 282.7 ppm and the error variance predicted by the random function model is 341.9 ppm^2. If we are willing to make a further assumption that the distribution is Normal, then we

Table 20.3 Comparison of weights calculated by accumulating local ordinary kriging weights and those calculated by the polygonal method.

Easting	Northing	V (ppm)	Accumulated Kriging Weight	Polygonal Weight
⋮	⋮	⋮	⋮	⋮
71	29	673.31	0.00070	0.00056
70	51	252.57	0.00193	0.00218
68	70	537.46	0.00263	0.00300
69	90	0.00	0.00168	0.00166
68	110	329.15	0.00168	0.00171
68	128	646.33	0.00069	0.00065
69	148	616.18	0.00060	0.00058
69	169	761.27	0.00078	0.00082
70	191	917.98	0.00055	0.00055
69	208	97.42	0.00167	0.00185
69	229	0.00	0.00433	0.00451
68	250	0.00	0.00369	0.00402
71	268	0.00	0.00367	0.00346
71	288	2.43	0.00439	0.00447
91	11	368.26	0.00386	0.00440
91	29	91.60	0.00222	0.00234
90	49	654.66	0.00054	0.00044
91	68	645.47	0.00059	0.00053
91	91	907.16	0.00060	0.00043
91	111	826.33	0.00067	0.00073
⋮	⋮	⋮	⋮	⋮

can turn this into a 95% confidence interval by calculating the standard deviation and doubling it. Using this traditional approach, our estimate of the global mean would be reported as 282.7 ±37 ppm.

Using the exhaustive Walker Lake data sets, we are able to check if this statement of uncertainty is truly representative of the actual fluctuations due to resampling. We can discard our 470 samples and draw new sample data sets with similar configurations. For this study we have chosen a particularly simple approach to producing sample data sets whose configuration is similar to that of our actual 470 samples: the sample data locations are simply translated slightly so that their locations relative to one another remain identical. With all of the sample locations being at least 7 m from the borders of the exhaustive data set, we can produce 225 alternative sample data sets by moving the existing sample grid in 1 m increments from 7 m south to 7 m north of its actual location and from 7 m east to 7 m west of its actual location. For each of these 225 sample data sets, we have repeated the estimation of the global mean using the same weights that we calculated earlier when we accumulated the local ordinary kriging weights. For each of these 225 estimates we can calculate the error or the difference between the estimate and the true exhaustive mean of 277.9 ppm and compare these to the errors we obtained using the modeled estimation variance. The standard deviation of these 225 estimation errors is 7.1 ppm which is considerably lower than the 18.5 ppm we predicted using the modeled estimation variance.

The discrepancy between the actual fluctuation of the estimates and the predicted fluctuation is not due to the adverse effects of preferential clustering. We have rescaled the sill of our variogram model so that it corresponds more closely to the overall variance of the data set. Some of this discrepancy may be explained by the inadequacies of the variogram model for small separation distances. We have already noted that our model has a higher nugget effect than does the corresponding exhaustive variogram. Even if we used the exhaustive variogram, however, our confidence intervals would still be too large. The problem, as we discussed earlier, is that we are using a result of our conceptual random function model, $\tilde{\sigma}_R^2$, which is based on an averaging over all other possible outcomes of the random function—outcomes that have no counterpart in reality. The fluctuations that we see in our single unique outcome are less than those that our model sees in its entire ensemble of outcomes.

This one case study does not discredit the other interpretation of global confidence intervals. If there are other areas that have similar statistical characteristics—other reservoirs within the same basin or other pollution sites from the same source—then these other areas may correspond to the other outcomes of our random function model and the global confidence interval may be an accurate reflection of the estimation errors averaged over all of these related areas.

A Dubious Use of Cross Validation

A dubious (but regrettably common) procedure in geostatistical estimation is to use the results of a cross validation study to adjust the variogram model. In this section we will present the traditional procedure and show how it can often lead to erroneous results.

The belief that cross validated residuals can help to improve the variogram model is clearly a tempting one. After all, a cross validation study does provide us with observable errors. In addition to doing the estimation at locations where we have samples, if we also predict the error variance, then it seems that a comparison of our predicted error variance and our observed error variance should be useful in building a better variogram model.

The traditional procedure for using cross validated residuals to improved the variogram model begins by defining a new variable r', often called the *reduced residual*:

$$r' = r/\tilde{\sigma}_R \tag{20.8}$$

The reduced residual is the actual residual, r, divided by the standard deviation of the error predicted by the random function model.

For any distribution of values, rescaling each value by the standard deviation produces a new set of values whose standard deviation is one. If $\tilde{\sigma}_R^2$ is an accurate estimate of the actual error variance, then the standard deviation of the variable r' should be close to 1.

The statistics of the reduced residuals from a cross validation study are often used as indications of how well our variogram model is performing in practice. If the standard deviation of r' is greater than 1, then our errors are actually more variable than the random function model predicts, and the sill of the variogram model should be raised. On the other hand, if the standard deviation of the reduced residuals is less than 1, then our actual errors are less variable than our random function model predicts and we should lower the sill of our variogram model. In fact, the precise magnitude of the adjustment is given by the variance of the reduced residuals. If a variogram model $\gamma(\mathbf{h})$ is used in a cross validation study and the variance of the resulting reduced residuals is $\sigma_{R'}^2$, then a second cross validation study with the variogram model $\gamma(\mathbf{h}) \cdot \sigma_{R'}^2$ will produce reduced residuals whose standard deviation is exactly 1. The temptation of this procedure lies in the fact that it is automatic—a computer program can easily perform the necessary calculations and readjust the user's model accordingly. Unfortunately, it often does not have the desired effect and can do more harm than good.

This automatic correction of the variogram model usually ignores the fact that adjustments to the variogram should include not only its magnitude, but also its shape. The use of cross validation to automatically readjust the shape of the variogram has been studied and various procedures have been proposed. Most of these, however, simply adjust the relative nugget effect until the reduced residuals behave as desired [4]. While the adjustment of the relative nugget effect will certainly have a big impact on the reduced residuals, the procedure is somewhat arbitrary. The spacing between available samples imposes a fundamental limitation on the improvement a cross validation study can make to the definition of the short scale structure of the variogram model. The simultaneous adjustment of all of the model parameters, including the type of function used, has not yet been proven successful in practice.

Research continues on the use of cross validation to automatically improve a variogram model. In our opinion, such an effort misses a fundamental limitation of the clever idea of cross validation. Even if an estimation procedure can be adapted to produce cross validated results that are encouraging, we still do not know if conclusions based on estimation at locations where we already have samples are applicable to the estimation that is the actual goal of the study. Conclusions based on cross validation should be used to modify an estimation procedure only after a careful consideration of whether the conditions under which cross validation was performed are truly representative of the conditions of the final estimation.

When we looked at cross validation in Chapter 15, we noticed several misleading conclusions. Most of the locations at which we could perform cross validation were in high valued areas while the estimation that really interests us covers the entire area. Though the cross validation may have been revealing features of the high valued areas we had preferentially sampled, the entire area did not have these same features and the results of the cross validation study were somewhat misleading.

We can see a further instance of this fundamental problem if we try to use our earlier cross validation results to improve our global variogram model. In Chapter 15 we presented a cross validation study

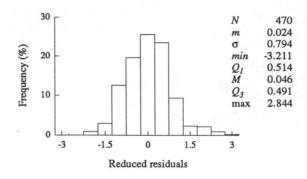

N	470
m	0.024
σ	0.794
min	-3.211
Q_1	0.514
M	0.046
Q_3	0.491
max	2.844

Figure 20.7   The histogram of the reduced residuals from the cross validation study presented in Chapter 15.

comparing ordinary kriging estimates and polygonal estimates. We have repeated this study and included the calculation of the ordinary kriging variance. The distribution of the reduced residuals is shown in Figure 20.7. The standard deviation of the reduced residuals is 0.79, suggesting that our actual errors are less variable than our random function model predicts. Were we to use this result in the automatic manner described earlier, we would decrease the sill of our variogram model by about one third. If we then use this adjusted model to produce estimates over the entire area, we would discover that our assessments of uncertainty were, in fact, now too optimistic. By basing the adjustment on observed errors at locations that are preferentially located in high-valued areas, we do not necessarily produce a more reliable variogram model for the estimation study we are actually interested in.

It is important to reiterate that the rescaling of the variogram by some constant does not change the estimates, only the estimation variance. So although the adjustment of the sill based on cross validated residuals is certainly easy to automate, it does not change the estimates themselves. As the previous example pointed out, it may not even make the estimation variance a more reliable index of uncertainty.

Realizing that the adjustment of the sill does not affect the estimates, some rather sophisticated variogram adjustment programs try to adjust other variogram model parameters. Since the nugget effect has a significant impact, it is often a natural target for automated "improvement." Starting with the variogram model that we fit to the sample variogram, we can experiment with different relative nugget effects (adding whatever we take away from the nugget to the other structures so that the sill remains constant) until we find the magic one which produces the reduced residuals whose distribution is closest to a Normal distribution with a mean of 0 and a variance of 1. For the 470 V values, a model with a relative nugget effect of 4% does the best job. Unfortunately, if we repeat the estimation over the entire area and base our comparison on actual errors at unsampled locations on a 10 x 10 m^2 grid, we discover that our original model performs better than the one whose nugget effect was automatically adjusted on the basis of cross validated residuals. The 470 preferentially located samples are not representative of the actual estimation we intend to perform, and the results of cross validation based on these samples should be used very cautiously.

Local Confidence Intervals

As we noted earlier when we first discussed various possible approaches to reporting uncertainty, a probabilistic statement such as a confidence interval has meaning only in the context of the population that we have chosen to group together. In the first case study in this chapter, we chose to group together all 780 block estimates. The confidence intervals we discussed earlier describe the possible range of errors when all similar configurations over the entire area are considered. Certain decisions require an assessment of uncertainty that is more locally customized.

Unfortunately, the most common method for reporting local uncertainty is the same $\pm 2\sigma$ 95% confidence interval that we explored earlier. It is unfortunate since the conditions that may make it a useful global measure of uncertainty rarely hold locally. There were two assumptions that were required for this approach. First, we had to assume that the error distribution is Normal. Second, we had to assume that we could predict the variance of the actual errors.

Though global error distributions are often symmetric, local ones are not. In low-valued areas, there is a high probability of overestimation and, in high-valued areas, a corresponding tendency toward underestimation. Particularly in the extreme areas that have the greatest importance, local error distributions are often asymmetric, with the sign of the skewness changing from one area to another. At a local level, an assumption of Normality is very questionable.

There are certain applications for which an assumption of Normality at the local level is acceptable. For geometric properties, such as the thickness of a coal seam, or the depth to some relatively flat-lying stratigraphic marker, the symmetry and low coefficient of variation of the original data permit an assumption of Normality of the error distributions at the local level. For most applications, however, this is not the case.

Even if we are willing to make the assumption that our errors are approximately Normally distributed, we still have the problem of predicting the variance of this distribution. Earlier, when we first looked at global confidence intervals, we discussed the problem of modifying the variogram model so that it is truly representative of the global area. This same problem arises, and much more severely, at the local level. If we intend that our $\pm 2\sigma$ 95% confidence intervals have local significance, then we have to make sure that the variogram model that we use to calculate the error variance is truly representative of the local pattern of spatial continuity. At the global level, we had a rather meager ability to adjust the sill to some reasonable level, but we had to admit that the modification of the shape was really beyond our ability. There is little hope, then, that we will be able to do much in the way of local modifications to the variogram model.

Perhaps the simplest approach to this problem is to assume that the shape of the variogram is the same everywhere but that its magnitude changes from one area to another. This assumption is based less on reality than it is on the convenient fact that it permits us to use a single variogram model and to rescale the error variance according to our estimate of the local variance.

We can define and model $\gamma_R(\mathbf{h})$, a relative variogram with a sill of 1, whose shape describes the pattern of spatial continuity. Like the traditional variogram, such a relative variogram can be flipped upside down to provide a relative covariance that can be used in the ordinary kriging equations. If such a model is used to calculate $\tilde{\sigma}_R^2$ using Equation 20.1, then the result is an error variance that is relative to the local variance. To predict the actual variance of the local errors, we must rescale this relative value to some estimate of the local variance:

$$\tilde{\sigma}_R^2 = \hat{\sigma}^2 \left[\tilde{C}_{AA} + \sum_{i=1}^{n} \sum_{j=1}^{n} w_i w_j \tilde{C}_{ij} - 2 \sum_{i=1}^{n} w_i \tilde{C}_{iA} \right] \qquad (20.9)$$

where the Cs are the relative covariances derived by subtracting the relative variogram from its own sill of one and $\hat{\sigma}^2$ is some estimate of the local variance.

The method used to estimate the local variance depends largely on what factors affect the local variance. This is the reason that moving neighborhood statistics are often an important part of exploratory spatial data analysis. Knowledge of the factors that influence the local variability permits a more reliable customization of the local variance. If the local variability can be related to other features, then these can be used later to help predict the appropriate rescaling of a relative variance.

Typically, the local variance is related to the local mean. A scatter plot of local means versus local variances will reveal if such a relationship exists and will also provide the data required for developing an equation that predicts local variance from the local mean.

Case Study: Local Confidence Intervals from Relative Variograms

To illustrate the use of relative variograms, we return to the block kriging case study shown in Chapter 13. In that case study, we estimated the average V value over 10 x 10 m^2 blocks. If we repeat the same estimation with a relative variogram model whose sill is one, the resulting error variance, $\tilde{\sigma}_R^2$, will be relative to the local variance. To obtain a measure of uncertainty that is customized to the local variability, we will have to multiply this relative error variance by an estimate of the local variance.

For this study, we will obtain our relative variogram by rescaling our earlier variogram model. In practice, relative variogram models are usually obtained by directly modeling one of the sample relative variograms discussed in Chapter 7. Since the sill of our earlier variogram model was 107,000 ppm^2, we can produce a model whose sill is 1 by simply dividing each of the coefficients of the model by this sill value:

$$\gamma(\mathbf{h}) = 0.2056 + 0.3738 \, Sph_1(\mathbf{h}_1') + 0.4206 \, Sph_2(\mathbf{h}_2') \qquad (20.10)$$

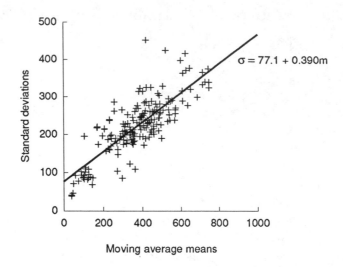

Figure 20.8 A scatterplot showing the regression line of moving window standard deviations versus moving window means from the sample data set of V.

This variogram model will provide block estimates identical to the ones obtained in the study in Chapter 13 since the shape of the variogram model has not been changed.

The most common way of estimating the local variance is to predict it from the estimate of the local mean. When we presented exploratory spatial data analysis in Chapter 4, we introduced the idea of moving window statistics. In Chapter 6 we used overlapping 60 x 60 m^2 windows to produce the scatter plot shown in Figure 20.8. This scatterplot showed a clear relationship between the local mean and the local variance. With a rather strong correlation coefficient, $\rho = 0.81$, this relationship is quite well described by the following linear regression line shown along with the cloud of points in Figure 20.8:

$$\hat{\sigma} = 77.1 + 0.390 \ m \qquad (20.11)$$

We can use this equation to predict the local standard deviation if we already know the local mean. In fact, we do not know the local mean; we have to estimate it from whatever samples are available. In practice, the block kriging estimate is typically used as an estimate of the local mean. If we are trying to assign local confidence intervals to a point estimate, it is not advisable to use the point estimate as an

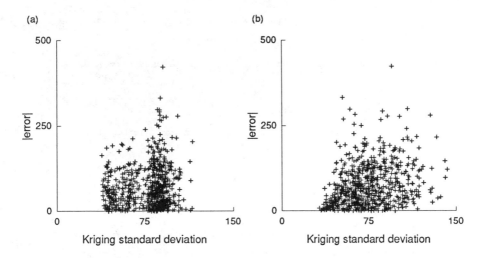

Figure 20.9 Scatterplots of the magnitude of the actual block kriging error of V versus the kriging standard deviations using the absolute variogram in (a) and a relative variogram with a rescaled local variance in (b).

estimate of the local mean. It is preferable to estimate separately the local mean by performing block estimation over a larger area centered on the point being estimated.

For each of the 780 block estimates, the local error variance will be estimated by multiplying the estimated local variance (obtained from Equation 20.11) by the relative error variance (obtained by using the relative variogram model in Equation 20.3). Once the local error variance has been obtained, the Normal 95% confidence intervals can be assigned by calculating the corresponding standard deviation and doubling it.

This procedure will produce good local confidence intervals only if the local error variances are good. To check whether or not our error variances are indeed locally relevant, we begin with a plot of the actual error versus the predicted local error variance. As we discussed earlier, we should not expect a strong relationship between these two; however, we should expect that, on average, the magnitude of the actual errors increases as the predicted local error variance increases.

Figure 20.9a shows the magnitude of the 780 actual errors versus the corresponding values of $\tilde{\sigma}_R$ given by ordinary kriging with the vari-

Table 20.4 A comparison of the actual spread of the errors to the spread predicted by ordinary kriging with an absolute variogram model.

	100 Lowest $\tilde{\sigma}_R$	100 Middle $\tilde{\sigma}_R$	100 Highest $\tilde{\sigma}_R$
Predicted	80	149	177
Actual	73	97	92

Table 20.5 A comparison of the actual spread of the errors to the spread predicted by ordinary kriging with a relative variogram model and a proportional effect correction.

	100 Lowest $\tilde{\sigma}_R$	100 Middle $\tilde{\sigma}_R$	100 Highest $\tilde{\sigma}_R$
Predicted	43	71	112
Actual	44	96	111

ogram model given in Equation 20.6. Figure 20.9b shows the same plot with $\tilde{\sigma}_R$ given by ordinary kriging with the relative variogram model given in Equation 20.10 and subsequently rescaled by the estimated local variance.

It is clear that in the presence of a proportional effect the value of $\tilde{\sigma}_R$ obtained by the ordinary kriging system with an absolute variogram model bears little relationship to the actual local errors as demonstrated in Figure 20.9a. The kriging standard deviations shown in (b), however, exhibit a more reasonable relationship with the absolute kriging error. Table 20.4 shows a comparison of the average predicted standard deviation of the errors versus the actual standard deviation of the errors for three groups of estimates: the 100 estimates for which $\tilde{\sigma}_R$ was the lowest, the 100 estimates for which $\tilde{\sigma}_R$ was in the middle, and the 100 estimates for which $\tilde{\sigma}_R$ was the highest.

Table 20.5 shows the same comparison when the local uncertainty is calculated by rescaling the relative variance given by ordinary kriging with a relative variogram. The improvement over the results in Table 20.4 is quite remarkable. If a proportional effect exists, it must be taken into account when assessing local uncertainty. Since the ordinary kriging variance does not depend on the actual magnitude of the sample data values but only on their locations, it does not take into account the possibility that estimates are more uncertain simply because they are in areas with higher values. If exploratory data analysis has revealed that the phenomenon becomes more erratic as the magnitude of the values increases, then the magnitude of the sample values is an important feature of uncertainty that the ordinary kriging variance does not account for. By rescaling the variogram to a sill of one and locally correcting the relative kriging variance, one can build confidence intervals that reflect local conditions[6].

Notes

[1] If we decide to use ordinary kriging, which minimizes the error variance, Equation 12.14 provides a quicker way than Equation 20.1 to calculate the error variance.

[2] Though the formula for error variance can be applied to any weighted linear combination in which the weights sum to 1, for the purposes of ranking data configurations, it is preferable to use the ordinary kriging weights. The use of other weighting methods may lead to results whose meaning is unclear since a high error variance may be due to a poor estimation method and not to the sample configuration.

[3] It is important to use the same discretization to calculate the average point-to-block covariances and the average covariance within the block. If different methods are used to calculate these, the resulting error variance may be slightly negative.

[4] This adjustment of the relative nugget effect changes the shape of the entire variogram and therefore leads to different cross validation results. There is another procedure, quite commonly used, in which the only part of the variogram model that is changed is the short-range structure between $h = 0$ and the first point on the sample variogram. The aim of this procedure is to use cross validation to improve the nugget effect. Unfortunately, this procedure cannot work since the value of the variogram for separation distances greater than 0 and less than the closest sample spacing are never used in cross validation.

[5] Though we are actually interested in the relationship between the local mean and the local variance, the local standard deviation will suffice since the local variance is easily obtained by simply squaring the standard deviation.

[6] Much of the wisdom regarding relative variograms and their use in producing better assessments of uncertainty can be found in David, M. , *Handbook of Applied Advanced Geostatistical Ore Reserve Estimation.* Amsterdam: Elsevier, 1988.

Further Reading

Journel, A. , "Non-parametric geostatistics for risk and additional sampling assessment," in *Priciples of Environmental Sampling*, (Keith, L. , ed.), pp. 45–72, American Chemical Society, 1988.

Srivastava, R. , "Minimum variance or maximum profitability," *CIM Bulletin*, vol. 80, no. 901, pp. 63–68, 1987.

21

FINAL THOUGHTS

Some 20 chapters ago, we began our presentation of geostatistics. Since this presentation has not been a conventional one, and since geostatistics has suffered in the past from not making clear its limitations, it is appropriate that we review the tool kit we have assembled. In this final chapter we will discuss the correct application of the various tools and their limitations. We will briefly discuss some of the important topics that we have chosen not to include in this introductory book. Finally, we will discuss the important new contributions of geostatistics.

Description and Data Analysis

The first seven chapters were devoted to description and data analysis. In many applications, this description and analysis is, by itself, the goal of the study. In other applications, it is an important first step toward the final goal of estimation. It is regrettable that in many geostatistical studies little attention is paid to this initial step. A good understanding of the data set is an essential ingredient of good estimation, and the time taken to explore, understand, and describe the data set is amply rewarded by improved estimates.

In exploratory spatial data analysis, one should not rigidly follow a prescribed sequence of steps but should, instead, follow one's instinct for explaining anomalies. Imagination and curiosity are the keys to unravelling the mysteries of a data set. If one tool uncovers something slightly bizarre, dig deeper, perhaps with other tools, until the reasons for the initial surprise are clear. This process not only leads to the discovery of the errors that have inevitably crept into a data set, but it also provides the necessary background for detecting errors that creep in during the course of a study. Estimation is prone to simple blunders particularly when computers become involved; a thorough exploratory data analysis often fosters an intimate knowledge that later warns of bogus results.

In Chapter 2, we looked at several ways of summarizing a univariate distribution. While the most commonly used summary statistics, the mean and the variance, do provide measures of the location and spread of a distribution, they do not provide a complete description. Furthermore, their sensitivity to extreme values makes other summary statistics, such as those based on quantiles, more useful for many descriptive tasks.

In Chapter 3 we presented methods for bivariate description. The scatter plot and its various summaries provide not only a good description of the relationship between two variables, but they also form the basis for the tools we use to analyze spatial continuity. Though the correlation coefficient is by far the most common summary of a scatter plot, it should be emphasized once again that it is a measure of the linear relationship between two variables and may not adequately capture strong nonlinear relationships. Furthermore, the correlation coefficient is strongly affected by extreme pairs. A pair of extreme values can produce a strong correlation coefficient that is not indicative of the generally poor correlation of the other pairs; it can also ruin an otherwise good correlation coefficient. The Spearman rank correlation coefficient is a useful supplement to the more common Pearson correlation coefficient. Large differences between the two often provide useful clues to the nature of the relationship between two variables.

One of the most important aspects of earth science data is its spatial location, and in Chapter 4 we presented several tools for describing the important features of a data set in a spatial context. While automatic contouring is an indispensable part of spatial description, it should be used judiciously since it can make very erratic phenomena appear quite smooth and can mask or blur anomalies. With recent advances in computer graphics software and hardware, there are many alternatives that reveal the detail of a data set without overloading the eye with too much information.

Moving window statistics are a good way of exploring the possible subdivisions of a spatial data set. All estimation methods involve an assumption that the data used in the weighted linear combination somehow belong in the same group; it is useful to explore the validity of this assumption through moving window statistics.

The most time consuming part of the data analysis and description is typically the description of spatial continuity. Though the variogram is the tool most commonly used by geostatisticians, it often suffers in practice from the combined effect of heteroscedasticity and the preferential clustering of samples in areas with high values. In such cases, there are many alternatives that may produce clearer and more interpretable descriptions of spatial continuity. Of these alternatives, the relative variograms are already quite commonly used by practitioners. Once quite common in the early days of spatial data analysis, the covariance function and the correlogram have largely been ignored by the mainstream of geostatistics and are only now becoming commonly used once again.

As the analysis of the Walker Lake sample data set showed, the description of spatial continuity typically involves a lot of trial and error. The various tools for univariate and bivariate description offer a broad range of innovative alternatives to the classical variogram. Rather than take the mean of the squared differences, why not take the median of the squared differences? Or the interquartile range of the differences? Or the mean of the absolute differences?

For the problem of describing the spatial continuity of strongly skewed and erratic values, one alternative that has not been discussed here is the possibility of using some transform of the original variable. Rather than calculate variograms of the original data values, one could choose to calculate variograms of their logarithms, for example, or of their rank. By reducing the skewness of the distribution, such transforms reduce the adverse effects of extreme values.

Even with robust alternatives to the variogram or with transformed variables, there may still be problems with describing the spatial continuity from the available samples. One tool that should be used more in exploratory spatial data analysis is the h-scatterplot. The erratic behavior of sample variograms may be revealed by a careful study of the paired values within each lag.

Estimation

In Chapter 8 we turned our attention to estimation and discussed three important variations of estimation problems:

- estimation of global parameters versus estimation of local parameters

- estimation of a single mean value versus estimation of an entire distribution of values

- estimation of unknown values whose support is the same as that of the available sample data versus estimation of unknown values whose support differs from that of the available samples

Before beginning an estimation exercise, one should consider which of these variations apply, and then choose an appropriate combination of the various estimation tools.

Global Estimation

If global estimates are required, then the important consideration is declustering. The presence of additional sampling in areas with extreme values can produce severely biased estimates if all samples are weighted equally. The polygonal and cell declustering methods presented in Chapter 10 are both quite simple and easily implemented methods that attempt to account for the effects of preferential sampling. In addition to these, one might also consider the method presented in Chapter 20 in which local kriging weights were accumulated to produce a declustering weight for each sample.

Local Estimation

For local estimation, the distance to nearby samples becomes an important consideration. The various methods presented in Chapters 11 and 12 all have certain advantages and drawbacks.

The polygonal method is the easiest to understand and does handle the clustering problems posed by irregular data configurations quite well. It is also easy to implement without a computer. By assigning all of the weight to a single nearby sample, however, it does not take advantage of the useful information contained in other nearby samples. In comparison to other methods, local estimates derived from polygonal weighting tend to have larger extreme errors.

While slightly more complex than polygonal weighting, triangulation is still relatively simple and has the advantage of using more of the nearby sample data. It is not easily adapted to the problem of extrapolation beyond the available sampling.

While polygons and triangulation are easy to implement manually, this is not a realistic possibility with data sets containing hundreds or thousands of samples. With such data sets, computers become necessary and computationally intensive methods then also become interesting. Inverse distance methods work well with regularly gridded data. The smoothness of estimates derived by inverse distance methods is desirable if contouring is the final goal of estimation. The biggest drawback of the inverse distance methods is their inability to account directly for clustering. If they are being used with irregularly gridded data or, worse, preferentially clustered data, it is advisable to use a search strategy, such as quadrant or octant search, that accomplishes some declustering of the nearby samples.

Though ordinary kriging is certainly the most computationally intensive and mathematically tedious of the point estimation methods discussed in this book, it does combine many of the desirable features of the other methods. It accounts both for the clustering of nearby samples and for their distance to the point being estimated. Furthermore, by considering statistical distance, through the variogram model, rather than euclidean distance, it offers tremendous possibilities for customizing the estimation method to the particular problem at hand. If the pattern of spatial continuity can be described and adequately captured in a variogram model, it is hard to improve on the estimates produced by ordinary kriging.

Ordinary kriging is not a panacea. The quality of estimates produced by ordinary kriging depends on the time taken to choose an appropriate model of the spatial continuity. Ordinary kriging with a poor model may produce worse estimates than the other simpler methods. Ordinary kriging is most successful when the anisotropy is properly described and when the variogram is locally customized. The description of anisotropy and local customization both depend heavily on good qualitative input and a good understanding of the genesis of the data set.

In addition to its other advantages, the ordinary kriging system shown in Chapter 12 can easily be extended from point estimation to block estimation. As was shown in Chapter 13, the replacement of the point-to-point covariances in the right-hand side **D** vector by average point-to-block covariances is all that is required to estimate block averages.

Accommodating Different Sample Support

A trick similar to that used for block kriging can be used to adapt the ordinary kriging system so that it can handle samples of different support. In many applications, the available sample data are representative of differing volumes. For example, in a petroleum application the porosity measured from a core plug is representative of a smaller volume than is the porosity inferred from a well log. If some samples have a support large enough that they cannot be adequately treated as point samples, then it is possible to incorporate them into the ordinary kriging equations and account for their support. As with the block estimation problem, the only adaptation that is needed is the replacement of point-to-point covariances with average point-to-block covariances.

In the left-hand side **C** matrix, the covariance between any two samples is replaced by the average covariance between the two sample volumes; in the right-hand side **D** matrix, the covariance between the sample and the point being estimated is replaced by the average covariance between the sample and the point being estimated.

A more general form for the ordinary kriging system, one that accounts for the possibility of samples with differing supports and for the possibility of yet another support for the arithmetic average value being estimated is given below:

$$
\underbrace{\begin{bmatrix} \tilde{C}_{11} & \cdots & \tilde{C}_{1n} & 1 \\ \vdots & \ddots & \vdots & \vdots \\ \tilde{C}_{n1} & \cdots & \tilde{C}_{nn} & 1 \\ 1 & \cdots & 1 & 0 \end{bmatrix}}_{(n+1)\times(n+1)} \cdot \underbrace{\begin{bmatrix} w_1 \\ \vdots \\ w_n \\ \mu \end{bmatrix}}_{(n+1)\times 1} = \underbrace{\begin{bmatrix} \tilde{C}_{1A} \\ \vdots \\ \tilde{C}_{nA} \\ 1 \end{bmatrix}}_{(n+1)\times 1} \qquad (21.1)
$$

$$ \mathbf{C} \qquad\qquad \cdot \qquad \mathbf{w} \qquad = \qquad \overline{\mathbf{D}} $$

The minimized error variance from the solution of this system is given by

$$
\tilde{\sigma}^2_{OK} = \tilde{C}_{AA} - \sum_{i=1}^{n} w_i \tilde{C}_{iA} - \mu \qquad (21.2)
$$

Search Strategy

In Chapter 14 we discussed the search strategy, an area that often receives inadequate attention. A well-conceived search strategy will improve any estimation procedure. In practice, the pattern of spatial continuity of the data set may fluctuate so wildly from one locality to the next that one is unable to build meaningful variogram models. In such cases, a customized search strategy may be the only hope for good estimates.

An important point that has been previously camouflaged in the geostatistical literature is that inverse distance methods will perform almost as well as ordinary kriging if the search strategy incorporates a good declustering procedure. The quadrant search that we discussed here is only one of several commonly used procedures. Rather than divide the neighborhood into quadrants, one could divide it into any number of sectors. These sectors need not be identical; if there is a strong anisotropy, one might prefer to have narrower sectors along the direction of major continuity. The criterion used to keep or to reject the samples within a particular sector need not be based on the euclidean distance. Some practitioners prefer to use the statistical distance, as measured by the variogram, to determine which samples should stay and which should go.

Incorporating a Trend

One of the considerations we posed in the choice of a search strategy was the question, "Are the nearby samples relevant?," which led to a discussion of stationarity. Many practitioners are rightly troubled by the assumption of stationarity. Their intuition is that their ore deposit or their petroleum reservoir is not well modeled by a stationary random function model. This good intuition does not mean that the geostatistical approach is inappropriate.

First, it should be emphasized that other procedures implicitly make the same assumption. The inappropriateness of the stationarity assumption is no justification for abandoning ordinary kriging in favor of, say, an inverse distance method. It is, however, a justification for attempting to subdivide the data set into separate populations. If there is sufficient information to support the intuition that stationarity is inappropriate, then this information can be helpful in subdividing the data set into smaller regions within which stationarity is more appropriate.

Second, the stationarity assumption applies not to the entire data set but only to the search neighborhood. While a large earth science data set nearly always contains interesting anomalies that seem to contradict the assumption of stationarity, it may still appear reasonably homogeneous within smaller regions the size of the search neighborhood. The *local stationarity* assumed by all of the estimation methods we have discussed is often a viable assumption even in data sets for which global stationarity is clearly inappropriate.

There are data sets in which the uncomfortableness with the stationarity assumption does not come from the belief that there are distinct subpopulations, but rather from the belief that there is a gradual trend in the data values. There is an adaptation of the ordinary kriging system that allows one to accommodate a trend. This procedure, known as *universal kriging* can be used to produce good local estimates in the presence of a trend especially in situations where the estimate is extrapolated rather than interpolated from the local sample values. Universal kriging can also be used to estimate the underlying trend itself, and is therefore interesting not only as a local estimation procedure, but also as a gridding procedure prior to contouring.

Though universal kriging can calculate a trend automatically, one should resist the temptation to use it as a black box, particularly when it is being used to extrapolate beyond the available data. It is always important to check the trend that an automatic method produces to see if it makes sense. If there is a trend, then it is likely that there is some qualitative understanding of why it exists and how it can best be described. Though automatic methods exist for finding the trend that is best in a statistical sense, this trend may not have the support of common sense and good judgement.

In many situations in which a trend exists, it is wiser to choose a trend based on an understanding of the genesis of the phenomenon, subtract this trend from the observed sample values to obtain residuals, do the estimation on the residuals, and add the trend back at the end. This procedure has the advantage of making the trend an explicit and conscious choice thereby avoiding the pitfall of bizarre behavior beyond the available sampling.

Cross Validation

In Chapter 15 we discussed the method of cross validation, a useful trick that gives us some ability to check the impact of our many choices about the estimation methodology. The results of a cross validation study give us a kind of dress rehearsal for our final production run. Though the success of a dress rehearsal does not guarantee the success of the final performance, its failure certainly raises serious doubts about the success of the final performance. The real benefit of a cross validation study is the warning bells that it sets off. In studying cross validation results, one should concentrate on the negative aspects such as the worst errors, the areas with consistent bias, or the areas where misclassification occurs.

It is dangerous to dwell on the positive aspects of set of cross validation residuals. Our ability to produce good estimates at sample locations may have little relevance to the final estimation study we intend to do. In particular, it is foolish, in our opinion, to use cross validated residuals for automatic improvement of the variogram model. As we showed in Chapter 20, this procedure can lead to an "improved" model that actually produces worse results. We are not saying that the models fit to sample variograms should never be adjusted. Indeed, there are many cases in which the model fit to a sample variogram is definitely flawed for the purpose of estimation. If the variogram model is flawed, it should be corrected and such corrections should be supported by qualitative information. For example, in the Walker Lake case studies it is reasonable and correct to argue that the sill of the sample variogram is too high due to the preferential sampling in areas with higher variability. A good declustered estimate of the global variance provides a more suitable sill.

In many fields of application, good variogram models depend on access to similar data sets that have been more densely sampled. For example, in petroleum applications the practice of inferring the shape of the horizontal variogram from the sample variogram of a related outcrop is becoming more common. Often, the anisotropy is difficult to determine from sample variograms due to the fact that well-behaved sample variograms often need large angular tolerances. If there is a similar data set that contains more data and permits a sharper definition of the directional variograms, it is reasonable to import the anisotropy evident from this related data set.

There are certainly situations in which the variogram model fit to the sample variogram is not appropriate for estimation, but adjustment through the use of qualitative information and variograms from related data sets is preferable to automatic adjustment through cross validation.

Modeling Sample Variograms

In Chapter 16 we looked at the practical detail of fitting a positive definite model to directional sample variograms. To the newcomer, the fitting of variogram models often seems difficult, a bit of a black art. Through practice, however, it loses its mystery and becomes more manageable. Initially, there is usually a tendency to overfit the sample variogram, using several structures to capture each and every kink in the sample points. Such complicated models do not usually do better than simpler models with fewer structures that capture the major features of the sample variogram. Simplicity is a good guiding principle in variogram modeling. For example, if a single exponential model fits as well as a combination of two spherical structures, then the simpler exponential model is preferable.

In deciding whether or not to honor the kinks in a sample variogram, it is wise to consider whether or not there is a physical explanation for them. If there is qualitative information about the genesis of the phenomenon that explains a particular feature, then it is worth trying to build a model that respects that feature. If there is no explanation, however, then the feature may be spurious and not worth modeling.

The fitting of a model to a particular direction is simply an exercise in curve fitting in which there are several parameters with which to play. A good interactive modeling program will make this step quite easy.

The simultaneous fitting of several directions calls for a little more care. By making sure that each direction incorporates the same type of basic model, with the same sills but different ranges, it is always possible to combine the various direction models into a single model that describes the pattern of spatial continuity in any direction.

Fitting models to auto- and cross-variograms for several variables requires even more care. Again, the use of the same type of basic model for all auto- and cross-variograms permits the linear model of coregionalization to be used and allows the positive definiteness of the entire set to be checked through the use of the determinants, as was shown in Chapter 16.

Using Other Variables

There are many applications in which estimates can be improved if the correlation between different variables is exploited. In Chapter 17 we discussed cokriging and showed how the ordinary kriging system could be adapted to include information contained in other variables. The case study in Chapter 17 showed how the information in the V variable could be used to improve the estimation of U. Cokriging need not be limited to only two variables; any number of additional variables can be incorporated into the estimation. The addition of a new variable calls for more variogram analysis and modeling. For each new variable included one needs its variogram model and also cross-variogram models between it and all of the other variables in the system. For example, with 10 variables one needs 10 autovariogram models and 45 cross-variogram models.

The improvement of cokriging over ordinary kriging with the primary variable alone is greatest when the primary variable is undersampled. With sample data sets in which the primary and all secondary variables are sampled at all locations, the improvement of cokriging over ordinary kriging is less dramatic. In practice, there are two reasons why the cokriging system is often unable to realize its full potential. The first is the requirement that the set of auto- and cross-variogram models be positive definite. The second is the unbiasedness condition.

The need for positive definiteness imposes many constraints on the models that can be chosen for the set of auto- and cross-variograms. In practice, the linear model of coregionalization provides a relatively simple way of checking whether large sets of auto- and cross-variogram models are indeed positive definite. Unfortunately, the use of this linear model of coregionalization also probably deprives the cokriging system of some of its potential improvement. A problem that has not yet received much attention is that of building less restrictive models of the spatial continuity and cross-continuity.

The unbiasedness condition that is most commonly used is that shown in Equation 17.8 in which the primary weights are made to sum to 1 and the secondary weights are made to sum to 0. This condition may be unnecessarily restrictive. As was shown in Chapter 17, the use of other unbiasedness conditions may improve the estimation procedure.

Estimating Distributions

The declustering procedures used for estimating the global mean are also appropriate for estimating a declustered global distribution. By applying the declustering weights to an indicator variable, which simply counts the data that are below a particular threshold, one can construct a cumulative probability distribution from which other declustered global statistics can be extracted.

The concept of an indicator variable also offers a clever way of adapting the ordinary kriging procedure so that it can be used to estimate a cumulative probability distribution. In using the indicator method, however, one does not obtain a complete description of the distribution, but obtains instead a estimation of particular points on the cumulative distribution curve. The implication of this is that once the indicator kriging procedure has done its job, there is still work to be done interpolating the behavior of the distribution between the estimated points.

The parametric approaches avoid this drawback, at the cost of not permitting as much detail about the pattern of spatial continuity to be injected into the estimation procedure. Parametric methods make rather sweeping assumptions about the nature of the indicator variograms at various cutoffs. For example, the multi-gaussian method implicitly assumes that the indicator variograms for the most extreme values show the least structure; in some applications, this assumption is not appropriate. For example, in petroleum reservoirs it is the most extreme permeabilities—the highs in the fractures and the lows in the shale barriers—that are the most continuous.

Other Uses of Indicators

Another common use of indicators is to separate populations. Earlier, we drew attention to the fact that some populations may not be separated by a sharp geometric boundary, but may be intermingled instead. In such situations, indicator kriging can be used to estimate the proportion of different populations within a certain block or local area.

For example, this is typically done in mining applications when there is a large spike of barren or completely unmineralized samples in a data set. Including these essentially constant values in the estimation is often undesirable. Their presence will make the spatial continuity appear to be greater than it is and this, in turn, may lead to severe misclassification problems around the edges of the anomalies. If the barren values are treated as one population and the positive values as another, one can perform separate estimations for the two populations. To solve the problem of what proportion of each population is present in a particular block or locality, indicator kriging can be used.

Indicators can also be used to directly estimate a conditional probability distribution. In Chapter 20, we discussed the desirability of estimating such a distribution and saw that the traditional approach of combining the kriging variance with an assumption of normality may not be appropriate in all situations. If one interprets the indicator as a probability, with a value of 1 revealing that at a particular location the value is certainly below the cutoff and the value 0 revealing that it certainly is not below the cutoff, then the ordinary point kriging of these indicators will produce estimates of the conditional probability distribution for particular cutoffs.

BIBLIOGRAPHY

Box, G. and Jenkins, G. , *Time Series Analysis forecasting and control.* Oakland, California: Holden-Day, 1976.

Brooker, P. , "Kriging," *Engineering and Mining Journal,* vol. 180, no. 9, pp. 148–153, 1979.

Buxton, B. , *Coal reserve Assessment: A Geostatistical Case Study.* Master's thesis, Stanford University, 1982.

Castle, B. and Davis, B. , "An overview of data handling and data analysis in geochemical exploration," Tech. Rep. Technical report 308, Fluor Daniel Inc., Mining and Metals Division, 10 Twin Dolphin Drive, Redwood City, CA., 94065, March 1984. Association of Exploration Geochemists short course at Reno, Nevada.

Chatterjee, S. and Price, B. , *Regression Analysis by Example.* New York: Wiley, 1977.

Cleveland, W. , "Robust locally weighted regression and smoothing scatterplots," *J. American Statistical Association,* vol. 74, pp. 828–836, 1979.

Cressie, N. and Hawkins, D. M. , "Robust estimation of the variogram, I," *Mathematical Geology,* vol. 12, no. 2, pp. 115–125, 1980.

David, M. , *Geostatistical Ore Reserve Estimation.* Amsterdam: Elsevier, 1977.

David, M. , *Handbook of Applied Advanced Geostatistical Ore Reserve Estimation*. Amsterdam: Elsevier, 1988.

Davis, J. C. , *Statistics and Data Analysis in Geology*. New York: Wiley, 2 ed., 1986.

Edwards, C. and Penney, D. , *Calculus and Analytical Geometry*. New Jersey: Prentice-Hall, 1982.

Friedman, J. H. and Stuetzle, W. , "Smoothing of scatterplots," Tech. Rep. Project Orion 003, Department of Statistics, Stanford University, CA., 94305, July 1982.

Gardner, W. A. , *Introduction to Random Processes*. New York: Macmillan, 1986.

Hayes, W. and Koch, G. , "Constructing and analyzing area-of-influence polygons by computer," *Computers and Geosciences*, vol. 10, pp. 411–431, 1984.

Isaaks, E. H. , *Risk Qualified mappings for hazardous waste sites: A case study in distribution free geostatistics*. Master's thesis, Stanford University, 1985.

Isaaks, E. H. and Srivastava, R. M. , "Spatial continuity measures for probabilistic and deterministic geostatistics," *Mathematical Geology*, vol. 20, no. 4, pp. 313–341, 1988.

Johnson, R. A. and Wichern, D. W. , *Applied Multivariate Statistical Analysis*. Englewood Cliffs, New Jersey: Prentice-Hall, 1982.

Journel, A. G. and Huijbregts, C. J. , *Mining Geostatistics*. London: Academic Press, 1978.

Journel, A. G. , "Non-parametric estimation of spatial distributions," *Mathematical Geology*, vol. 15, no. 3, pp. 445–468, 1983.

Journel, A. G. , "Non-parametric geostatistics for risk and additional sampling assessment," in *Priciples of Environmental Sampling*, (Keith, L. , ed.), pp. 45–72, American Chemical Society, 1988.

Koch, G. and Link, R. , *Statistical Analysis of Geological Data*. New York: Wiley, 2 ed., 1986.

Matern, B. , "Spatial variation," *Meddelanden Fran Statens Skogs-forskningsinstitut, Stockholm*, vol. 49, no. 5, p. 144, 1960.

Matheron, G. F. , "La théorie des variables régionalisées et ses applications," Tech. Rep. Fascicule 5, Les Cahiers du Centre de Morphologie Mathématique de Fontenebleau, École Supérieure des Mines de Paris, 1970.

Matheron, G. F. , "Estimer et choisir," Fascicule 7, Les Cahiers du Centre de Morphologie Mathématique de Fontainebleau, Ecole Supérieure des Mines de Paris, 1978.

McArther, G. J. , "Using geology to control geostatistics in the hellyer deposit," *Mathematical Geology*, vol. 20, no. 4, pp. 343–366, 1968.

Mosteller, F. and Tukey, J. W. , *Data Analysis and Regression*. Reading Mass: Addison-Wesley, 1977.

Mueller, E. , "Comparing and validating computer models of orebodies," in *Twelfth International Symposium of Computer Applications in the Minerals Industry*, (Johnson, T. and Gentry, D. , eds.), pp. H25–H39, Colorado School of Mines, 1974.

Olkin, I. , Gleser, L. , and Derman, C. , *Probability Models and Applications*. New York: Macmillan, 1980.

Omre, H. , *Alternative Variogram Estimators in Geostatistics*. PhD thesis, Stanford University, 1985.

Parker, H. , "The volume-variance relationship: a useful tool for mine planning," in *Geostatistics*, (Mousset-Jones, P. , ed.), pp. 61–91, McGraw Hill, New York, 1980.

Parker, H. , "Trends in geostatistics in the mining industry," in *Geostatistics for Natural Resources Characterization*, (Verly, G. , David, M. , Journel, A. G. , and Marechal, A. , eds.), pp. 915–934, NATO Advanced Study Institute, South Lake Tahoe, California, September 6-17, D. Reidel, Dordrecht, Holland, 1983.

Rendu, J. , "Kriging for ore valuation and mine planning," *Engineering and Mining Journal*, vol. 181, no. 1, pp. 114–120, 1980.

Ripley, B. D. , *Spatial Statistics*. New York: Wiley, 1981.

Royle, A. , "Why geostatistics?," *Engineering and Mining Journal*, vol. 180, no. 5, pp. 92–102, 1979.

Shurtz, R. , "The electronic computer and statistics for predicting ore recovery," *Mining Engineering*, vol. 11, no. 10, pp. 1035–1044, 1959.

Silverman, B. , "Some aspects of the spline smoothing approach to nonparametric regression curve fitting (with discussion)," *J. Royal Statistical Society, Series B*, vol. 47, pp. 1–52, 1985.

Srivastava, R. M. , "Minimum variance or maximum profitability," *CIM Bulletin*, vol. 80, no. 901, pp. 63–68, 1987.

Srivastava, R. M. and Parker, H. , "Robust measures of spatial continuity," in *Third International Geostatistics Congress*, (Armstrong, M. e. a. , ed.), D. Reidel, Dordrecht, Holland, 1988.

Srivastava, R. , *A Non-Ergodic Framework for Variograms and Covariance Functions*. Master's thesis, Stanford University, 1987.

Strang, G. , *Linear Algegbra and Its Applications*. New York: Academic Press, 1980.

Sullivan, J. , "Conditional recovery estimation through probability kriging- theory and practice," in *Geostatistics for Natural Resources Characterization*, (Verly, G. , David, M. , Journel, A. G. , and Marechal, A. , eds.), pp. 365–384, NATO Advanced Study Institute, South Lake Tahoe, California, September 6-17, D. Reidel, Dordrecht, Holland, 1983.

Tukey, J. , *Exploratory Data Analysis*. Reading Mass: Addison-Wesley, 1977.

Verly, G. , David, M. , Journel, A. G. , and Marechal, A. , eds., *Geostatistics for Natural Resources Characterization*, NATO Advanced Study Institute, South Lake Tahoe, California, September 6-17, D. Reidel, Dordrecht, Holland, 1983.

Verly, G. and Sullivan, J. , "Multigaussian and probability krigings-application to the Jerritt Canyon deposit," *Mining Engineering*, pp. 568–574, 1985.

A
THE WALKER LAKE
DATA SETS

This appendix provides detailed information on origin of the Walker Lake data sets. Should one wish to repeat some of the case studies presented in the text or experiment with new ideas, the Walker Lake data sets can easily be reproduced using publicly available digital elevation data.

The Digital Elevation Model

The Walker Lake data set is derived from digital elevation data from the Walker Lake area near the California-Nevada border. These elevation data were obtained from a digital elevation model (DEM) of the National Cartographic Information Center (NCIC)[1]. The NCIC provides DEMs in three formats; the data used throughout this book were obtained from a DEM that was in the Defense Mapping Agency digital or planar format. This format consists of digital elevation data for an area covering 1 degree of latitude by 1 degree of longitude; two contiguous blocks correspond to the area covered by one standard 1 : 250,000 topographic map sheet. The ground distance between adjacent points is approximately 200 feet, so each 1° x 1° quadrangle contains about 2.5 million elevation points. The eastern half of the Walker Lake topo-

---

[1]NCIC, U.S. Geological Survey, 507 National Center, Reston VA 22092, U.S.A.

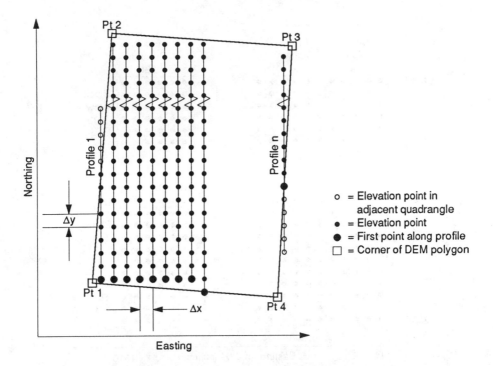

Figure A.1 Typical structure of a digital elevation model. The Defense Mapping Agency format consists of elevation points on a regular grid with Δ x and Δ y equal to 0.01 inches on a 1 : 250,000 map sheet.

graphic map sheet forms the basis for the exhaustive data set discussed in Chapter 5.

Figure A.1 illustrates a typical layout of digital terrain data; note that the profiles of elevation data are not necessarily all the same length nor do they necessarily begin at the same latitude. The first point on each profile is identified with map X and Y coordinates in one-hundredths of an inch. A perfectly rectangular subset was obtained from the DEM corresponding to the eastern half of the Walker Lake topographic map sheet. The coordinates of the corners of this rectangle are (in map units of 0.01″):

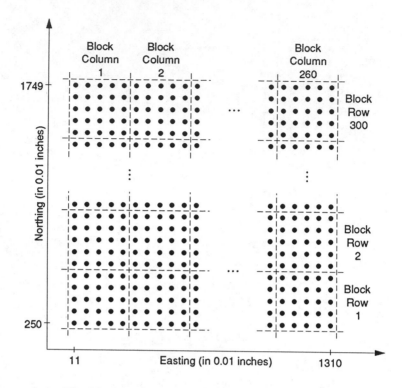

Figure A.2 The blocking pattern superimposed on the grid of digital elevation data.

southwest corner	(11,250)
northwest corner	(11,1749)
southeast corner	(1310,250)
northwest corner	(1310,1749)

This rectangle consists of $1,300$ north-south profiles, each of which contains $1,500$ elevation points. These 1.95 million data were grouped into 5×5 blocks (see Figure A.2) providing a regular grid, 260 blocks east-west by 300 blocks north-south. The original elevations were rescaled by subtracting $4,000$ from each elevation and dividing each result by 3.28. The subtraction shifts the minimum elevation from $4,000$ to 0 feet while the division by 3.28 converts the measurements from feet to meters.

The Exhaustive Data Set

The variable referred to as U in the Walker lake data set is the variance (in meters squared) of the 25 values within each block:

$$U = \sigma^2 = \frac{1}{25} \sum_{i=1}^{25} (x_i - \bar{x})^2$$

where $x_1, x_2, \ldots, x_{25}$ are the elevation values (in meters) of the 25 points within a block. In flat terrain, the elevation values will be very similar and U will be quite low; in very rugged terrain, the elevation values will be much more variable and U will be quite high. U can be seen, therefore, as an index of roughness of the topography.

The variable referred to as V is a function of the mean and variance of the 25 values in each block:

$$V = [\bar{x} * ln(U + 1)]/10$$

The type variable T records whether the average elevation within a block is above or below $5,000$ feet:

$$type = 1 \qquad \text{if } \bar{x} < 5,000'$$
$$type = 2 \qquad \text{otherwise}$$

Artifacts

The 1.95 million original elevation data contain some peculiarities that are most likely artifacts of the digitizing process. The DEM was created by digitizing contour lines and using a bicubic spline to interpolate these irregularly spaced values to a regular grid. One of the appeals of a bicubic spline as an interpolating function is that it produces a smooth and eye-pleasing interpolation. Unfortunately, if the original data points are quite widely spaced, many of the interpolated values are identical to the nearest available digitized elevation. The result of this is that in flat lying areas the elevations of the digitized contour lines tend to dominate the grid of interpolated values.

With a class size of 1 foot, a histogram of the original 1.95 million elevation data contains many little spikes at elevations corresponding to the contour lines of the original 1:250,000 topographic map. Since the contour interval is 200 feet, the spikes occur at elevations that are evenly divisible by 200. These spikes of similar values form small plateaus along the north-south profiles of the DEM model at elevations corresponding to the original contour line and are most noticeable in relatively flat lying areas.

The plateaus of similar elevation values along the DEM profiles are responsible for the curious features of the V and U values. Since secondary variable U is an index of the roughness of the topography, it is very low (often 0) near these plateaus. In relatively flat lying areas, where the 200-foot contour lines are quite far apart, the U data set shows bands of low values that track the original contour lines. The northwest corner of Figure 5.10c provides a good example of this banding.

The primary variable V is also related to the roughness, though less directly than U; it also shows a banding in flat-lying areas. The northwest corner of Figure 5.9c shows a good example of this banding; the southwest corner of the Walker Lake data set, which covers Mono Lake and its shoreline, provides another good example in Figure 5.9b.

The interpolation of the elevations digitized from contour lines to a regular grid involves a search strategy that selects nearby values within a prescribed neighborhood. In relatively flat lying areas, with the original digitized elevations quite far apart, the interpolated values show small discontinuities as particular elevation values are included or excluded by the search strategy. Though not very visible on a contour map of the interpolated elevations, these discontinuities become very noticeable when an index of roughness is contoured. These discontinuities are most pronounced when there are few available digitized elevations; these are the flat lying areas that tend to produce very constant interpolated values. In such areas, an index of roughness, such as the secondary variable U, is usually very low but becomes abnormally high above a discontinuity.

On the indicator maps in Figure 5.9 and 5.10, the prominent features that dangle like icicles in the area southeast of Walker Lake itself are due to these discontinuities in the grid of interpolated elevation values.

The spikes in the histogram of the original elevation values are responsible for the banding that is visible on the scatterplot of V versus U (Figure 5.8). From its definition given earlier in this appendix, it is clear that for a particular value of U, the variable V is simply the average elevation multiplied by some constant. The conditional distribution of V given a particular value of U is therefore a rescaling of the distribution of the average elevation. With the histogram of the original elevation values having spikes at multiples of 200 feet, the histogram of average elevation will also contain spikes. Due to the fact that averaging has been performed over 5 x 5 blocks, the spikes on the histogram of original elevation values will be somewhat blurred and will appear as modes on the histogram of average elevation. The banding of the exhaustive scatterplot is a result of these many small modes in the histogram of average elevation.

B

CONTINUOUS RANDOM

VARIABLES

This appendix provides the continuous analogy of the presentation of discrete random variables given in Chapter 9. Though the use of random function models can be understood through discrete random variables, most of the geostatistical literature uses continuous random variables.

The use of continuous random variables requires a few changes to the discrete approach. Rather than a set of possible outcomes and an associated set of probabilities, a continuous random variable has a probability distribution function that describes the relative probability of occurrence for a range of possible values. Rather than expressing the parameters of a random variable with summation signs, the continuous version uses integrals. Despite these conceptual and notational differences, continuous random variables behave very much like discrete ones.

The Probability Distribution Function

Random variables need not be limited to a set of discrete values; in the earth sciences we usually deal with random variables that have a continuous range of possible outcomes. For example, we could choose to view the depth to a particular stratigraphic horizon as a random variable D that can take on any positive value. Some outcomes of D are more probable than others, and some are extremely improbable.

It is not feasible to define a continuous random variable by listing all of its possible outcomes and their associated probabilities; not only are there an infinite number of possible outcomes, but any particular outcome has a probability of 0. It is easier to define continuous random variables with a function that gives the cumulative probability of all outcomes below a certain value:

$$F(v) = P\{V \leq v\} \tag{B.3}$$

It is certain that the outcomes of a random variable are between $-\infty$ and $+\infty$, which gives us the following limiting values of $F(v)$:

$$F(-\infty) = 0 \quad \text{and} \quad F(+\infty) = 1$$

This function $F(v)$, usually referred to as the *cumulative density function* or *cdf*, allows us to calculate the probability that the random variable will take on values within any interval:

$$\begin{aligned} P\{a < V \leq b\} &= P\{V \leq b\} - P\{V \leq a\} \\ &= F(b) - F(a) \end{aligned}$$

The first derivative of the cdf is called the *probability density function* or *pdf*, and is usually denoted by $f(v)$:

$$\frac{dF(v)}{dv} = f(v) \qquad F(v_0) = \int_{-\infty}^{v_0} f(v)dv \tag{B.4}$$

The pdf is similar to the set of probabilities $\{p_1, \ldots, p_n\}$ that we used in Chapter 9 to define a discrete random variable in the sense that it provides relative probabilities of occurrence of particular outcomes. These relative probabilities are scaled in such a way that the total probability of all outcomes is 1:

$$\int_{-\infty}^{+\infty} f(v)dv = 1$$

Parameters of a Continuous Random Variable

As with a discrete random variable, which is completely defined by its set of outcomes and the set of corresponding probabilities, a continuous random variable is completely defined by its cdf or its pdf. From the complete definition of a random variable, one can calculate many parameters that describe interesting features of the random variable.

Expected Value. The expected value of a continuous random variable is given by either of the following equations:

$$E\{V\} = \tilde{m} = \int_{-\infty}^{+\infty} v \cdot f(v)dv \qquad (B.5)$$

$$= \int_{-\infty}^{+\infty} v \, dF(v) \qquad (B.6)$$

The use of either of these equations usually requires an analytical expression for $f(v)$ or $F(v)$. There are many well-studied distributions whose probability distribution functions can be written analytically. Even without an analytical expression for $f(v)$, it is always possible to tabulate the cumulative probability distribution $F(v)$ for several possible values $v_{(1)}, \ldots, v_{(n)}$ and to approximate the expected value by replacing the integral in Equation B.6 by a discrete sum:

$$E\{V\} = \tilde{m} \approx \sum_{i=1}^{n-1} \frac{v_{(i)} + v_{(i+1)}}{2} \left[F(v_{(i+1)}) - F(v_{(i)}) \right] \qquad (B.7)$$

This approximation will improve as n, the number of values for which $F(v)$ is tabulated, increases. In this approximate form, the expected value of a continuous random variable appears very much like that of a discrete random variable, as given in Equation 9.1.

As with discrete random variables, the expected value of a continuous random variable need not be a possible outcome; for continuous random variables with more than one mode, the expected value may fall between modes.

The mean value of a finite number of outcomes of a continuous random variable need not be identical to the expected value. As the number of outcomes increases, however, the sample mean usually converges to the expected value.

Variance. Equations 9.3 and 9.4, which defined the variance solely in terms of expected values, are applicable to both discrete and continuous random variables. If the pdf is known and has an analytical expression, the variance of a continuous random variable can be expressed as:

$$Var\{V\} = \tilde{\sigma}^2 = \int_{-\infty}^{+\infty} v^2 \cdot f(v)dv \quad - \quad \tilde{m}^2 \qquad (B.8)$$

or, if the cdf is known, as

$$Var\{V\} = \tilde{\sigma}^2 = \int_{-\infty}^{+\infty} v^2 \, dF(v) \quad - \quad \tilde{m}^2 \qquad (B.9)$$

As with the expected value, $\tilde{m}$, the calculation of the variance of a continuous random variable ideally requires an analytical expression for $f(v)$ or $F(v)$, but can be approximated using the form given in Equation B.9 with the increment $dF(v)$ taken from a tabulation of cumulative probabilities.

Joint Random Variables

Joint random variables may also be continuous. Returning to our earlier example of the depth to a particular stratigraphic horizon, we can add a second random variable, T, the thickness of the horizon. There may be some statistical relationship between the depth to the horizon, D, and the thickness of the horizon, T; a complete definition of these two random variables should include a statement of how they jointly vary.

A pair of continuous random variables (U, V), can be completely defined by their joint cumulative density function:

$$F_{UV}(u,v) = P\{U \le u \text{ and } V \le v\} \tag{B.10}$$

Like a univariate cdf, the bivariate cdf gives a probability that something might occur. While the univariate cdf, $F_V(v)$, reported the probability that the outcome of a single random variable was less than a particular cutoff, v, the bivariate cdf, $F_{UV}(u,v)$, reports the probability that the outcomes of two random variables are both less than two particular cutoffs, u and v.

We can also define a bivariate probability density function, $f_{UV}(u,v)$, which, as before, is the derivative of the corresponding cdf:

$$F_{UV}(u_0, v_0) = \int_{-\infty}^{u_0} \int_{-\infty}^{v_0} f_{UV}(u,v) \, dv \, du$$

Like its univariate counterpart, the bivariate pdf gives the relative probability of occurrence of any particular pair of outcomes (u, v), and is scaled so that the total volume beneath its surface is 1:

$$F_{UV}(+\infty, +\infty) = \int_{-\infty}^{+\infty} \int_{-\infty}^{+\infty} f_{UV}(u,v) \, dv \, du$$

Marginal Distributions

The knowledge of the joint cdf of two random variables allows the calculation of the univariate or marginal distribution of each random variable. Given the joint cdf of U and V, $F_{UV}(u,v)$, the cdf of U by itself, $F_U(u)$, is calculated by simply accepting all possible values of V:

$$F_U(u) = F_{UV}(u, +\infty)$$

Conditional Distributions

The knowledge of the joint distribution of two continuous random variables also allows us to calculate the univariate distribution of one variable given a particular outcome of the other random variable. The cdf of U given a particular range of values of V, is given by the following equation:

$$F_{U|V \leq v_0}(u) = \frac{F_{UV}(u, v_0)}{F_V(v_0)} = \frac{F_{UV}(u, v_0)}{F_{UV}(+\infty, v_0)}$$

The denominator is the marginal probability that V is below a given cutoff and serves to rescale the probability given in the numerator so that it serves as a proper cdf (i.e., so that $F_{U|V \leq v_0}(+\infty)$ is 1).

Parameters of Joint Random Variables

Equations 9.9 and 9.10, which gave the definition of the covariance solely in terms of expected values, are applicable for either discrete or continuous random variables. The continuous form of Equation 9.11 is

$$Cov\{UV\} = \int_{-\infty}^{+\infty} \int_{-\infty}^{+\infty} uv \cdot f_{UV}(u,v) \; du \; dv \quad - \quad \tilde{m}_U \cdot \tilde{m}_V \quad \text{(B.11)}$$

where $\tilde{m}_U$ and $\tilde{m}_V$ are the expected values of the two marginal distributions.

With these definitions, the remainder of the presentation of probabilistic models in Chapter 9 and the development of ordinary kriging in Chapter 12 can be applied either to discrete or to continuous random variables.

INDEX